Tower of Babel

Tower of Babel

The Evidence against the New Creationism

Robert T. Pennock

A Bradford Book
The MIT Press
Cambridge, Massachusetts
London, England

First MIT Press paperback edition, 2000
©1999 Massachusetts Institute of Technology

This book was set in Sabon by Crane Composition and was printed and bound in the United States of America.

Library of Congress Cataloging-in-Publication Data

Pennock, Robert T.
  Tower of Babel : the evidence against the new creationism / Robert T. Pennock.
    p.   cm.
  "A Bradford book."
  Includes bibliographical references and index.
  ISBN 0-262-16180-X (hc. : alk. paper), 0-262-66165-9 (pb)
    1. Evolution (Biology). 2. Evolution (Biology)—Religious aspects—
  Christianity. 3. Creationism. 4. Religion and science. 5. Historical
  linguistics. 6. Science—Philosophy. I. Title.
  QH366.2 .P428   1999
  576.8—dc21   98-27286
                                                          CIP

Contents

Preface: Creationism's Tower

Creationists are building a tower to heaven, and they are raising the banner of antievolution upon its ramparts. They see themselves as participants in a holy war against forces that would undermine the foundations of true Christianity, and they see "evolutionism" as the godless philosophy that unites the enemy. A poster given out by the Institute for Creation Research (ICR) depicts a cartoon of creationists (all wearing clerics' collars) firing cannons from a tower at evolutionists (depicted in pirates' costumes), who are firing their cannons in return while gleefully sending aloft balloons upon which are written "Atheism," "Abortion," "Homosexuality," and other supposed elements of the philosophy of evolutionism. The poster illustrates one danger of the creationist movement: the way in which it mixes issues in religion, philosophy and science and presents them in a manner that is biased, simplistic, and polarizing. The foundation of the Creationist Tower is the Book of Genesis. Though not always strictly Fundamentalist, creationists typically hold a literal reading of the Bible. They believe that there is only one correct interpretation of scripture—their own—and only one true stairway leading up the Tower to God. In public debates creationists often reject any explicit discussion of religion, insisting that they present only an alternative scientific position; but in their own literature, they make it clear that they are engaged in a religious battle for the hearts and minds of those who have fallen away from God and who would lead their children away from the true path to salvation.

Secular humanists and a few atheistic scientists are the most easily identified among the enemy because they wear their non-Christian colors openly. But creationists also believe that the tower to heaven is put at risk by people who may think of themselves as Christian but who have compromised the integrity of their faith by accepting the scientific, evolutionary history of life, in opposition to the record of creation revealed in Genesis. The question of "origins," as creationists term the issue, lies at the very center of their theology. I will examine the complexities of this view in some detail, but for the moment I can put their position simply: If we were not specially created in just the manner that a "plain reading" of the Bible says but rather have been produced by evolutionary processes, then Scripture loses its authority as revealed truth and with it crumbles the ground of religion, morality, and the possibility of salvation after death through Jesus Christ. On their view, the teachings of Christianity and the theory of evolution are strictly incompatible. What this means is that those who accept evolution are either deceived or are not "true" Christians. The creationism controversy is not just about the status of Darwinian evolution—it is about a clash of religious and philosophical worldviews.

If creationism were simply an incorrect view about biology, then the problem would be of concern mostly to scientists, philosophers of science, and those who want their children to receive a good education. But it is clearly more than this, and Christians whose beliefs do not turn them against science should also be concerned. The Catholic Church and many mainline Protestant denominations have issued explicit statements that evolution and Christianity are not incompatible, but according to creationists such accommodation to evolution is seriously misguided at best and heretical at worst. They have specific views about what is required for one to be a Christian, and most people who think of themselves as Christian simply do not qualify. One reason that creationism is gaining adherents despite its fringe theology is that creationists have positioned themselves as part of the so-called Religious Right and so typically promote a conservative political agenda, especially on social issues. However, political terms like "liberal" and "conservative" do not adequately differentiate the camps because although creationists certainly would reject liberal Christianity, they also would reject the theological views of many politically conservative Christians. Some creationists now try to portray themselves as advocating only a tolerant ecumenical notion of "mere creation," echoing C. S. Lewis's notion of "mere Christianity"—but scratch this surface and one finds a strict orthodoxy.

One problem with religious orthodoxy, however, is that it has an inherent drive to refine its boundaries to distinguish itself more clearly from the heterodox, and this process inevitably leads to schisms. Creationists try to portray the creationism controversy as a battle between dogmatic anti-Christian scientists and fair-minded believers, but if we look carefully we see that the lines are far more complex. Obscured by the noisy confrontation with scientists in the public fields, a quieter, private battle is being waged within the ramparts of the Creationist Tower. Creationists disagree vehemently among themselves about the theological details of what counts as a proper interpretation of Scripture, and since biblical interpretation drives their physical picture of the universe, we find a fascinating array of conflicting views. Chapter 1 takes a look inside the Tower and describes some of these internal struggles, showing how they are transforming creationism.

One of the most common creationist argumentative strategies against evolution is to quote statements scientists make when they are criticizing one or another element of evolutionary theory, and then to suggest that we should reject evolution as a whole because scientists themselves find fault with it. Would not turnabout be fair play here? In the battle over the teaching of creationism in school science classes, creationists unite against their common foe—but what about the great confusion within the Tower? I will call attention to this confusion but I do not mean thereby to follow the creationists' own fallacious strategy and urge that we should reject creationism simply because creationists find fault with the details of each other's views. After all, we philosophers are no less known for our disagreements. Rather, I want to suggest that if we listen carefully to their internal quarrels, we will notice how greatly these differ from debates about evolution within science. The point is to look at what each side—creationists and scientists—offers in the way of arguments and evidence for its particular point of view and against its rivals. Viewed in this way, creationism functions as a rather nice case study for an examination of some basic issues in philosophy of science, such as questions about the nature of scientific theories, scientific explanation, and especially scientific evidence. Developing this case study by examining creationist attacks upon evolutionary theory and other sciences is one of the principal tasks of this book. Chapter 2 provides an introduction to the elements of evolutionary theory and reviews some of the evidence upon which the theory is based, beginning with Darwin's own studies that led him to reject his earlier creationist views. It also begins to consider some of the major arguments that creationists give against evolution—that it supposedly violates the second law of thermodynamics, and that chance processes could not produce the world's complex and useful biological structures.

Creationists have a quiver of such arrows that they let fly to try to poke holes in evolutionary theory. I will address several of the most well-known of these, showing why the scientific account is not in any danger of losing out to the familiar creationist view that God created the world, its animals, and the first human beings in a week some six thousand years ago, but many others have been answered previously by other philosophers and scientists and do not warrant a second look, so in chapter 3 I take a new tack. Rather than discussing the evolution of organisms I talk about the evolution of languages. Linguistic evolution has strong theoretical parallels with biological evolution both in content and in the sort of evidence scientists use to draw conclusions about it; but it is also pointedly relevant to creationism, in that Genesis tells us that languages did not evolve but were specially created by God in the great confusion of tongues at the Tower of Babel. To my knowledge, no one has drawn out this important parallel before. Looking at the creationist view in this context will give a better sense of how much of science creationists are willing to reject to maintain their preferred conclusions—biological evolution is only the beginning of the scientific findings they oppose—and, I hope, will make it easier to recognize the weaknesses of their arguments. The Creationist Tower of Babel will be a recurring image throughout the book.

Of course, academics are often accused of residing in a tower of their own, an ivory tower out of which they do not deign to descend into the real world. I do not accept this view of academia, but it is true that theoretical discussions of the philosophy of science can seem rather remote from the ordinary concerns of those outside of the academy. As we will see, however, philosophy of science has played and continues to play a significant role in the creationism controversy. It is also true that most academics, especially in the sciences, have ignored creationist activism because they feel it is a waste of time, and because it has had negligible impact on their work. Until just a few years ago, I probably would have expressed a similar sentiment.

My high school biology teacher avoided the controversy by following an all-too-common strategy of saying little about evolution at all, and so if someone had asked my view at that time I would have said that I thought it rather unbelievable that a creature like the long-necked giraffe could have evolved. When I now think back to that opinion I formed from that vague acquaintance with evolutionary theory, I feel a bit more understanding of people who are convinced of creationism; few of us receive an adequate education in even the basics of evolution, and most are totally ignorant of the evidence for it. As an undergraduate at Earlham College, however, I studied biology under the guidance of a superb group of professors who made sure we understood, whether we were studying ecology, plant taxonomy, genetics, or dung beetle behavior, how everything fit into the framework of evolutionary theory. By the time I completed my joint biology and philosophy degree I had a good grounding in the science as well as the critical thinking skills that philosophy develops. When I subsequently encountered creationist arguments, I could recognize easily that they were mostly based on ignorance or misunderstandings. In the mid-1980s, while I was working on my doctorate in history and philosophy of science at the University of Pittsburgh, the First International Conference on Creationism was held in the city, and out of curiosity I attended some of the sessions. I was surprised by the large number of participants and their sometimes heated exchanges and I was intrigued by some of the people I met, but at the time the event did not strike me as anything more than a curiosity. After I started teaching, I was interested in creationism only as a pedagogically useful case study—until a series of events stimulated me to begin to take the subject more seriously.

The first event was a creationist talk sponsored by a campus Christian group at the University of Texas at Austin. As I recall, I attended because I was teaching my course on the relationship of science and religion, "God and the Scientist," which included a section on creationism, and I wanted to hear if any of the standard arguments had changed. Indeed they had. What I heard was the argument of one Phillip Johnson, a professor from the University of California at Berkeley, who rehearsed just a few variations of the standard complaints but then launched into an indirect attack upon evolution by way of an attack on philosophical naturalism. The edifice of evolutionary theory had no supporting evidence, he claimed, but was merely scientific dogma propped up by a speculative philosophy. Recognize and knock out that prop, he argued, and evolution would come crashing down, leaving creationism the victor. Johnson had an affable manner on stage and seemed to have much of the audience of several hundred students in the palm of his hand. Several of his subsequent speeches that I have attended have drawn even greater numbers of spellbound attendees. I also attended a symposium held at Southern Methodist University (S.M.U.) in March 1992 which featured a "debate" between Johnson and Michael Ruse, a renowned philosopher of biology who had played a key role as an expert witness in the important 1982 Arkansas trial that had overturned legislation mandating "balanced-treatment" of creationism and evolution in the state's public schools. The day-long symposium was titled "Darwinism: Scientific Inference or Philosophical Preference?" and here I encountered for the first time a group of creationists—Michael Behe, William Dembski, Stephen Meyer and others—who followed Johnson's lead in attacking evolution as flimsy naturalistic philosophy and in advocating a new creationism, euphemistically termed "intelligent-design theory."

I began to write on this topic in 1993 while I had a National Endowment for the Humanities fellowship to attend a Summer Institute on the topic of naturalism, led by Robert Audi at the University of Nebraska. My research that summer led to an article criticizing Johnson's view in his book *Darwin on Trial*. In 1994 my article, "Naturalism, Creationism and Evidence: The Case of Phillip Johnson,"[1] was accepted for publication in *Biology and Philosophy* and material from it forms the core of chapter 4, which discusses the new creationists' attack on scientific naturalism, and Johnson's attempt to resuscitate purely negative argumentation. In it I dismiss Johnson's claim that evolution rests on a dogmatic metaphysical naturalism, and I show that science only makes use of naturalism methodologically.

Having completed that original article, I did not plan to pursue the matter further, but in the world of academic publishing it can take a year or more before an article appears in print, and in the meantime I was invited to give a talk at Ohio State University, sponsored by the biology department. There I mentioned the forthcoming paper. Within two days Johnson had called Michael Ruse, who edits *Biology and Philosophy*, to say he had heard about my paper from someone in that audience and wanted to write a rebuttal. In his response Johnson criticized me for not having read his second book, *Reason in the Balance*. (Since that had only appeared in 1995 I felt I could not have been expected to have had the foresight—literally!— to read it two years in advance of its publication.) When I did read it for my reply[2] I realized that Johnson's project was not just to attack evolution, but to take arms against what he called the "modernist naturalist worldview" in general for what he saw as its inherent immorality. Like the creationists at ICR, though in a less cartoonish manner, he was arguing that evolutionary naturalist thinking gives rise to ethical relativism, and was causing the breakdown of traditional male and female roles, the acceptance of homosexuality, and what he took to be other forms of cultural and moral decay. I suddenly recognized that this was part of a common underlying pattern of all creationist views, namely, a fear that evolution undermined the basis of morality and the basis of purpose in life. This led me to write a follow-up article, "Naturalism, Creationism and the Meaning of Life: The Case of Phillip Johnson Revisited,"[3] material of which forms the core of chapter 7, which discusses what creationists take to be at stake in the battle. The relationship between factual and moral issues in the controversy over creationism and evolution—the points of contact between philosophy of science and ethics—is one of the most interesting aspects of the issue that has drawn me into the debate.

A second series of events also made me begin to change my attitude, as I saw renewed creationist activism in the schools. I remember in particular when one of my best undergraduate students, now pursuing a joint M.D./ Ph.D. program, came into my office, upset about news from her hometown of Plano, Texas, where the school board was considering a proposal to adopt a creationist textbook for use in science classes. This led me to my first look at *Of Pandas and People*, and I was surprised to find it had been put together by many of the same "intelligent-design theorists" who had been involved with the conference at S.M.U. It was a surprise because this group, led by Johnson, had claimed that they were only interested in gaining acceptance for their view from the "top down" by arguing their position at the university level rather than trying to push it in the schools. Around this time, creationists were making news around the country with similar new proposals to incorporate creationism into the science curriculum. These proposals were introduced, often with advice from the Institute for Creation Research, through local school boards to which a religious conservative majority had been elected, in towns from Vista, California, to Merrimack, New Hampshire. This led me to write an article arguing for the wisdom of keeping private religious beliefs separate from public scientific knowledge. Elements of that article—"Creationism in the Science Classroom: Private Faith vs. Public Knowledge"[4]—are incorporated in chapter 8. Behind the arguments in this and the earlier papers is the view that there are good reasons for science to rely on naturalism as part of its methodology. Science imposes severe constraints upon itself to ensure that its conclusions are intersubjectively testable, constraints that require that it not appeal to supernatural hypotheses or allow the citation of special (private) revelations as evidence. The new creationists, including Johnson and philosophers such as Alvin Plantinga, reject these constraints and share the view that supernatural explanations should be admitted into science. My article on this issue, "The Prospects for a Theistic Science," which appeared in *Perspectives on Science and Christian Faith,*[5] is incorporated here as a section of chapter 6.

The final impetus that made me take the new creationism more seriously was seeing the size and renewed power of the movement. Talks given by Johnson and other new creationists such as Plantinga and Walter Bradley regularly draw an audience of four hundred to eight hundred people. Creationists are also becoming financially well-supported; the Center for Renewal of Science and Culture, which funds intelligent-design creationists, was established in 1996 as a branch of Seattle's Discovery Institute upon receipt of a million dollars in grants. Intelligent-design creationists are beginning to catch up in influence to the venerable ICR, which itself continues to draw wide support. I attended a three-day *Back to Genesis* seminar led by the ICR at a Baptist church in Austin, and estimated an attendance of close to a thousand people. During one of the breaks, attendees gathered in the front to form a local creationist activist group that would meet and prepare to challenge evolutionists. Since I began research for this book, I have heard from biology professors in various parts of the country who are finding students coming to class with mental defenses prepared so they will not be "brainwashed" into accepting evolutionary theory, or with antievolutionary tracts in hand to pass out.

One such tract that was distributed at my university has a cover illustration of an ape chomping on a banana and is entitled "Big Daddy?"[6] In comic book form we are shown a biology professor, depicted as an overweight, bug-eyed Lenin with a goatee and a thin mustache, who browbeats a polite Christian student who stands to express his disbelief in evolution. ("You can GET OUT of MY class!! After you've apologized to everyone for your rudeness and ignorance, we MIGHT let you back in! . . . On second thought . . . I think I will systematically tear your little beliefs to shreds in front of the entire class!" "Thank you, sir!") However, the student is prepared with rebuttals to every argument and has "data" at the ready to challenge paleontologists' reconstruction of human evolution, geologists' use of radioisotopes to date the Earth, biologists' understanding of the origin of living molecules, and physicists' understanding of the forces that hold the atom together. He soon has reduced the arrogant professor to a sweating, defeated figure who confesses he has no answers. ("I don't know!" "I'm sorry sir, but I can't hear you." "I said—I don't know. You tell me!" "Sir, may I quote from the Bible?" "YES, YES, YES!!") As the professor slinks away to resign from his job, the other students realize what has happened ("[W]e didn't evolve! The establishment has been feeding us THE BIG LIE! We really do have a soul!"), and they are warned by the triumphant Christian student of the hellish consequences of refusing to acknowledge God as the Creator. ("What if I don't believe this and die— what then?" "Then you'll die in your sins—and be eternally lost.")

The comic book form is prone to hyperbole, but by the time I read this tract I was not surprised by creationism's new aggressive stance. Several biology professors I have talked to have expressed concern about the vehemence of the antievolutionary views of some of their students in introductory biology classes, and they struggle to find ways to teach their subject so as to reach those who have been told by creationists to view evolution with fear and suspicion. They are typically unaware of how creationism has evolved into a variety of disparate forms and so do not recognize or know how to deal with what they encounter. Moreover, too few scientists are taught anything about the philosophical basis of scientific methods, so most are unprepared to explain why science investigates the world as it does, or why evolutionary theory does not fall to the various "philosophical" criticisms that the new creationists level against it. In part, this book is a field guide to help science teachers understand who is attacking them and why; it is a primer to help them know how to respond to the intellectual chaos that creationists would create with their confused arguments.

Let me say, however, that although I will argue against creationism as a *science* and do what I can to counter its divisive rhetoric, I do not mean to attack the sincerity or intentions of creationist believers. Also, none of what I write should be taken as an attack upon religion in general or Christianity in particular. Indeed, as a member of the Society of Friends (Quakers), a minority sect that historically was persecuted for its unorthodox views by larger Christian denominations, I believe strongly in the freedom of religious belief (or unbelief), and I hope that others who do as well will recognize that the broad creationist program actually poses a real danger to that freedom. I will argue that "creation-science" (or "intelligent-design theory," or however it is euphemistically termed) is antithetical to science and certainly does not belong in science classrooms. Fundamentalist and evangelical Christians are not the only creationists who want their religious views to take precedence over science. We will also meet Native American tribes, Hindus, and others who have their own Creation stories that conflict with evolution, as well as a religious group that rejects evolution because its members believe that life on Earth was created by aliens who arrived in UFOs. To turn science classes into a clash of such so-called alternative theories to biological evolution is a recipe for disaster because creationists of each religious stripe are enemies of every other. I hope that the reader will come to understand the wisdom of leaving creationists to build and fight within their tower and of not allowing their battles to spill into the schools.

Acknowledgments

In writing this book and the articles that led up to it I have benefited in numerous ways from many people including Laurie Anderson, Katherine Arens, Tim Berra, John Beatty, Richard Dawkins, Edwin Delattre, Wesley Elsberry, George Gilchrist, William Jefferys, Philip Kitcher, Leon Long, Clifford Matthews, Risto Miikkulainen, Michael Ruse, Eugenie Scott, Kelly Smith, Elliott Sober, and two anonymous reviewers. I am indebted to the National Endowment for the Humanities for their funding in 1993 which assisted my research on naturalism. I also want to take this opportunity to express belated gratitude to my Earlham professors, especially Bill Stevenson, Dorothy Douglas, and Jerry Woolpy (on the biology side) and Len Clark and Richard Wood (on the philosophy side) for what I now, as a professor myself, appreciate all the more as their exemplary teaching. I would also like to thank students from my graduate seminars on the philosophy of biology, especially Matt Brauer, Dan Brumbaugh, Andy Czebieniak and Jason Fosson, for stimulating discussions. I must give special thanks to my wife Kristin, who patiently and lovingly supported my work on this project, and to my sister Mary, who combed the entire manuscript several times as my informal copy editor and advisor. And finally I want to express my gratitude to my parents Roger and Elizabeth Pennock, to whom I dedicate this book.

Tower of Babel

1

Creation and Evolution of a Controversy

It has often and confidently been asserted, that man's origin can never be known: but ignorance more frequently begets confidence than does knowledge: it is those who know little, and not those who know much, who so positively assert that this or that problem will never be solved by science.
—Charles Darwin

Speciation in Progress?

Creationism is evolving. Several new varieties of creationism have appeared recently and are competing to stake out a niche in the intellectual landscape. Someone who last looked in on the creationism debate in the 1980s would today still find much that is familiar but would also be struck by the significant changes the controversy has undergone. Creationism is no longer the simple notion it was once taken to be. Of course, the transformation should not be surprising if, using a slightly modified Darwinian model, we think of creationism as a cluster of ideas that reproduces itself by spreading from mind to mind and struggling with competing ideas for a home among a person's beliefs.[1] Sometimes it loses out to more powerful rival ideas, but sometimes it finds receptive mental soil, takes root and waits to be passed on again. But mental environments themselves differ, and occasionally, it takes a change in the original cluster—a conceptual mutation, if you will—for a view to survive and successfully reproduce in the dynamic world of ideas. After a time, such changes add up to a point at which the conceptual landscape is peppered with distinguishable varieties. The differences now emerging within creationism have reached that far, and perhaps farther. We have reached a critical juncture, for when a population diverges significantly in its distribution of character traits we must ask whether we still have just different varieties of the same species or whether the varieties have become new and distinct species. This is an exciting time for creationism-watchers, for we may be observing a conceptual speciation event in progress.

Probably in most people's minds the archetypal, or in biological terms the "wildtype," creationist is characterized by beliefs close to the following: God dictated the Bible word for word and so we must take it as literally true. From the Book of Genesis we know He miraculously created the world and all life, including our original ancestors, Adam and Eve, during a six-day work week just six-thousand years ago. He subsequently destroyed the world in a global flood, allowing only Noah and his family to survive in a huge Ark into which they had herded pairs of every kind of animal. All current people and animals are descendants of those on the Ark. And, oh yes, evolution did not happen, and we definitely are not related to monkeys.

Today, however, one may find self-proclaimed creationists who modify or reject almost every element of this cluster. Moreover, there are other aspects of creationism that to date have received little attention. For most of us, our knowledge of the history of the creationism controversy begins with the trial of John Scopes, a substitute biology teacher in the little town of Dayton, Tennessee, in 1925, who had been charged with violating a state law against the teaching of evolution.

A Brief History of Creationism

The Scopes Trial and Other Cases
The Scopes trial had actually been provoked by the American Civil Liberties Union to challenge the Tennessee law as unconstitutional. As in many legal battles, the particular case that gets used to test a law is less important than the larger issues that it exemplifies, and in this instance even the major constitutional issue wound up taking a back seat to larger cultural issues. The "Monkey Trial" was widely seen as a clash between science and religion, with the larger-than-life attorneys—Clarence Darrow and William Jennings Bryan—as the battling gladiators representing, respectively, evolution and creationism. Scopes himself was a bit player in this dramatic

contest. In his memoirs he says he does not even remember for sure whether he actually taught evolution or not in his class.[2] It also turned out to be insignificant in the big picture that Scopes lost the case; the Scopes trial was the first case that was really fought in the media. The national and international press covered the trial in great detail and the public followed the proceedings with the same fascination that it more recently followed the O. J. Simpson trial. Bryan was a well-known politician who had run for President, and saw the trial in part as an opportunity to raise awareness about the immorality that he thought was bringing American society to ruin—moral decay that he blamed on scientific materialism in general and evolution in particular for making people question biblical authority. Bryan himself took the stand to defend the Genesis account, though he came to regret this hubris. Under Darrow's pointed questioning, he quickly floundered. That he officially won the case only accentuated what was a very public humiliation. Having heard the evidence itself, the public mostly ignored the court's ruling and concluded on its own that evolution had triumphed. Though Fundamentalists were able to pass three more antievolution bills in other states in the four years following the Scopes trial, for the most part they abandoned their legislative efforts against the teaching of evolution. On the other hand, with the statute officially upheld, most textbook publishers chose to avoid potential problems by simply deleting mention of evolution or Darwin in new and revised editions of their science texts until nearly the end of the 1950s, so in fact the teaching of evolution in science classes lost further ground.

The Tennessee statute under which Scopes was convicted remained in force until 1967, when another science teacher, Gary Scott, who had been fired from his job for violating it, successfully challenged the law. By this time, largely because of the push to upgrade American science education that had begun following the Soviet Union's launch of Sputnik in 1957, evolution had finally begun to become integrated into the science curriculum, at least in the biology textbooks put together by the Biological Sciences Curriculum Study. In 1973, however, the Tennessee legislature passed a new law that required that any textbook that discussed the origins of man and the world had to give equal emphasis to the Genesis account. The explicit reference to Genesis made it a straightforward matter for the U.S. Court of Appeals to overrule the new law in 1975 as unconstitutionally giving preferential treatment to the biblical view of creation. But creationists were working to win back lost territory and, in the wake of the latest Tennessee defeat, they adopted a new strategy of introducing legislation (in some twenty states) that promoted creationism without making any explicit reference to the Bible. This led to a series of important cases that made their way in the 1980s through the U.S. courts, the most significant of which was the 1982 *Rev. Bill McLean et al. v. Arkansas Board of Education* case. Creationist activists had gotten the State of Arkansas to pass Act 590, legislation that required public schools to give "balanced treatment" to what they called "creation-science" and "evolution-science," and it was the constitutionality of this Act that was at issue in the case. The idea of creation-science had arisen at about the same time as the Sputnik launch, and can be dated from the publication in 1961 of *The Genesis Flood,* by John C. Whitcomb Jr., a Protestant theologian, and Henry M. Morris, who held a doctorate in engineering from the University of Minnesota and who subsequently went on to found the Institute for Creation Research (ICR), which remains the largest and most influential creationist organization. It is no exaggeration to credit this book as the impetus for the revival of the movement and the contemporary image of the wildtype creationist. We will see more of the details of the position shortly, but the basic thrust of creation-science was (and is) that creationism qualified as an alternative model to evolution and that it is verifiable—and indeed verified—scientifically. Act 590 legislated that the public schools incorporate creation-science into the biology curriculum alongside evolution and treat the two models on a par. The McLean case challenged that law.

The plaintiffs called upon scientific luminaries such as Francisco Ayala and Stephen Jay Gould as expert witnesses on evolution and the fossil record, Harold Morowitz on the second law of thermodynamics, and G. Brent Dalrymple on radiometric and other methods of geological dating. Their combined testimony devastated the pseudoscientific arguments of creation-science. Reading the post-trial write-up in the journal *Science,*[3] one gets the impression that the case was won solely on the basis of the scientific testimony, but this assessment misunderstands a key feature of the case. It would not have been enough to show that creation-science was bad science, because the suit sought to overturn Act 590 on the grounds that creation-science was not a science but, rather, represented a disguised *religious* view of origins and thus still violated the Establishment Clause of the Constitution that "Congress shall make no law respecting an establishment of religion." To establish this conclusion, the more important testimony was that given by the witnesses whose expertise dealt with religion and philosophy of science. Among these were the Reverend Kenneth W. Hicks, Methodist bishop of Arkansas; Father Bruce Vawter, a De Paul University biblical scholar; George Marsden, a professor of American Religious History at Calvin College; Langdon Gilkey, professor of theology at University of Chicago Divinity School (who also was a consultant for the IRS, helping them determine whether particular groups qualified for religious tax exemptions); and Michael Ruse, philosopher of science from Guelph University. Judge William R. Overton's final opinion on the case makes little reference to the detailed scientific arguments refuting creation-science's claims but focuses more on the testimony that dealt with its status as religious or scientific. Overton defined science descriptively as "what scientists do" and "what is accepted by the scientific community," but he also incorporated what he called the "essential characteristics" of science that he culled from Ruse's testimony, namely:

1. It is guided by natural law;
2. It has to be explanatory by reference to natural law;
3. It is testable against the empirical world;
4. Its conclusions are tentative, i.e., are not necessarily the final word; and
5. It is falsifiable.

Judge Overton concluded that creation-science "fails to meet these essential characteristics" and thus "is not science."[4] Furthermore, based on the theological testimony, he concluded that creation-science was religious and thus that Act 590 did indeed violate the Establishment Clause. I will look more closely at these characteristics of science and the issue of distinguishing science from religion in later chapters. Here, I just want to note that the defeat of creation-science in the Arkansas case was not primarily a victory of evolutionary theory itself but rather, because of the way the case was tried, depended upon deeper philosophical issues about the nature of science and religion and their relation to the law.[5]

Cases that followed *McLean v. Arkansas* in the 1980s dealt with creationists' subsequent attempts to find a crack in its ruling to push creationism through. Louisiana's Act for Balanced Treatment of Creation-Science and Evolution, for example, did not *require* teaching creation-science, but simply prohibited the teaching of evolution in public schools except when it was accompanied by instruction in creation-science. The U.S. Supreme Court ruled that Act unconstitutional in the 1987 *Edwards v. Aguillard* case, on the grounds that the Act impermissibly promoted religion by advancing the view that a supernatural being created humankind, and that comprehensive science education would be undermined if schools were forbidden to teach evolution.

The New Creationists
Entering the last decade of the millennium, a new generation of creationists began to reevaluate the old approach and to recast themselves in order to try new avenues of attack upon evolution. The textbook *Of Pandas and People*,[6] for example, looks as though it was hand-tailored to try to slip between the lines of the law as drawn in the cases mentioned above. Also, at this time, some creationists began to avoid using the term "creation-science" altogether in favor of one or another euphemism, such as "abrupt appearance theory" or "initial complexity theory." The *Pandas* textbook was put together by the most significant group of new creationists, and the term that they use is "intelligent-design theory," or sometimes "theistic science." At the ICR, a generational transition was also underway, as Henry Morris turned over the reins to his son John Morris in the mid-1990s and stepped back to concentrate on writing. The present is a critical period for the ICR's leaders: Although they continue to expand their operations and seem more successful than ever, they find that new creationist groups with different theological commitments are challenging their leadership of the movement. There is a struggle going on in the Creationist Tower, and this is what has sparked the interest of creationism-watchers.

In nature, when closely related groups of organisms (varieties of a species, say) occupy the same ecological niche, their competition for the same resources puts pressure on the populations and forces a transformation. Sometimes one group loses the battle and becomes extinct as the winner takes the spoils. Other times the battle forces the groups to diverge in their characteristics over generations so they eventually no longer compete for the same resources. This latter possibility, known as *character displacement*, may turn out to be the first step of a speciation event, the evolutionary creation of a new species. Could a similar process be occurring now in the conceptual environment as varieties of creationism struggle for dominance? We shall have to watch and see what happens. Think of this next section as a field guide to the varieties of creationism that are locked in battle not just with evolution, but also with each other.

Factions within the Tower

Terminological Issues

How shall we name and characterize the new varieties of creationism? Terminological issues will turn out to be very important as we sort through the creationism controversy. This will be especially clear once we get into questions dealing with the philosophy of science, for words that have one meaning in everyday circumstances are often given special definitions in philosophical and scientific contexts; but the point holds generally. If we are to try to understand the way people think who do not share our views, we must be careful not to assume that we mean the same thing even when we are using the very same terms. Often in this controversy, the disputants simply talk past one another because they are unaware that they are using the same words in quite different ways. This can easily happen, for instance, when a nonscientist misinterprets a scientific term by thinking of it as it is used in ordinary speech rather than its technical definition, but it can also happen when one is thinking about unfamiliar theological viewpoints. This problem will make it difficult to keep a steady eye on what is going on in the internal creationist battles, for the creationist factions find serious theological differences among themselves that outsiders may find inconsequential or even indistinguishable. As far I can, I will try to describe people's views using the terms as they themselves use them, but once we begin to compare different views this will sometimes become difficult. Readers should stay alert to the changing meanings of terms when we shift from one perspective to another, especially when one or another side is being quoted.

Related to the problem of definition is another important terminological problem to watch out for. Once people do finally realize that disputants are using words in different ways they may simply judge that the conflict is "just a semantic issue." That is, they may conclude that there is nothing going on but a quarrel about the meaning of words, and that if all we are arguing about is whose definitions will be used, then the dispute really is without substance. One of the most difficult lessons that new students of philosophy must learn is how to move past pointless semantic quarrels about words and into the deep and substantive arguments about concepts. As we will see, several of the points of dispute in the creationism controversy are indeed little more than terminological differences, but many more have to do with very basic concepts such as God, Christianity, evidence, truth, and scientific knowledge.

With these warnings about terminological pitfalls, I now turn to characterizing the new creationists.

Fundamentalists, Evangelicals, and Biblical Inerrancy

It is not quite correct to say that creationists are Fundamentalists, for within the spectrum of their religious views this term has a special theological meaning that is narrower than the way it is used in everyday contexts. Many creationists explicitly disavow the Fundamentalist label and describe themselves rather as Evangelicals. This sometimes reflects simply a denominational division—affiliation with the National Association of Evangelicals rather than with the American Council of Christian Churches. More often it reflects a difference of theological emphasis, with Fundamentalists being more likely to refer exclusively to the King James version of the Bible, to hold to the imminence of Christ's Second Coming, and to separate themselves from the modern world to prepare for the "rapture" of true believers into heaven.[7]

For our purposes, the most important difference is in their stances on biblical hermeneutics. Evangelicals are more comfortable acknowledging the difficulties of interpretation, provided that one retains faith in "biblical inerrancy." This is the view that the Bible, as the revealed word of God, is itself without error in its original writings. However, not all Evangelicals believe that each and every word of the Bible comes directly from God as though the biblical writers were just taking dictation, a view that Matthew Arnold, the nineteenth-century essayist, described as "divine ventriloquism." While the most common hermeneutic among Evangelicals is *plenary verbal inspiration* (which holds that every part of the Bible is equally inspired, and that biblical writers were directed by God in their choice of subject matter and words, though they used their own style), a smaller percentage would hold the weaker view of *inspired concepts* (whereby God gave the thoughts to the writers, but permitted them, perhaps years later, to express these in their own words as they remembered them).[8] While the inerrancy view does favor a literal reading, it recognizes possible vagaries of translation and allows that occasionally biblical language may be figurative, as in the parables. From without, these small differences of degree may seem inconsequential, but from within they are highly significant, thus setting up an environmental difference that profoundly affects the evolution of the creationism meme.

As we lean in closer to examine the details of creationist beliefs, it will be easy to lose sight of the fact that their views stand as but fringe positions in the vast woods of Judeo-Christian theology (let alone the broad forests of religious belief generally). So, before we begin to look at the ways in which creationists see their antievolutionary positions as arising out of the doctrine of biblical inerrancy, we must keep in mind that the majority of Christian theologians see no conflict between their faith and the findings of evolutionary biology. Even holding that the Bible is inerrant does not require that we think of it as giving a *plainly* literal account, especially with regard to scientific matters. Indeed, some argue that faith in biblical inerrancy *requires* that we not use a literalist hermeneutic, because taking all biblical statements at face value leads to dozens of explicit internal self-contradictions. Episcopal Bishop John Shelby Spong is one of many who have developed this argument. Spong is highly critical of biblical literalists, charging that by flying in the face of modern science their view is "destroying Christianity."[9] Few theologians are as critical as Spong, but most are adamant that Christians should understand that believing in the Christian God and taking the Bible seriously does not mean that one must accept the creationists' antievolutionary view. Many offer compatibilist theological views, holding that it is perfectly reasonable to be a theist and an evolutionist. With that final reminder, let us peer into the Tower.

Young-Earth Creationism

Even creationists disagree about precisely how to understand the notion of biblical inerrancy, but the dominant view among them is that the Bible is meant to be read literally not only on matters of faith and spirit but also on all matters about the physical world that are mentioned. The creation of the world took six twenty-four hour days. God did not create human beings using physical, evolutionary processes, but formed Adam directly from the dust of the ground and Eve from Adam's rib. Jonah really was swallowed and lived for several days in the belly of a great fish. Methuselah in fact lived nine-hundred years.

As we will see, creationists make a wide range of claims about matters of physical fact, and for them, these are inextricably tied to strong theological commitments. In their public activism, however, they try to disguise the theology and also to limit the physical claims to a smaller core set. In the Arkansas balanced-treatment act they pared down their explicit commitments to six theses, and these will serve as a useful initial characterization.

(1) Sudden creation of the universe, energy, and life from nothing; (2) The insufficiency of mutation and natural selection in bringing about the development of all living kinds from a single organism; (3) Changes only within fixed limits of originally created kinds of plants and animals; (4) Separate ancestry of humans and apes; (5) Explanation of the earth's geology by catastrophism, including the occurrence of a worldwide flood; and (6) A relatively recent inception of the earth and living kinds.[10]

It is this last thesis especially that has come to characterize creationism in most people's minds. Though the vague wording obscures the content of the specific thesis, most people are aware that biblical literalists calculate the date of "inception" of the world by reference to the generations and chronology found in Genesis.

The original calculation that the world is about 6,000 years old was made in the seventeenth century by James Ussher, Irish Archbishop of Armagh, and it was refined a few years later by Cambridge University's John Lightfoot who concluded that Creation took place on October 18, 4004 B.C. Adam was created on October 23, at 9 a.m. Of Lightfoot's estimate, historian E. T. Brewster wryly commented: "Closer than this, as a cautious scholar, the Vice Chancellor of Cambridge University did not venture to commit himself."[11] Such misplaced precision seems amusing now but at that time their conclusion was taken as authoritative and was included as a margin note or sometimes as a section heading in early editions of the King James Bible.

Current literalist creationists usually retain the 6,000 year figure but accept that there may be a margin of error in this interpretation of the Genesis chronology, and so they concede that the universe could perhaps be up to 10,000 years old. Though it is conceptually possible to endorse thesis (6) and reject the previous five points, in fact no creationist holds such a view. In the standard parlance, a "young-earth creationist" is someone who endorses all six theses. Sometimes these creationists refer to themselves as "recent creationists" (to emphasize that they believe Creation happened recently, not that they just recently became creationists) or, because the calculation refers to the date of Creation of *everything ex nihilo* and not just the creation of the earth, as "young-*universe* creationists." In the telegraphed language of the Internet, disputants commonly use the acronym "YEC" as a convenient label for this view (for example, "As a YEC, I believe that. . ."). It would actually be helpful to be able to keep "YE creationism" and "YU creationism" distinct, and to use the latter term for the current dominant view described above and to drop "recent creationism" as ambiguous, but I will follow the well-established convention and use them all interchangeably.

Today, the young-earth view is advanced by a veritable army of activist creationist groups. On the front lines and still in command of most of the Tower is the venerable Institute for Creation Research. Headquartered on the outskirts of San Diego in Santee, California, ICR describes itself in its publications catalog as engaging in a wide variety of "activities and ministries, all promoting the truths of scientific creationism and inerrant biblical authority in all fields of study and in all areas of life." As a publishing venture, ICR's output is prodigious; its catalog includes over a hundred and fifty titles, and its free monthly periodical "Acts and Facts" and quarterly devotional Bible study booklet "Days of Praise" are sent to hundreds of thousands of recipients on their mailing list. Because ICR is the largest, one of the oldest, and still by far the most influential of all the creationist organizations, I will be looking at its official views in detail later; here let me mention several of the less well-known young-earth groups.

The Answers in Genesis (AIG) organization with headquarters in Florence, Kentucky, was founded by Ken Ham and Dr. Gary Parker. Describing itself as "a nonprofit, nondenominational evangelical organization dedicated to the urgent task of spreading the creation message," AIG holds seminars and publishes the *Creation Technical Journal* which includes papers by "leading creation scientists" and *Creation ex nihilo*, a colorful magazine that is aimed at the whole family. One recent article considered the question "Did Adam Have a Belly-button?" The Bible-Science Association, Inc. (BSA), recently renamed Creation Moments, is based in Minneapolis, Minnesota and publishes *Bible-Science News*. According to its promotional materials, BSA "[s]eeks to analyze the arguments used in various science disciplines to support origins viewpoints, especially those arguments which are destructive to the Christian world view." In Ashland, Ohio is the Creation Research Society (CRS) that publishes the journal *Creation Research Society Quarterly*, and was founded to give a "complete re-evaluation of science from the theistic viewpoint." The United States is the home to by far the largest number of creationist groups and creationism for the most part remains a distinctly American phenomenon. However, one does find a growing number of creationists in other countries, including Australia and New Zealand, as the Fundamentalist movement continues to spread beyond its historic roots in the rural American Bible Belt. One of the most well-established overseas groups is the Biblical Creation Society (BCS) in Great Britain. It used to publish the journal *Biblical Creation*, but in 1987, beginning a trend, BCS changed the name of its periodical to the less obviously religious-sounding *Origins*. Among its articles one finds titles that range from "Where was Eden?" to "Dinosaurs—Designer Made?" There are many other smaller organizations that promote young-earth creationism.[12] Several defer to the ICR, but others promote some unique variation of the young-earth view. A brief description of one of these will give a flavor of this sort of subfactionalism.

The Center for Scientific Creation (CSC), located in Phoenix, Arizona, was founded by Air Force colonel Dr. Walt Brown, upon his retirement from the service. On its web site, from which the following quotations are taken,[13] CSC describes itself as a

non-profit organization . . . dedicated to the research of the origins of humankind as well as the questions listed below. Have you ever wondered about the evidence for creation and against evolution? Have you ever considered the possibility that the earth is less than 10,000 years old? Was there a worldwide flood during Noah's lifetime? Where did the water come from? Where did it go?

Brown, whose doctorate was in mechanical engineering, travels around the country giving seminars on creationism, and he wrote a book—*In the Beginning: Compelling Evidence for Creation and the Flood*[14]—that purports to answer these and dozens of other questions that he says baffle scientists. Most of the questions involve the issue of a global flood and illustrate the importance of thesis (5) for YECs. Brown is typical in wanting to explain all major terrestrial features (from the Grand Canyon and ocean trenches to major mountain ranges and the jigsaw fit of the continents) by reference to the catastrophic Noachian flood, but his special contribution to this topic is his unique "hydroplate theory," which he claims also explains how the continental plates move.

Perhaps the most perplexing question in the earth sciences today is barely verbalized in classrooms and textbooks: "What force moves plates over the globe and by what mechanism?" What is the energy source? The hydroplate theory gives a surprisingly simple answer. It involves gravity, the Mid-Atlantic Ridge, and water—lots of it.

Geologists would be very surprised to hear that they are perplexed by this question. Geology students cramming for exams and trying to memorize the textbook illustrations and the many details of the mechanisms of plate tectonics from their professors' lectures would certainly complain that Brown's statement does not match their experience. The way creationists handle continental drift and plate tectonics not only illustrates the great rift that separates their approach from that of scientists, but also reveals a few of the fault lines that divide their own ranks. Some creationists want to claim continental drift as a point in favor of the predictive ability and truth of "Biblical Science," claiming that it was already recognized and described in the Old Testament. In particular, they cite Genesis 10:25, which says "In the days of Peleg the earth was divided." Others, such as Wayne Frair (whose research specialty is herpetology) and Percival Davis argue that creationists must reject continental drift because it would undermine the recent chronology.

If the usual geological time scale is accepted, continental drift would have occurred at a rate of inches per year, which is reasonable. But if the much shorter chronology consistent with biblical relevation [sic] is accepted, the rate would have had to be many miles per year to produce the present location of the continents.[15]

For YECs like Davis and Frair the brief chronology that they believe is divinely revealed in the Bible takes precedence over the wealth of geological evidence supporting continental drift and the standard geologic time-scale. Brown's hydroplate view also tries to avoid this purported conflict with Genesis by rejecting the scientific account and imagining the continents breaking apart and zipping, like surfers who have caught the curl of a monster wave, to their present locations on the subsumed waters of Noah's Flood.

As we shall see, Brown's hydroplate theory is just one of many telling examples of how the creationist attack is aimed not just at evolution, but at science on many different fronts. We will look in some detail at the volleys of arguments that creationists hurl down upon the field of science from their Tower. These rain most incessantly upon evolutionary biology and geology, but they also fall upon chemistry, astronomy, physics and beyond. To hold that the world was created at most ten thousand years ago requires that YECs reject most of the scientific chronology—and everything that goes with it. The common public perception that the creationist controversy is just a fight about evolution seriously underestimates the scope of their attacks. However, to really understand creationists one must pay attention not just to the way they attack biology and the other sciences but also to the way they wage their internal battles. Young-earth creationism is no longer securely in control of the Tower. Although it remains by far the dominant form, the young-earth view is showing a few signs of weakening as some evangelical Christians promote competing creationist positions.

Old-Earth Creationism

The most obvious fault line dividing creationists has to do with the age of the earth. Opposing the young-earth camp are the old-earth creationists (OECs) who understand that the geological record simply cannot be squared with a recent creation. The following quotation from Fred Heeren gives a sense of how the line between the views is drawn.

Some of the great Bible teachers who I respect most happen to be recent Creationists, though I disagree with their position on creation's timing. I just don't think they're aware of all the facts. Science isn't their field. . . .The *fact* of creation is critical, but God's timing is not something to break fellowship over.[16]

Heeren supports the scientific Big Bang theory and the view that the universe is around fifteen billion years old. Heeren explains that he used to be a YEC until he looked at the evidence for the Big Bang and saw that it proved that there was a beginning to the universe, and that the fine-tuning of the laws of the universe was a good argument for God. One of the main points of commonality among creationists is a shared desire that science be seen as providing scientific evidence of the Creator as depicted in the Bible. Young-earth creation-scientists do this by arguing that science supports the revealed claims of plain Scripture, and thereby proves that the Bible had it exactly right all along. Heeren, it seems, was able to break out of his YEC worldview when he decided that with but the small price of a change in interpretation, he was able to reconcile Scripture with the scientific age of the earth and get what he took to be a passel of new scientific support for the biblical Creator in the bargain. Despite having accepted this part of the scientific picture, he remains opposed to evolution.

Old-earth creationists typically hold fast to the biblical inerrancy thesis, and many still think of themselves as Fundamentalists. Heeren is representative of OECs in this way; in general, they too would prefer something close to a literalist reading of Scripture, though there are disagreements about what this amounts to when confronted with specific passages. So how do OECs reconcile Genesis and an ancient earth? There is no simple answer, for as one looks closer one finds more fault lines, and factions within the factions, each with its own interpretation of the text. This is where the story really begins to get interesting.

Probably the largest old-earth faction holds that the days of creation are not to be thought of as twenty-four hour human-sized days, but rather as God-sized days. Each "day" is like an age on a human scale, and each may have been millions or even billions of years long. For theological justification this "day-age interpretation" may appeal to God's omnipresence. We sometimes hear that God exists "simultaneously" at all points of space and at all times throughout eternity, or that that space and time exist "within" God, though such locutions are metaphorical since God is not taken to be part of the physical space-time continuum. In any case, because God exists outside of time as we know and experience it, we should not expect that our ordinary temporal notions could even apply to God's creative action. On this reading, Genesis speaks of days in the broad sense that we might speak of something happening "in Moses' day." Furthermore, proponents point out that this understanding of "day," though not strictly literal, is not unusual and clearly occurs in other sections of the Bible. On this reading we need not think that the six "days" of Creation would be short or even that they be the same length as each other. Perhaps they could even overlap one another. This sort of flexibility gives the day-age advocates plenty of room to accommodate as extended a geological chronology as might be needed to avoid a conflict with the scientific view.

In fact, the day-age view has a much longer history than the YEC view described above, which originated as part of the "Flood Geology" of George McCready Price, who first developed the notion in his 1902 book *Outlines of Modern Christianity and Modern Science*.[17] Price was a tireless crusader, but his view initially attracted little support outside his own Seventh-day Adventist sect. Before the young-earth creationists rose to power in the 1960s, bringing Price's marginal view for the first time to the forefront of the theological dispute in the Tower, the day-age interpretation was by far the predominant creationist view. Moreover, its advocates point out, this interpretation of Scripture was not concocted as an accommodationist concession to Darwin but may be traced back to the earliest Jewish and Christian thinkers; Augustine, Philo, Josephus, Irenaeus, Justin Martyr, Hippolytus, Origen, and Eusebius are among those frequently cited. Today the day-age view is most thoroughly defended by Hugh Ross, an astronomer who founded the "Reasons to Believe" ministry in 1986, and is endorsed by many others, such as Davis Young and many other members of the American Scientific Affiliation (ASA). The ASA was founded in 1941 to "promote and encourage the study of the relationship between the facts of science and the Holy Scriptures," and its membership originally included both young-earth and old-earth creationists. However, over the course of its first two decades, ASA members who worked in fields related to radioactive dating and who were not tied to a literalist hermeneutic were able to shift the view of the organization against young-earth Flood Geology. The Creation Research Society was founded by unyielding young-earthers who split from the ASA, believing that it had abandoned the doctrines upon which it had been founded.

A second common way that Christians have harmonized a literal reading of Genesis with the scientific picture of an ancient earth is known as the "gap interpretation." According to this reading of the "gappists," there is a time-gap of indeterminate length between Genesis 1:1 and 1:2. All of geological history including the ages of the dinosaurs may be fit into these "pre-Edenic" times. This view is also known as the "ruin and restoration" interpretation, because it holds that God destroyed this original ancient world in a great cataclysm, perhaps as a judgment on Lucifer's rebellion, and then re-created the world in six literal twenty-four hour days as Genesis describes. C. I. Scofield was one important conservative theologian who promoted the gap interpretation in his *Reference Bible* writing: "The first creative act refers to the dateless past, and gives scope for all the geologic ages."[18]

Until the rise to dominance of the young-earth creationists beginning in the 1960s, the day-age and gap interpretations represented the standard creationist understanding of Genesis. In his masterful history of scientific creationism, *The Creationists,* historian Ronald Numbers shows how these were the commonly accepted creationist views in the previous century and through the first half of this century. He points out that William Jennings Bryan, whose defense of a literalist interpretation of Genesis at the Scopes trial made him an easy target for Darrow's ridicule, was no young-earth creationist but rather believed in an ancient earth under a day-age interpretation.[19] Though many old-earth creationists today continue to support one or another of these two interpretations, in their recent debates on the Internet and elsewhere some have begun to promote several other ways that one may understand the Genesis account so that it does not conflict with the scientific view of the age of the world.

One newly popular alternative, though with a history that goes back to the 1850s, is the "revelatory" or "visionary day interpretation." The description of the Creation in Genesis is from the point of view of some observer, but of course no one but God was present to have seen the formation of the universe *ex nihilo*. So how was Moses able to narrate this history? Since the Bible is taken to be directly inspired by God, the story goes, it is not unreasonable to suppose that God showed visions of the Creation to Moses. Certainly the descriptions of Creation in Genesis cannot be from God's perspective, given that an omnipresent God would know things "all at once" and from all points of view rather than seeing them in a temporal sequence from a particular point of view. Thus, the six days of Genesis 1 do not refer to the period of God's actual Creation, which may have begun long ago and continued for eons, but to days during which God revealed aspects of the Creation to Moses so that he could act as a virtual observer and witness.

Another recently promoted view holds that the days of Genesis 1 are actual days, but that they are not consecutive. Rather, they are days, separated perhaps by millions of years, on which God *proclaimed* the next phase of creative activity. This "days of proclamation interpretation" is defended by Glenn Morton in his *Foundation, Fall and Flood*.[20] Morton is a geologist who had been a young-earth creationist until the overwhelming evidence for an old earth he encountered in his work in petroleum geology forced him to reject that view and to make an agonizing reassessment of his Christian faith. He reports that this new interpretation has allowed many other YECs to become open to the scientific view.

The Battle between YECs and OECs

The young-earth and old-earth creationist factions are quite distinct. Though both take up arms against biology, holding that evolution must be false, they rarely are found on the field together because they are simultaneously sniping at one another in a battle for power within the creationist Tower. Heeren may plead that since both sides agree on the fact of Creation, the differences about timing are not worth breaking fellowship over; but in the end his view depends upon linking the details of Big Bang cosmogony to his preferred interpretation of the biblical account of God's creation of the world, so he is as adamant about rejecting the young-earth view as are scientists. For their part, YECs are unable to reach a truce with the old-earth camp's acceptance of the scientific account of the evolution of the universe for the same reason they cannot come to terms with the scientific account of the evolution of organisms, because they believe the OEC view to be in direct opposition to scriptural chronology as well as to central matters of theology.

In their literature and presentations, each faction usually just ignores other positions and acts as though theirs is the only Christian view, and concentrates upon attacking the "anti-Christian" evolutionary view. For instance, in his book *Creation Scientists Answer Their Critics* (1993), ICR's veteran debater Duane Gish responds to scientists like Stephen Jay Gould, Ilya Prigogine, and Niles Eldredge, and philosophers of science such as Philip Kitcher and Michael Ruse whose work has rebutted his arguments, but he does not mention proponents of opposing forms of creationism who also have criticized his young-earth position, let alone other Christian views that criticize the project of creation-science generally. Furthermore, his bibliography of creation-science literature sticks almost exclusively to explicit proponents of the young-earth view (heavily weighted to works by fellow members of ICR) and omits any reference to Ross, Heeren, or other old-earth creationists. In this manner the reader may be led to believe that the argument is simply with biology and that there is just one correct Christian view. Phillip Johnson, who tries to rede-fine creationism in his own generic terms, never even mentions young-earth creationists in most of his writings and presentations. On the Internet, web sites devoted to one or another creationist viewpoint may provide links to pages of like-minded creationists, but rarely provide links to rivals' sites.

When different factions do confront one another, each side quotes its favored biblical passages and marshals its exegetical forces to show that its view is the correct Christian position, and attacks the opposing side as being ignorant, misguided or damaging to the Faith. The dispute between the young-earth and old-earth camps hinges on how one should under-stand the Hebrew term that is translated as "day" in Genesis. This is an ancient debate. St. Augustine set an early precedent for the old-earth view's figurative interpretation when he weighed in on the issue in the fifth cen-tury, pointing out in *De Genesi ad Litteram* that these could not be ordi-nary solar days if only because Genesis itself tells us that the sun was not made until the fourth "day." For their part, young-earthers can cite the opinion of Martin Luther, who took issue with Augustine on this point in his *Lectures on Genesis*, writing: "So far as this opinion of Augustine is concerned, we assert that Moses spoke in the literal sense, not allegorically or figuratively, i.e., that the world, with all its creatures, was created within six days, as the words read."[21] As one might imagine, to come down on one side or another of this debate requires that one make substantive decisions about how to approach scriptural interpretation, and such issues of herme-neutics quickly lead the disputants into deep theological waters.

Hugh Ross defends the day-age interpretation in great detail in his 1994 book *Creation and Time: A Biblical and Scientific Perspective on the Creation-Date Controversy*. Ross takes special care to trace its theological history and to try to show that its acceptance does not imperil Christian faith. He is distressed that young-earthers put so much weight on the relatively minor issue of the date of Creation and pit their naïve interpretation of words of the Bible against the facts of nature. As an astronomer, Ross is well aware of the strong evidence for the scientific view of an ancient universe, and he, like Heeren, is much more impressed by the way that the scientific cosmological theory supposedly meshes with the distinctly Christian story of Creation. To get this supposed theological payoff, however, he and other OECs have to try to show why the YEC interpretation of Scripture is doubtful.

Old-earthers often begin by pointing out that Ussher's and Lightfoot's calculations were based upon an erroneous reading of the biblical genealogies, since they did not realize that the Hebrew words for father (*'ab*) and son (*ben*) can also mean forefather and descendant. Furthermore, they failed to recognize that Scripture occasionally "telescopes" the genealogies to emphasize the more important ancestors. Of course, even allowing for this sort of error will not get one very far in making up the billions of years of geologic history, so the argument quickly returns to the interpretation of *yom*. OECs admit that *yom* may mean a twenty-four hour solar day, but they argue that it may also refer to an indefinitely long period of time. It is in that sense that the term appears in Genesis 2:4, speaking of the "day" that all of God's creative activity took place in the course of the "days" of creation. Some translations render this sense using the English "when." For example, the New International Version uses the translation "when they were created," rather than the more transliteral "in the day that the Lord God made" as the King James version renders it. Furthermore, OECs note that in the Hebrew, Genesis 1 omits the definite article before each of the creation days (i.e., not "the first day" but "day one") and that in Hebrew prose the definite article could only be omitted in poetic style, which indicates that the use in Genesis is figurative rather than literal.

One might think that it makes more sense that God rested after billions of years of creative labor than after a mere six ordinary days, but then again it is odd that an omnipotent being would have to rest in either case.

Rather than give arguments from God's limitations, old-earth creationists focus on Adam's. They take a line advocated by Gleason Archer, an Old Testament scholar, who wrote that "it would seem to border on sheer irrationality to insist that all of Adam's experiences in Genesis 2:15–22 could have been crowded into the last hour or two of a literal twenty-four-hour day."[22] In particular, could Adam really have named all the animals that God brought to show him in that short time, as Genesis tells us? Old-earther Walter Bradley, responding to a recent creationist's challenge about the proper interpretation of *yom* following one of his talks, ridiculed this implication of the twenty-four hour view, wondering if somehow the animals could have galloped and zoomed past Adam at top speed in the few hours that would have been open in his schedule that day, giving him perhaps a fraction of a second or so to name each.

At this point the interpretive contest becomes increasingly arcane. Morris responds, first, that Adam could have completed the task because he was much more intelligent at that time just after his creation than we are today. This is because human beings supposedly became stupider after the Fall, the perfect state of their minds having been darkened by sin. Ross admits the loss of mental and physical perfection after the Fall, but says that in Adam's earlier perfect state he would have been all the more "meticulous"[23] in performing the naming task, and so would have taken care to observe even those animals unique to distant continents, which would have actually slowed him down all the more, so the task might have taken years. Van Bebber and Taylor question Ross's scenario on the grounds that "the kind of God we know" would not have given Adam a task that would have required him to be alone for so long a time.[24]

As a second reply, YECs point out that Genesis does not say that Adam named all the animals, but only all of those kinds included among the livestock, the birds of the air, and the beasts of the field, which is a smaller group. Be that as it may, estimating the number of these kinds from information in the Bible is more difficult even than estimating the age of the earth (though perhaps a bit easier than figuring out the number of angels that could dance on the head of a pin). The calculation is complicated by the fact that the created "kinds" may not correspond to the contemporary notion of species. Ross assumes there were many thousands of kinds for Adam to name. Archer had thought Adam's task impossible even though he estimated Adam had to name only hundreds of kinds of animals. Henry Morris estimated perhaps 3,000 kinds, but Van Bebber and Taylor come to his defense, arguing that "all" may not mean "each and every" but could refer to "the collective nature of God's creation."[25] Lest one think that this leaves us with an even trade of a literal reading of "day" for a figurative reading of "all," they suggest that the actual number of animals that Adam had to name was really just that number needed for him to realize God's purpose, which they say was to "get the point" that he was alone and that there was no helper among the animals for him. They think it "probable" that he could have realized this in a matter of hours.

Mark Van Bebber and Paul Taylor, who are both directors of the young-earth creationist company Films for Christ, have written a book with the same main title as Ross's book, *Creation and Time*, that attacks Ross's brand of old-earth creationism point for point. They say they believe that Ross is a saved Christian, and they praise his opposition to evolution, and his desire to evangelize, as well as "his ability to remain relatively cool and self-controlled under pressure."[26] Nevertheless, they believe that his old-earth teachings "are leading people down a wrong and dangerous path— a trail trod by many in the past that has repeatedly led ultimately to even more serious theological problems and loss of faith in God's word."[27] In this they sound a warning about the slippery slope to atheism which we will hear echoed again and again by creationists of all stripes: If you can't trust that Genesis is literally true, then how can you trust the rest of the Bible?

Furthermore, the age of the earth is not a minor or peripheral doctrinal point, they argue, because: "Ultimately, all biblical doctrines of theology are based directly or indirectly on the book of Genesis. If the foundation (Genesis) is damaged, the structure (Christianity) is in peril."[28] Creationists of all sorts are sympathetic to this general argument, but YECs have a particular theological interpretation of Genesis and its relation to Christ's message in the New Testament that makes the age of the earth (and other points about Creation) not open to compromise. It would be a fascinating excursion to look into this doctrinal debate but this would take us too far afield, so here I just mention the most important among these theological points, which involves the issue of death. Van Bebber and Taylor say that

Ross's old-earth creationism is in the same boat as evolution in that it accepts physical death and suffering for long ages before Adam's sin. Death before the Fall, they say, is not only clearly contradicted by the Genesis account of Creation, but to accept it would be to obviate the need for Christ's sacrifice. Jesus' crucifixion atoned for the sin that Adam brought into the world for which death was the punishment, and it is through Christ alone that one may be saved from sin and thus from death. If there was death and suffering before sin entered the world when Adam disobeyed God's command not to eat the fruit of the tree of knowledge, then this would undermine what they take to be the whole message of the Bible and the very meaning of Christ's own death and resurrection. YECs believe that to accept an old-earth view, be it creationist or evolutionist, is tantamount to digging the grave of Christianity.

But what about the scientific evidence that supports an ancient universe? Ross claims in his writings that he is only showing how the Bible fits with the facts of nature. Van Bebber and Taylor attack Ross with the same charge creationists make against evolution, namely, that what he calls the "facts" of cosmogony and cosmological evolution are no more than theories, hypotheses, and assumptions.

It is apparent that Dr. Ross's teachings concerning the length of the Creation Week and small extent of the Flood has little to do with real study of the Hebrew syntax or literary genre. He came to his conclusions based upon his assumptions about scientific "fact." His arguments do not change the Bible's statements.[29]

Citing Calvin as their authority, Van Bebber and Taylor state the standard creationist view that our understanding of the world must be based upon Scripture and that we must interpret the world through the spectacles of God's special revelation. If there is a conflict with secular science then secular science must give way to God's revealed truth. Ross, they say, has the relationship backwards.

Dr. Ross clearly believes that we must put on the eyeglasses of modern science if we are to fully understand the Bible. Dr. Ross's view . . . stands in complete and total opposition to that of John Calvin, one of the most influential theologians of all time. . . . Sound interpretations of the Bible (built upon a literal, historical, grammatical hermeneutic) should not be invalidated by the ever-changing whims of secular origins science. . . . Due to human depravity, the "facts" of science can be misrepresented; even the experimentation of man can be twisted to sinful ends. True scientific facts will always complement God's Word.[30]

But of course, the scientific facts often do not fit the plain reading of the Bible. Scientists looking at the empirical evidence find that it clearly supports an ancient evolving universe and an evolving biological world, in contradiction to the YEC view on both these points. Ross argues that we may learn from God's Creation directly, and YECs agree but only up to a point. When push comes to shove on the testimony of the book of nature versus the testimony of the Book itself, YECs say that we must follow the latter. They will then try to show why the physical evidence is doubtful, or, if finally forced to it, they may adopt the solution proposed by Henry Philip Gosse in 1857 in his *Omphalos*. In this work Gosse introduced what is known as the "appearance of age" thesis, namely, that the universe is as recent a creation as Genesis tells us it is, but that God created it so that it *looked old*. Embedded in deep layers of rock we find what appear to be ancient fossils of long-extinct creatures, but that is because God placed them there when he laid down those beds six-thousand years ago. We see light from stars millions of light-years away because God neatly created it already in transit. God simply created a "mature" earth. YECs who accept this view are called "mature-earth creationists."

Of course, it is only fair to acknowledge that many people find it troubling to think that God would deceive us by creating the world in such a way that the physical evidence disguised the truth. Mature-earth creationists regularly have to counter the worry that their view makes God a deceiver. In the end the issue perhaps turns on the question of whether or not Adam had a belly-button. Since Adam was not "born of woman" but was formed directly from the dust he would not have had an umbilical cord. But would it have been deceptive of God to nevertheless form Adam with its mark? Gosse had considered this an important enough issue to refer to it in naming his book; *Omphalos* means "navel." Creationists stand ready to debate this central point in the controversy. But let us not delve into it any deeper. I here leave the issue of the navel for the reader to contemplate.

I have discussed the conflict between the YECs and the OECs in considerable detail . . . perhaps in greater detail than many readers would have wished. Viewed from the sidelines, the hermeneutical battle between the YECs and the OECs may appear to many to be like the Great Butter Battle described by Dr. Seuss. Does it really matter so much on which side of the bread they spread the theological butter? To someone not engaged in the battle, the outcome may seem inconsequential; but I believe that it has been worth laying out the lines of this one aspect of the struggle for power in the Tower with some care, to illustrate a few important points.

The first lesson to be learned is that creationism is not a single conceptual species but has significant distinguishable varieties and subvarieties. Recognizing this will alert us to watch out for differences when observing and confronting creationists in the wild. Many scientists still seem to think that all creationists are young-earthers and this leaves them unprepared to respond to new threats from unfamiliar quarters.

A related second point is that, from the creationists' own point of view, it makes a great deal of difference indeed which side of the bread gets the butter. Holding to the proper theology—the True Christianity—is absolutely essential. Everything else is to be judged by the orthodox standard, and seemingly minor points of difference may be just those defining beliefs that differentiate the right from the wrong. Creationism is a pure form of religious orthodoxy and, as such, it defines itself by those it excludes. As we will see in a later chapter, creationists portray their struggle with evolution as a battle for truth, purpose, morality, and the very survival of the Christian Faith. One important lesson we need to take from the review above is that creationists have narrow views of what constitutes the true Christianity . . . so narrow that they end up factionalizing and excluding even one another, not to mention the vast majority of other religious and nonreligious believers.

The third point I hope that the review illuminated is the way in which these theological commitments drive creationists' stand on scientific matters. In one sense this is a familiar point for it was always patently obvious that the young-earth "creation-science," even though stripped of explicit biblical references and advocated using "scientific" terminology, is based on and motivated by a literalist reading of the book of Genesis. What is new is the recognition of the several distinct creationist theologies and that these determine what part or parts of science the different factions are willing to accept.

Finally, I hope the reader has begun to get a sense of what creationists take to be evidence and how they go about evaluating it. Creationists do cite scientific evidence when it supports their view, especially in public forums, but among themselves they quickly return to theological arguments, because they take the Word of God to be the true source of all knowledge and the basis upon which empirical evidence must ultimately be judged. In her superb insider sociological study of Fundamentalism, Nancy Ammerman notes that for Fundamentalists, the Bible "provides not only practical and ethical guidance but also scientific information about the physical universe and the history of the world."[31] We noticed that evangelical OECs were a bit more lenient in their interpretation of Scripture, which allows them to be open to the scientific evidence for an ancient universe, but in the end they too believe that empirical evidence must pass theological muster. This last point about how we should evaluate empirical evidence will be of particular importance to our story for it is where philosophy of science enters the picture and where the new creationists are pressing the attack.

Having used the battle between the YECs and OECs to highlight these points, I will not continue to describe in detail the other divisions among the creationist factions in the Tower, though one could easily do this on other characteristic points, such as whether Noah's Flood was global or localized. Having looked in on one struggle, one can fairly well predict how these other skirmishes go. There are, however, a few other important parties, so let me introduce them briefly.

Progressive Creation vs. Theistic Evolutionism
The battle between the YECs and the OECs is representative of internal creationist struggles and its battle lines are probably the most clearly drawn since they involve a specific issue—the timing of Creation—and because the incompatibility of the scientific account with a plain reading of Scripture is so clear. Other divisions are fuzzier, especially between those positions that accept more and more of the scientific picture. I want now to mention two positions—progressive creationism (PC) and theistic evolutionism (TE)—that are especially important because their interface marks the edge of creationism.

Progressive creationism accepts much of the scientific picture of the development of the universe, assuming that *for the most part* it developed according to natural laws. However, especially with regard to life on earth,

PCs hold that God intervened supernaturally at strategic points along the way. On their view, Creation was not a single six-day event but occurred in stages over millions of years. George McCready Price dismissed this idea as a "burlesque"[32] of creation, and many YECs continue to hold this opinion of progressive creation. The PC view tends to overlap with other views, particularly with old-earth creationism. Hugh Ross is a progressive creationist and is attacked by YECs for that view as much as for his view regarding the age of the earth.

Just across the border from progressive creationism lies theistic evolutionism, which is the most common term used to describe theists who accept the scientific evidence for Darwinian evolution. I should point out that some people take TE to be the view that God directly guided the process of evolution, but this latter view seems more often to be known as "evolutionary creationism" (EC) and I shall keep these views distinct. (It is hard to see how to distinguish EC from PC, except that the latter has a longer history and seems to have a somewhat more specific set of commitments.) Another term that is used on occasion as a synonym for TE is "providential evolutionism" (PE), though it too seems occasionally to be used inconsistently. Because there do not yet seem to be commonly accepted definitions for PE and EC, for the most part I will avoid them. There are many people one could cite as TEs and several, such as Oxford University biologist and theologian Arthur Peacocke,[33] who have articulated the view in considerable detail, but I will briefly discuss just one— Howard van Till—because he seems to brush right up against the border.

Van Till is a scientist and a Christian of Calvinist and Kuyperian background, who assumes that the biblical doctrine of creation is essential to Christian faith, as he writes, "To know God as Redeemer, we must first know him as Creator."[34] This may make van Till sound like a creationist, but he holds that we must carefully distinguish two categories of questions about the material world. The first category has to do with its "internal affairs" and includes its properties, behavior, and history; and the second category has to do with its "external relationships" and includes its status, origin, governance, value, and purpose. Although we should look to the Bible to answer the second category of questions, the "appropriate source of answers for the Christian" to the former category is "the created cosmos itself, which is constituted and governed in such a way that it is amenable to empirical investigation and is intelligible to the human mind."[35] Opposing the creationist view while still accepting both God and science, van Till promotes what he calls a "creationomic" view. On the issue of evolution he concludes:

> I see no reason, either scientific or theological, to preclude the possibility that the temporal development of life-forms follows from the properties and behavior of matter. . . . I believe that the phenomenon of biological evolution, like any other material process, is the legitimate object of scientific investigation. . . . I would be terribly surprised to discover that we live in a universe that is only partially coherent, a universe in which the temporal development of numerous material systems proceeds in a causally continuous manner while the history of other systems is punctuated by arbitrary, discontinuous acts unrelated to the ordinary patterned behavior of matter.[36]

Because theistic evolutionists do not reject evolutionary theory, they may not enter the Tower itself, and even though they believe in God they regularly find themselves under attack by creationists. It is in the debates between creationists and evangelical TEs that the arguments become most vitriolic. Creationists accuse TEs of collaborating with the enemy, and TEs accuse creationists of giving Christianity a bad name; though both sides may begin with charitable intentions the debates often degenerate into a verbal slugfest. The animosity among the various factions is especially evident in on-line creationist discussion groups, where, to give just a few examples, TEs charge PCs with being "destructive to dialog," YECs say OECs are "accommodationists," and members of every faction regularly charge advocates of others with being "deceitful," "naïve," or "unchristian." One long-time discussant, frustrated with the "personal attacks and patronizing suggestions" he said he had endured, vowed that he would "not pull my punches any more." Though on-line messages appear with the sender's moniker, I'll leave these comments anonymous. One finds similar accusations leveled at one time or other from and toward members of all the various factions.

Intelligent-Design Creationism

The OECs' battle against the YECs and the other factional disputes are important developments, but such infighting is only one aspect of the evolution of the new creationism. Now we come to what may be the most significant recent development in the conceptual evolution of creationism.

A more powerful movement is gaining strength within the Tower and is beginning to take the lead in the battles against evolution in the field. This is the group of creationists that advocates "theistic science" and promotes what they call "intelligent-design theory." Creationism-watchers have called the advance guard of intelligent-design creationism (IDC) the "upper tier" of creationists because, unlike their earlier counterparts, they carry advanced degrees from major institutions, often hold positions in higher education, and are typically more knowledgeable, more articulate, and far more savvy.

There are a dozen or two names that appear most frequently in association with the ideas of intelligent-design and theistic science, but because this variation of creationism is still relatively new and its advocates have not all published or explicitly identified themselves under these labels it is not yet clear whom to list among its leaders. Walter Bradley, Jon Buell, William Lane Craig, Percival Davis, Michael Denton, Mark Hartwig, J. P. Moreland, Hugh Ross, and Charles B. Thaxton are important figures. Another is John Angus Campbell, a University of Memphis rhetorician, and he mentions Nancy Pearcey, Del Ratzsch, Tom Woodward, John Mark Reynolds, Walter ReMine, and Robert Koons (who is a colleague of mine in the philosophy department at The University of Texas at Austin), as being among the "key players" of "our movement."[37] Among the more well-known names to sign on to the crusade are Michael Behe (Lehigh University) and Dean Kenyon (San Francisco State University) on the scientific side, and Alvin Plantinga and Peter van Inwagen (both of Notre Dame) on the philosophical side. Perhaps more significant, however, are the younger members of the group—William Dembski, Paul Nelson, Stephen C. Meyer, and Jonathan Wells. These "four horsemen" have dedicated their lives to the creationist cause and have been collecting multiple graduate degrees (Dembski in mathematics, philosophy and theology; Meyer and Nelson in philosophy; and Wells in religious studies and molecular and cellular biology) so they will be fully armored and ready to ride forth. (Nelson, in particular, has an impressive creationist pedigree; he is the grandson of Byron C. Nelson, who helped organize one of the earliest creationist organizations—the Religion and Science Association—and who authored several classic YEC books.)[38] This young group has garnered significant financial backing, and in 1996 they transformed the creationist journal *Origins Research* into *Origins & Design*, which, as part of the Access Research Network, is now the official platform for intelligent-design creationism. We will see more of these warriors and others who wear their colors,[39] but there is one more new creationist champion to introduce.

The most influential new creationist and unofficial general of this elite force is Phillip Johnson, of the University of California at Berkeley. Johnson is neither a scientist nor a philosopher nor a theologian, but is a professor of criminal law at Berkeley Law School. He joins the ranks of other lawyers like Norman MacBeth[40] and Wendell R. Bird,[41] who have taken up the creationist banner over the decades. Johnson burst onto the field of battle in 1991 with the publication of his book *Darwin on Trial*, which he has followed up with more books, including *Reason in the Balance* (1995c) and *Defeating Darwinism* (1997), as well as a barrage of articles.

In most of his writings and speeches, Johnson tries to avoid making specific commitments on the points of contention that divide the main creationist camps. This is one of the identifying characteristics of intelligent-design creationists (IDCs). In articles by IDCs one never sees anything about the Great Flood, and the issue of Noah's Ark is avoided like the plague. As far as possible, they shun even mentioning the Book of Genesis or its interpretation. Usually, one has to look carefully to find a veiled reference, let alone a forthright statement, that indicates a specific stand on the age of the earth. Trying to offer a banner under which the different factions could unite, IDCs, taking a cue from C. S. Lewis's ecumenical notion of a simple "mere Christianity," at times promote the idea of "mere Creation."

According to Johnson, the key elements of the creationist view are that there is a (1) Personal Creator who (2) is supernatural, and who (3) initiated and (4) continues to control the process of creation (5) in furtherance of some end or purpose. This generic definition simply ignores all the factional disputes and sets forth a common minimal set of principles. I will look more closely at Johnson's characterization of creationism in chapter 4, but here let me just note a sixth common element that is implicit in this definition, namely, (6) a rejection of theistic evolution. Other IDCs make this final commitment more explicitly. In a published talk he gave at the

Princeton Apologetics Seminar, William Dembski drew the line clearly in the sand:

Design theorists are no friends of theistic evolution. As far as design theorists are concerned, theistic evolution is American evangelicalism's ill-conceived accommodation to Darwinism.[42] [Emphasis in original]

As we proceed, we will have to compare this minimal notion of creationism to the more specific views we saw in the earlier definition of creation-science, but for the most part this will be the basic sense of creationism that I will take for granted and that we will return to again and again. There is much more to be said about the views of Johnson and the IDCs; evaluating their position in comparison with the views of YECs and other creationists will be one of the main tasks of this book. For the remainder of this chapter, however, let me finish laying out some of the other ways in which creationism has been evolving.

New Fields of Battle

One of the features of biological evolution is that it often involves the expansion of territory. It was not too many years ago that the creationism meme was transmitted primarily orally and through the print media in pamphlets, a few periodicals, and books. In the last decade, however, it has spread throughout the electronic media beyond radio to television, videotapes, and now the Internet. One cannot help but note the wry irony of the fact that creationists have become very adept at exploiting the technological fruits of science to proclaim their antiscientific message.

Anyone who has tuned into public access television knows that Fundamentalist and Evangelical proselytizers are a regular ingredient of its eclectic populist stew, and that criticizing evolution and the evils it purportedly spawns is one of their favorite topics. Even more significant are the national networks—the Christian Broadcasting Network and the Trinity Broadcasting Network—that are devoted to bringing the "Good News" directly and with high production values to the televisions and radios in the nation's living rooms. The John Ankerberg Show is a notable case in point. It has been among the best of these programs, having successfully adapted the format of syndicated talk shows, with each show devoted to a single topic as Ankerberg moderates and poses questions to the guests on stage from his position among the audience members. Most shows focus upon establishing the falsity of other religions or the errors of non-Fundamentalist Christianity, or on the proper Bible-based stance to take on controversial social issues. Ankerberg has taken on evolution as one such issue on his show and has published a small companion book on the subject—*The Facts on Creation vs. Evolution* (1993)[43]—that viewers can buy who want to see more details about the falsity of evolution and how it contradicts Scripture. Ankerberg is also one of the contributors to *The Creation Hypothesis* (1994),[44] an anthology that presents intelligent-design creationism, though in that forum he and co-author John Weldon keep their young-earth view under wraps.[45] Most programs, like Ankerberg's, bring up evolution as just one of their range of topics, but the TV series *Origins* is devoted entirely to promoting creationism. The *Origins* studio set is a cartoon version of a scientist's office: floor to ceiling book shelves (but very few books), a chalkboard, a hanging anatomical chart such as one might find in a doctor's office, and a couple of stools around a lab bench upon which are placed a number of flasks and beakers filled with blue and red liquids. On each show, the host brings in a guest speaker, and they discuss one or another way in which the scientific view of origins purportedly conflicts with the divinely revealed truth about origins found in the Bible. (The term "origins," as a nonreligious synonym for "Genesis," is a common code word for creationists, and they use the term "origins science" as another synonym for creation-science.)

Videotaped productions for use in classrooms and by study groups are a recent growth industry for spreading the message. Walter Brown sells a videotape explaining his views, and Hugh Ross's group Reasons to Believe puts out videos including *Darwinism on Trial*, a recording of one of Phillip Johnson's early speeches delivered at the University of California at Irvine shortly after the publication of his first book. I was surprised to find a professional film crew taping all the proceedings at what I had expected to be a typical academic conference at Southern Methodist University in March 1992, which featured a face-off between Johnson and philosopher of science Michael Ruse as well as talks by many of the IDCs mentioned above. Johnson has spoken several times at my own university at the invitation of campus evangelical groups, who tape and then broadcast videos of his speeches on community access TV. The Institute for Creation Research leads them all, however, with a growing collection of videotape programs in its large publication list. One can learn of scientific creationism from Henry Morris himself in his seven-tape *Basic Creation Series*, take a trip with Steve Austin through the Grand Canyon to hear his creationist interpretation of its formation, or hear John Morris discuss special topics such as *Natural Selection Versus Supernatural Design*, or *Evolution and the Wages of Sin*, as he defends creationism against "the typical [public school] classroom dogmatic propaganda for evolution" in the Velikovskian-titled *When Two Worldviews Collide*.[46]

The most significant new medium that creationists have successfully exploited, however, is the Internet. All the major creationist organizations and a slew of smaller ones have slick web sites. IDCs have a variety of sites, most currently under the umbrella of the Access Research Network (formerly known as Students for Origins Research). As mentioned, their main promotional organ is the journal *Origins & Design* (formerly known as *Origins Research*), which is available both in print and on-line under the editorial leadership of Paul Nelson (formerly known as Peter Gordon). Old-earth creationist Hugh Ross's Reasons to Believe organization has a large site that includes, among other information, transcriptions of his regular radio programs. The Institute for Creation Research site still leads the pack for its impressive professional graphic design and its comprehensiveness. By the time this book is out, ICR should have completed its "distance learning" creation study course on the Internet that it hopes "will reach millions worldwide."[47] I will not list any more creationist home pages on the web; a simple search on the keyword *creationism* will quickly turn up dozens. It will also bring up a smaller number of pages of creationism-watchers and defenders of evolution, the most prominent of which are the sites of the National Center for Science Education (N.C.S.E.) and the talk.origins archive, a depository for evolutionist-selected information culled from the Usenet newsgroup of the same name.

The talk.origins newsgroup is where the battle between creationists and evolutionists rages in its most chaotic form. The Internet is the new public square, and talk.origins is the corner of the square where many creationists first come to challenge evolutionists for their territory. On the Internet newsgroups, regular participants are very aware of the varieties of creationist positions, and their posts are sprinkled with identifying acronyms and abbreviations—"SciCre," "OEC" and so on. Neophytes who leap into the fray without first lurking in the background long enough to pick up the lingo are quickly referred to the Frequently Asked Questions (FAQ) page of the talk.origins web site or to the handy on-line glossary compiled by Wesley Elsberry, a graduate student in marine biology who founded a similar newsgroup on Fidonet back in 1992 and who continues to participate in and document talk.origins. As is often the case on the electronic frontier, in the rambunctious free-for-all of the Internet newsgroups the arguments often degenerate into "flame wars." This sort of invective does not bear repeating, but I mention just one mild bit of name-calling that has become part of the ongoing culture of the newsgroup. A few years ago a creationist on talk.origins derisively referred to the evolutionist contingent as "a tree full of howler monkeys." For their part, the evolutionary regulars on the newsgroup gleefully leapt on the phrase and adopted the moniker as a badge of honor. They still at times speak of themselves as "Howlers," as when some of them gathered at the Museum of Natural History in New York for a "Howlerfest" to meet one another in person and to view some of the "nonexistent evidence" for evolutionary descent.

If the Usenet newsgroups form the rowdy open town square, mailing lists are the just slightly more sedate saloons and clubs. To try to understand what was going on in the creationist Tower, for a couple of years I listened in on one creationist list, based in Calvin College, known as the "Reflector." The participants mostly identified themselves as Evangelical Christians, though there were a few Fundamentalists and other Christians, and they typically argued in great detail about the proper Christian stance to take toward one or another aspect of evolution or scriptural interpretation. On the Calvin Reflector, YECs, OECs, PCs, IDCs and TEs were all represented, and much of what went on involved attacks and defenses of one or another of these positions. The Reflector had a policy of Christian civility, but even so the discussion sometimes degenerated into personal attack or into something just short of an electronic brawl. It was here that I really first began to appreciate the struggle that is taking place within the Tower.

In their activism involving the public schools, however, all creationists are working toward the same ends, though here too their strategies of attack have evolved. Having been repeatedly defeated in their attempts to get major creationist legislation to stick, they have turned their attention to more local activism.

In 1997, I attended a public hearing held by the Texas State Board of Education on their proposed curriculum standards for the state schools, and I listened in amazement as creationists stood in turn to testify against inclusion of evolutionary terminology in the science curriculum. They claimed that biologists were abandoning the theory and that, in any case, it was "not that important in biology" and so students should not waste their time on it. If evolution had to be included, then at least teachers should be instructed to present the scientific evidence against it as well. Religious conservatives on the Board spoke in strong support of these proposals and urged that evolutionary concepts be omitted or put in "neutral language." Another proposal they recommended was to include discussion of "alternative theories" such as "design." This same scenario is played out in public hearings around the country and, with too few scientists taking creationism seriously enough to pay close attention, state boards of education have often compromised or given in to creationists' demands. In Alabama, the proposed curriculum was amended to water down statements on evolution. For example, a science requirement to "Explain how fossils provide evidence that life has changed over time" was changed to "Examine fossil evidence for change." Other states have included evolution "disclaimers." In Clayton County, Georgia, the school board directed science teachers to paste into biology textbooks a disclaimer that began as follows:

This textbook may discuss evolution, a controversial theory some scientists present as a scientific explanation for the origin of living things, such as plants, animals and humans. . . . No human was present when life first appeared on earth. Therefore, any statement about life's origins should be considered as theory, not fact.[48]

Across the country, communities from Vista, California to Plano, Texas to Merrimack, New Hampshire have discovered that they have inadvertently elected creationist "stealth candidates" to their local school board who then work to modify the science curriculum in such ways. Sometimes they are more forthright, as when twelve Fundamentalist candidates ran for the twelve school board seats in Fairfax Country, Virginia just outside Washington, D.C. They wanted to take sex education out of the public schools and put creationism in, following the curriculum of parochial schools. Speaking to the press in support of the candidates, Gil Hansen of the Fairfax Baptist Temple school said that it was wrong to teach the science of evolution as fact: "We certainly let the young people know that, hey, there's a lot of people out there that believe that Darwin's theory of evolution is the way things were created. We expose it as a false model." Of his own school's educational approach, Hansen says, "I make it very emphatically clear that what we do here is base everything on the Bible. This becomes really the foundation, the word of God is the foundation from which all academics really spring."[49]

One now finds explicit support of creationist views expressed at progressively higher levels of government. The Iowa Republican Party included a plank in their 1996 platform that stated:

We believe the theory of Creation Science should be taught in public schools along with other theories . . . [and] support the stocking of *Creationist* produced resources in *all* tax-funded public and school libraries. [Emphasis in original]

A study of twenty-two party platforms by People for the American Way found these as well as similar pro-creationist planks in Kansas, Missouri, Oklahoma, and Texas. Creationism even appeared as an issue at the national political level when presidential candidate Pat Buchanan included it prominently in his campaign. He put his position this way:

Look, my view is, I believe that God created Heaven and earth. . . . I think this: What ought to be taught as fact is what is known as fact. I don't believe it is demonstrably true that we have descended from apes. I don't believe it. I do not believe all that.[50]

What this implied for public education policy was that parents should "have a right to insist that Godless evolution not be taught to their children or their children not be indoctrinated in it."[51] Buchanan was clear about who he thought was to blame for the current pro-evolution policy: "I tell you," he opined in his campaign stump speeches, "we don't need some miserable secular humanist in sandals with beads at the Department of Education telling us how to educate America's children!"

Finally, while on the political front, I cannot fail to mention Alabama Governor Fob James who mocked evolutionary theory by physically mimicking evolution from monkeys at a meeting of the state's board of education, and then bought nine hundred copies of Phillip Johnson's *Darwin on Trial* and had them distributed to biology teachers around the state.[52] Such creationist antics and grandstanding would be amusing if their implications for science education were not so serious.

This brings us to another new field of battle, the universities. It was not too long ago that one could say with confidence that creationism had little or no impact in institutions of higher education. Though ICR's professional debater Duane Gish did appear on college campuses whenever he could find a scientist willing to share the stage with him, such events were regarded more as a carnival side-show than a serious intellectual exercise. Today, however, creationists have entered the halls of academe on two fronts. The first is that young-earth creationists have been successful in promulgating their views throughout Fundamentalist and Evangelical churches, so that many students of those religious backgrounds now enter university primed to resist evolution. Biology professors find more and more that they have to find ways to deal with students in introductory classes who have been taught in Sunday School that evolution is anti-Christian, and who view the subject and those who teach it with suspicion and profound misgivings. The second is that many of the new creationists, unlike their predecessors who operated out of private ministries, have acquired positions in colleges and universities and are leading an attack from within. They work in tandem with campus Christian student groups to hold creationist conferences, and they arrange for departments to sponsor anti-evolution speakers. Their academic credentials and affiliations give them entrance to broader public forums, so, for example, one can now find a few of their books published by academic presses instead of only by small Christian presses. They rightly see this move into higher education as an indication that their movement has achieved a new momentum. Creationism is poised to break into the mainstream.

The Nature of the Controversy

A Scientific Controversy?
As the new creationists portray the battle, they are now undermining the foundation of the Darwinian citadel, and evolutionary theory is about to

collapse in upon itself. From the point of view of biologists, however, such statements are absurd, and there is no controversy about whether or not evolutionary theory is true. The repeated claims by creationists that the theory is false are not taken seriously by scientists, except in the sense that they worry that the rest of the public might take them seriously, which would then be a problem. But that is a problem of education. As a problem for science, however, the "creationism debate" is basically a nonissue.

Creationists of course would be quick to challenge this assessment, bringing out lists of juicy quotations that they have culled from one or another biologist's text or talk that are critical of one or another aspect of evolutionary theory. They often carry around a pack of three-by-five cards with these purportedly damning quotations so they can disprove evolution using "the scientists' own words." Of course there are many unanswered questions about evolution, as there are in any other area of science, and scientists point them out all the time and often argue with great agitation about their solution. But within the big picture these are matters of detail; the core framework is firmly in place. Working biologists resent having to take the time from their research to rebut anew each revised creationist argument against well-established facts. They are much more interested in spending their time answering new questions and furthering biological knowledge. Certainly there are still significant and interesting *extensions* to be made to evolutionary theory, and they are eager to be the ones who will discover them.

Although there is no scientific controversy about creationism within biology, there is certainly a conflict in the sense that biology must deal with the regular attacks from without. In this sense, however, the scientific "controversy" goes well beyond evolutionary biology. As we will see, creationism calls into question not only the conclusions of biology, but also of many other specific sciences. More significant still, especially from my point of view as a philosopher of science, is that the new creationism also calls into question scientific methodology itself. Going beyond the classic "creation-science" that simply proposed that the *content* of evolutionary science was wrong, the new creationists' call for a "theistic science" is far more radical in that it would replace not only the content but also the methodological foundations of science. They would never put their position in such open terms, but the new creationists would generally agree with one Fundamentalist, quoted by Ammerman as holding a representative position, who said simply: "The Bible is the ultimate scientific approach."[53]

A Religious Controversy?

Because almost all of the conflict that reaches the level of public debate involves creationists attacking evolution and scientists defending the same, most people have the erroneous, though understandable, view that this is just a battle between Fundamentalist Evangelicals and scientists, and do not recognize that many mainstream Christian theologians are equally involved in opposing creationism. They are appalled that creationists presume to limit the means by which God's creative power can operate and to claim that their antievolutionary view is the only true Christian viewpoint. Theologically they object to thinking of Genesis as giving a literal description of Creation as though it were a science textbook, and they caution us not to forget the notorious earlier "conflict" between the scientific and religious views about the movement of the earth and the heavens, and the aphorism that was the lesson of "the Galileo affair," namely, that the Bible teaches how to go to heaven, not how the heavens go.

Many mainline religions and Christian denominations have explicitly declared that they find no conflict with evolution. In *Voices for Evolution* (1995), Molleen Matsumura compiled statements supporting evolution from a wide range of religious groups. Many more denominations implicitly hold the same position. Pope John Paul II, in an October 22, 1996 message to the Pontifical Academy of Sciences, explicitly endorsed the findings of evolutionary theory, stating that "fresh knowledge leads to recognition of the theory of evolution as more than just a hypothesis." The Pope's comments were carried by wire services around the world, though his reassurance that there was no opposition between evolution and the faith was old news given that Pope Pius XII had previously attested to their compatibility in his Encyclical *Humani generis* way back in 1950.

From another point of view, however, creationism is a classic religious controversy. Its defenders hold to an orthodoxy, and it is very important for them to draw lines that clearly determine the limits of what is to count as "the true" Christian view and differentiate that view from others. Theistic evolutionists regularly find themselves told that they are "accommodationists" and that their view is not really Christian. In return, they charge that it is rather creationists who depart from the Christian view in holding that the Creation hypothesis is not religious and that science may investigate a generic notion of "creator" or "designer" of the world as if this aspect of God were conceptually separable from God as Redeemer. At the Arkansas trial, theologian Langdon Gilkey testified that to the extent they held this view creationists come "very close, yes, very close indeed to the *first, and worst, Christian heresy!*" that had been committed originally by the Gnostics in the second century.[54] I will have little more to say about such theological arguments and controversies beyond noting how greatly they differ from what goes on in science and how unwise it would be to inject them into the science classroom.

A Philosophical Controversy?

There is good reason that the clash of creationists with evolutionary theory has been compared to the sorry affair between the Church and Galileo in the seventeenth century. Both dealt with emotion-laden questions that were central to our understanding of our place in the cosmos. Both are taken to represent major turning points in the relation between science and religion. More significantly, both deal with the same interconnected pair of issues: the truth of nature and the nature of truth. On the surface, the aspects of either debate that stirred passions supposedly depend upon the truth of specific theses about nature—in the one case the motion of the planets, in the other the origin of species. Whether Galileo's or Darwin's hypotheses are true or false is ostensibly the key point upon which the other issues of the debate turn. At a deeper level, however, is an equally significant debate about truth itself and how we can come to know it. Can we check the truth of such empirical matters (and related matters of purpose and moral value) by human experience and reason, or must we rely on divine revelation? It is in large part because of the intersection of these two sets of issues that the creationism case is of particular interest for the philosopher of science.

Some people might judge it a waste of time and effort to devote a book to a discussion of the many problems with creationism. There are, after all, plenty of goofy antiscientific views that float about at the margins of mainstream thought, so why is this one worth taking seriously? The reason is not just because it is mistaken, but because it is mistaken in a way that is potentially dangerous. Creationists' with-us-or-against-us arguments and their accompanying take-no-prisoners rhetoric is polarizing and divisive, and introducing their viewpoint into the school curriculum would be an educational disaster. Of intelligent-design creationism, Phillip Johnson says "We call our strategy 'the wedge.' "[55] The idea is that they will split the log of scientific naturalism and materialism, and sunder the ranks of theistic evolutionists and others who hold to a middle position that evolution is not incompatible with religion. They aim to force a choice not only between creationism and evolution but also, in one way or other depending upon the faction, on a range of other issues that they believe cleave along the same line:

The Bible is either inerrant or worthless
Christianity or Atheism
Certainty or Skepticism
Absolute Morality or Subjectivism (Relativism)

Readers will have to decide for themselves whether they agree that the conceptual choices are really so stark as creationists portray them. This all-or-nothing-at-all view is overly simplistic, and I will try to point out intermediate positions that are available.

A Political Controversy?

Scientific? Religious? Philosophical? Many people might conclude that in the end the controversy over creationism and evolution is none of these but is simply a political controversy . . . a struggle for power. Some would say that rational analysis of these issues is irrelevant and that it will be which position garners the most votes at the school board meeting or which candidate wins the gubernatorial election that will determine the outcome, at least regarding what will happen in science classrooms. But even when the controversy is considered in this "external" political mode there is still an important sense in which events will be determined by ideas. This takes us back to the original battle scene and the analogy of evolution in the intellectual landscape. The new creationists may or may not succeed in taking over the Tower, and in their struggle a speciation event may or may not take place. Whatever happens within the Tower, however, we can be sure that the new creationists will continue to press the war upon evolution.

Scientists have defended the truth of evolutionary theory from the attacks of creation-science on the merits and must continue to do so. The new attacks are against science as a way of knowing and here the defense must be taken up on the philosophical side. The truth of nature and the nature of truth: these are the issues. It would be wrong to allow such matters to be decided simply by appeal to political power. They must be resolved rather by the power of the evidence.

2

The Evidence for Evolution

It is naive to suppose that the acceptance of evolution theory depends upon the evidence of a number of so-called "proofs"; it depends rather upon the fact that the evolutionary theory permeates and supports every branch of biological science, much as the notion of the roundness of the earth underlies all geodesy and all cosmological theories on which the shape of the earth has a bearing. Thus anti-evolutionism is of the same stature as flat-earthism.

—Sir Peter Medawar

Descended from apes! My dear, let us hope it is not so; but if it is, let us hope that it does not become generally known.

—Wife of the Bishop of Worcester

Bonobos and Biologists

The bonobo mother reclined on a rock outcropping in the shade of a tree and cradled her baby to her breast. A couple of her older children played close by. They amused themselves with a stick, grinning gleefully while waving it about and tugging at it for possession. It was not long before one of them seemed to lose interest in the game and toddled over to his mother, plopped down beside her and prodded her for attention. She patiently indulged his playfulness, but did not allow him to disturb his infant sibling.

I was watching the bonobos at the San Diego Zoo with Shadrack Kamenya, a young African biologist, then a graduate student at the University of Colorado. Shadrack had observed chimps in the wild for years in Tanzania in the Gombe Stream Research Center, but bonobos, or "pygmy chimps," are a different species from the common chimpanzee found in

Gombe. He held his camera to the ready and snapped several photographs as we watched the behaviors and interactions of the little family.

It was only in the late 1920s that biologists recognized bonobos as a distinct species. On average they have longer legs, and a slightly smaller build than the common chimp. Recent analysis of their DNA shows that the two species share 99.3 percent of their genes.[1] Primatologists were surprised to discover that despite their physical similarities bonobos have strikingly different patterns of social behavior than common chimpanzees, and Shadrack was eager to point out to me some of these differences he saw. I found myself equally interested in the behavior of our fellow human spectators and could not help but listen in on their conversations as they observed the pygmy chimps with obvious fascination.

"Look at its arm muscle!" exclaimed a teenage girl, comparing the mother bonobo's well-defined biceps to that of her friend, who looked like she might be a gymnast. The girl who pointed out the arm muscle had probably not studied the details of comparative anatomy but was simply noting a similarity that would strike anyone who was a little observant. "And check out its legs!" she continued, laughing, "They look like our legs when we're not shaven, don't they?"

The previous evening Shadrack and I had attended a meeting of Sigma Xi, The Scientific Research Society, at which the Society presented the William Proctor Prize, its highest award, to Shadrack's mentor Jane Goodall.

Jane Goodall was the first person to observe tool-making behavior among nonhuman animals, thereby shattering the myth that it was a defining ability of human beings alone. When she reported the news to her mentor, Louis Leakey, he noted wryly that humans would now either have to change the definition of "tool" or "man," or else accept chimpanzees into the human race. In our history of self-definition we have faced this sort of problem again and again, especially in recent years as scientists have discovered more and more previously unknown animal behaviors. It should be humbling to us that every time we cite some feature that is apparently unique to ourselves as human beings, someone discovers the same feature, in a less or nearly equally developed form, elsewhere in the animal kingdom. To take another apropos example, one of the unexpected discoveries about the pygmy chimps had to do with their sexual behavior. Bonobos copulate in a variety of positions, including the eponymous missionary face-to-face position that, so the story goes, Christian missionaries recommended to their Polynesian converts as the proper, dignified approach for reproduction because it supposedly was unique to humans, distinguishing us from all the animals. Nor may we say any longer with assurance that we are alone in our ability to communicate through language; studies with a variety of animals ranging from the African gray parrot to the dolphin and especially the great apes suggest that animals have hitherto unappreciated conceptual and linguistic abilities. Goodall's research was sharply criticized in its early stages because she used names rather than subject numbers to label the chimps she observed, and because she described their behaviors in what was condemned as improperly "anthropomorphic" terms, but her careful and systematic observations sustained over more than three decades, eventually opened scientists' eyes to the complex social behaviors of chimpanzees and led others to begin to investigate what seems to be the rich emotional lives of animals. Many of these investigations are still controversial, not in the least because of the tricky philosophical issues that are involved, but we may someday have to admit that animals have feelings of happiness and anger, playfulness and sadness, loss and even grief that are comparable to emotions we ourselves feel.

Goodall gave a lecture accompanied by slides, reporting on the latest news about the children and grandchildren of Flo, the matriarch Gombe chimp who came to be known fondly by millions in the pages of *National Geographic* and in the Society's special series on television. She recounted her old and new findings about chimpanzee social behavior, but she also described the devastating effects that deforestation caused by human population growth is having on the chimpanzees' habitat, and the large numbers of chimps that are killed by poachers simply to satisfy tourists' desire for "monkey paw" ashtrays. Let it not be said that the scientific ideal of objectivity means that scientists are cold and unfeeling—as the lights came up, and Goodall brought her talk to a close with a plea that we not abandon our evolutionary cousins, many in the audience of scientists were wiping tears from their eyes.

How jarringly at odds with Goodall's inspiring story were the comments I heard the next day from a docent at the Museum of Creation and Earth History.

A Guided Tour of the Museum of Creation

The Museum of Creation and Earth History is located on the outskirts of San Diego, in Santee, in the same building that houses the Institute for Creation Research. ICR has expanded the museum over the years and they have clearly put a large investment into making it an impressive facility. Entrance to the exhibits is free, as is the regular tour, which I took with a group of about a dozen enthusiastic creationists. Our tour guide took every opportunity to stress the special status of human beings, how different humans are from the animals, and how absurd it is to think we could be related to apes. It was evolution theory, she said, that led people to think that we were animals, and when people think they are animals it is not surprising that they start behaving like animals. Evolutionist thinking is the reason for the terrible state our society is in today. Evolutionism leads people to think that animals have rights but that unborn babies do not. "I learned recently," she mentioned, nodding her head significantly toward a photo on the wall, "that *Hitler* believed in animal rights."

Most of these final comments came at the end of the tour in the exhibits that pressed home what was really at stake in the choice between evolution and creationism. The debate is not just about deciding between two views about the history of life, it is about a choice of worldviews that determines whether we will head on the upward or the downward path. We will hear this warning echoed and amplified by the new creationists. But this gets ahead of our story; we will consider these issues in the chapters to come. For now let us return to the origin of the controversy.

The tour began where it should begin, our docent had told us, namely at the beginning, which of course was "In the beginning . . . ," with the creation of the world by God on the first day of Creation. ICR interpreters put a scientific spin on the opening passages of Genesis. When the Bible says that God created the heavens and the earth on the first day they take this to refer respectively to three-dimensional space and to matter, and so they claim that the very first phrase tells us that God also started time at

Creation. On the second day God created the waters below and also above the firmament—a water vapor canopy that they claim would be the source of the forty days and nights of rain to come. They suggest, however, that until that time, this canopy would have made the world's climate consistently warm, thus allowing organisms to grow to large sizes and to live to hundreds of years of age after being created, starting from day three onward. This explains, of course, the "giants" that lived in those times and the remarkable longevity of Old Testament figures. The first plants that were created included complex fruiting plants, they claim, not simple ones that later evolved into such complexity. Furthermore, plants, and animals on subsequent days, were created "according to their kind," indicating that they did not descend from some other kind, as evolution holds. The guide emphasized that God created life only on the earth. Astronauts returning from the moon were kept in quarantine because evolutionists, thinking that life could have arisen elsewhere, worried about the possibility of hostile moon germs, but of course these did not exist. A park bench lets the visitors take a breather on the seventh "day of rest," as the guide asked us to consider how absolutely perfect the design of the world must have been at every level for God to have pronounced His Creation "very good." Only a supreme intelligent designer, not chance, could ever have produced such perfection and complexity.

The rooms we passed through as we walked through the days illustrated these and other points with large colorful murals, photographs of planets, nebulae, and comets, and cages of birds, rodents, and snakes. Many of the exhibits have a professional look to them. However, because of the antiscientific content of much of the material, the religious framework that is stressed, and the copious quotations from Scripture, there is no mistaking this for a science museum. That it is theology and not science that drives the picture being presented is especially clear in the next room, in which we learn the effects that sin had when it entered this perfect world upon Adam's and Eve's disobedience of God. Death, violence, disease, and mutations are all the result of the curse God set upon the world as punishment for their sin. God's curse also introduced into the world the second law of thermodynamics, which they call the "death principle," which says that everything without exception must fall inevitably into decay. Prior to the Fall, no animals died since all were vegetarians; the struggle among animals and their carnivorous behavior began only after the loss of Eden's peaceable kingdom. Henry and John Morris speculate that "God performed genetic engineering on animals [to give them sharp teeth for eating flesh] to forever remind Adam and Eve of the awful consequences of sin."[2] Eventually, though, God became so disgusted with the violence and sin of the world that he decided to wipe the slate clean with a global flood and to allow only a pair of representatives of each kind of being to survive in an Ark built by the righteous Noah.

The Noah's Ark room is crafted to make one feel as though one is below deck on the great ship. Sounds of wind, thunder, and creaking boards add to the effect, and a carefully painted mural one sees in perspective upon entering gives one the unmistakable impression of looking down an immensely long row of animal stalls. Is that the back of a stegosaurus we see in one of the stalls? Yes, our guide explained, dinosaurs were certainly on board the Ark, since the Bible tells us that Noah took two of every kind of animal. Another wall includes reproductions of dozens of sketches and descriptions from explorers who claimed to have glimpsed or uncovered portions of the remains of the Ark. Part of ICR's "scientific research" is to continue that search. At a "Back to Genesis" seminar I attended, John Morris said he had been on over a dozen expeditions to the Middle East looking for the Ark. "I haven't found it yet," he admitted. "When I get up to heaven I'm going to ask Noah where he parked that thing!"

This obsession with the Ark is understandable, given the key theological role that God's destruction of the world in the Noachian deluge plays for creationists, and because it is a central explanatory element of creation-science. Our guide explained that this global catastrophic flood was the cause of all the major geological features of the earth. Mountains arose by sudden upthrusting as the fountains of the deep broke open. The Flood itself scooped out valleys and canyons, deposited gravel and rocks in the layers as we find them, destroyed almost all life on earth, and was the origin of most fossils. As John Morris explains it in a videotaped tour of the museum, all the supposed evidence for evolution comes from looking at fossils in the layers of rock, but when we recognize that these were all laid down at once in the Flood then "there is no evidence left for evolution and an old earth."[3]

We will go further into the Museum of Creation in subsequent chapters, but this is an opportune time to discuss an overarching issue that frames the foregoing material and everything else we will consider, namely, the evidential basis of knowledge claims about evolution.

Proofs and Evidences

One of the points that our guide emphasized repeatedly was that there was no proof of evolution. Again and again she would give an argument against biological or cosmological evolution and then would refer to a passage from Scripture that supported the opposing Creation hypothesis. The only trustworthy guide to what was true was the Bible, she said, claiming that not a single statement in it had ever been shown to be false. Evolution, on the other hand, was just an assumption. One finds this sort of claim made regularly in creationist literature. In some cases the claim is even stronger, that there *could* be no proof for evolution. Narrating the videotape of the museum, John Morris says evolution and Creation are both "simply ideas about the past" and that "neither one can ever be proved or disproved."[4] He goes on to say "We don't even try to prove the Bible. We believe it."

Scientists find such statements that there is no proof of evolution to be baffling and bizarre, but one possible reason for such differences of view is that people seem to have different notions of what "proof" means. As we noted earlier, we must always take great care about terminology and definitions because there can be significant differences between the meaning of a word in a colloquial sense and as a technical term. Moreover, there can be a "technical" religious sense to a term in the same way that there can be a scientific or philosophical technical sense. The term "proof" has just such a special religious meaning for some creationists that may partially explain such comments.

Proof Texts

"Bible-believing Christians" have the notion of a "proof text," which determines the biblically correct answer to a question. Ammerman explains this special Fundamentalist notion of "supporting an argument by finding the 'texts' that are the 'proof' of God's answer."[5] A Bible-

believer with a question will "search the Scriptures" for a proof text—
typically a verse, a phrase, or even just a word—that will give God's answer
to the question he or she has in mind. Such "proofs" range from the sublime
to the ridiculous, as Ammerman describes. In one case a couple decides
that God wants them to tell their children the truth about a child their
father had sired while separated from their mother on the basis of the verse
John 8:32: "The truth shall make you free." In another case a man who is
wondering where to buy a tent checks the Bible before leaving for the
store and happens upon the fifth verse of Deuteronomy 14 that mentions
"roebuck" as a food the Jews are allowed to eat, and thereby concludes
that God approves his buying the tent from the Sears Roebuck Company.[6]

This notion of the proof text certainly seemed to play a role throughout
the Museum of Creation. Without exception, the exhibits that laid out the
various elements of creation-science were backed up by plaques that gave
supporting quotations from Scripture. For example, the second law of
thermodynamics that describes a tendency towards disorder was shown
to be correct by reference to a line from Psalms (102:25–26) that says the
earth and heavens "shall wear out like an old garment." Thus, it may be
that when creationists say that there is "no proof" for evolution, they
simply mean that it is not supported by Scripture.

Of course, even this last claim is contentious, for theistic evolutionists
are quick to point to a wide variety of scriptural texts that they say not
only show that the Bible is compatible with evolution but actually endorses
it. For instance, the phrase "Let the land produce . . ." appears twice in
Genesis (1:11 and 1:24) applied to plants and animals as part of the description of Creation, suggesting that the Creator's intention was not to create
these directly, but rather to have nature produce life. This indirect mode
of creation from "the land" also applies to the origin of human beings,
they say, shedding light on Genesis 2:7a: "The Lord God formed the man
from the dust of the ground." Given this understanding, the part that says
that "[The Lord God] breathed into his nostrils the breath [Hebrew for
'spirit'] of life, and the man became a living being" (Genesis 2:7b) may
thus indicate that what God did with respect to human beings was to
spiritually affect a preexisting body. This interpretation is purported also
to be supported in Genesis 1:27. Theistic evolutionists argue further that
the Bible even explicitly endorses the evolutionary view of the transmuta-

tion of one species into another, citing Isaiah 14.29b, which says: "From the root of that snake will spring up a viper, its fruit will be a darting, venomous serpent." Could Scripture be any clearer in telling us that species are not immutable, they ask? Is not this as literal a description as one could want that species evolve one into another? Evolution thus has its scriptural proof.

Creationists of course believe this interpretation is completely wrong, and the hermeneutical battle begins anew. In any case, irrespective of who is right about such questions of exegesis, this religious notion of "proof" is not relevant to the scientific or the philosophical senses of the term. Indeed, most creationists seem to have another notion in mind when they claim that there is no proof of evolution from a scientific point of view, or when they say that no proof of evolution is even possible. Again, it is hard to know how to understand such statements, but it appears that the problem stems from narrow and mistaken beliefs about scientific proof. Many creationists think that any scientific argument that provides less than complete certainty is not really proof (they are happy with nothing less than absolute truth), and they are confused about the sort of proof that science does offer. To see what is going on and to help us sort out the confusions, let us look briefly at a few other technical notions of proof.

Buckets and Searchlights

Philosophers distinguish two sorts of proof: deductive and inductive. In deduction an argument is either valid or not, and this depends solely upon the formal structure of the argument. Given that all men are mortal and that Socrates is a man, we may validly conclude that Socrates is mortal. If the premises of a deductively valid argument are true then the conclusion must be as well; the form of the argument guarantees it. In this sense, deductive proof does provide certainty. However, deductive certainty comes at a price; the reason a deductive argument can guarantee the truth of its conclusion is that that conclusion was already in a sense contained in the premises. What makes a deductive argument valid is just that it has a formal structure such that the conclusion can never say anything more than was already assumed from the start. In Wesley Salmon's terminology, deductive proofs are truth-preserving but "non-ampliative,"[7] meaning that they do not amplify or increase our information. Inductive proofs, on the other hand, are those that do increase our information. When the detectives dust for fingerprints on the murder weapon they are hoping to find evidence that will help them discover who pulled the trigger, a discovery which will increase their knowledge significantly. Induction is ampliative.

Deductive proof is found in the realm of logic and mathematics, but although mathematics is important in science, the amplifying feature of induction should make it clear why it is actually this second notion of proof that is most relevant to the empirical sciences. To say that a hypothesis has been "proven" in science is not to say that it is deductively certain but that it has been tested and supported well by the evidence. On this inductive notion of proof, hypotheses are not determined to be Truths with a capital "T"; scientific truths always must be written lowercase. To say this is not to be a relativist, but to understand that scientific truths are defeasible. This means that we must always stand ready to revise or even abandon them should new countervailing evidence be found. Judge Overton's ruling in the Arkansas "Balanced-Treatment" case had alluded to this as a feature of science in noting that scientific conclusions are always "tentative." Of course this should not be taken to mean that all scientific conclusions are *equally* tentative, for some are only weakly supported while others are well supported. This point highlights another distinguishing characteristic of inductive proof: deductive validity is "all or nothing," but inductive evidence comes in varying degrees of strength. The stronger the inductive evidence, the more confident we are in the hypothesis. Indeed, some scientific conclusions are so well supported that it becomes hard to imagine any realistic scenario of possible evidence that could overturn them. In such cases we may take them to be certain "for all practical purposes," but even so it is understood that because of the nature of induction they remain in principle subject to revision. For all practical purposes we are certain that all men are mortal, but if we were to seem to find Socrates (or Methuselah) still alive today we should begin to reconsider our generalization or perhaps our judgment that he was a normal man.

Because of the difference between deduction and induction, it can be confusing to speak generically about proof in a scientific context. Thus, to be clearer, when dealing with the inductive notion philosophers of science will speak less often of "proof" and more often of "evidence," and rather than saying that a scientific hypothesis is "proven" or "disproven" we will say that it has been "confirmed" or "disconfirmed."

Even these distinctions, however, do not resolve all the possible confusions, for there are various types of inductive evidence. Here I will just mention the two sorts that are most relevant to the topic at hand. The first kind of induction is known as "induction by enumeration." In this method one simply makes direct observations and enumerates or lists what one finds, drawing generalizations from strings of these observations. This method is often attributed to Francis Bacon, whose writings in the early seventeenth century helped inspire the scientific revolution. A second kind of induction is known as the *method of hypothesis*, or sometimes the *hypothetico-deductive method*. Philosophers Gil Harman (1965), Peter Lipton (1991) and others have discussed this as *the inference to the best explanation*.[8] In this method one assumes a hypothesis for the sake of investigation, asks what would follow empirically if it were true, and checks its probable consequences against the phenomena. One way to do this is to make a prediction based upon the hypothesis and then to see whether the prediction is borne out. Because it is no mean feat to correctly predict the unknown, if the prediction from the hypothesis is successful then this is good reason to infer that the hypothesis is likely to be true. On the other hand, if the prediction turns out to be incorrect then this is good reason to infer that the hypothesis is false. Actually, one does not really require a prediction of a future observation; what are called "retrodictions" or "postdictions" of past phenomena also work. The key feature of this form of inference is not whether the data occurs in the future or the past or the present, but whether it stands in the proper relation to the hypothesis. What we are looking for is that the hypothesis is able to adequately *explain* the observed pattern of data. Hypotheses that are inadequate must either be modified or else be rejected in favor of a better alternative. One may think of this as a kind of Darwinian process in which various hypotheses compete to explain patterns of data, weaker ones lose out, and the one with the most explanatory power gets selected. Putting this another way, when considering rival hypotheses one should infer the likely truth of whichever one provides the best explanation of the phenomena.

In introducing these two forms of induction to students, philosophers of science sometimes express the difference with a couple of metaphors. It is a mistake to think that scientists investigate the world as though they simply carried around a bucket into which they collected whatever facts they happened to stumble upon. Science is more than a collection of observations. Scientists are not passive observers but active researchers who seek out and bring new knowledge to light by following out the consequences of their hypotheses. We should thus think of scientists not as simply using a collection bucket, but as using a flashlight. One tests a hypothesis as one tests a flashlight—by turning it on and seeing whether and how well it can illuminate one's surroundings. If the light is dim one might have to twiddle the bulb or clean the contacts. If it provides no light at all one might have to put in some batteries or just get a whole new flashlight. Particularly powerful theories are like searchlights that shed a broad, bright, and sharply focused beam upon the world, allowing us to clearly see and distinguish its features.

Evolutionary theory is such a searchlight. Sir Peter Medawar, quoted at the head of this chapter, was using a slightly different set of metaphors in speaking of science's acceptance of evolution, but this is the sort of thing he had in mind. It is the great explanatory power of evolutionary theory—that it accounts for so much data so well—that testifies to its truth. The great geneticist Theodosius Dobzhansky was making the same point in a different way when he said that nothing in biology makes sense without evolution.

Scientists who daily test their hypotheses using the method of hypothesis are very experienced in thinking in terms of searchlights, but most creationists seem to think only of buckets. To "prove" that species are evolving into new ones, they want to see reptiles changing into birds or monkeys into men. The reason many think it is impossible for science to know that evolution occurred in the past is that no scientist was there, bucket in hand, to see that it happened. This narrow view of evidence has been a characteristic of creationism from the early history of the movement. In the mid-1940s Dobzhansky had a lengthy exchange of correspondence with Frank Lewis Marsh, the author of *Evolution, Creation, and Science* (1944), and also the only creationist who did not seem simply ignorant of biology. Marsh was one of the first creationists to accept the truth of microevolutionary change (and was dismayed at being branded an evolutionist by his "brother Fundamentalists" for doing so), though he continued to hold that this occurred only within created kinds. Dobzhansky labored to help Marsh take the next step, explaining how the evidential basis for large-scale evolution rested on inference from the observable data rather than from direct observation. But Marsh would never accept this notion of proof: "Alas! Inferential evidence again! Is there no *real* proof for this theory of evolution which we may grasp in our hands?" After a series of such exchanges Dobzhansky finally gave up in exasperation, writing "If you demand that biologists would demonstrate the origin of a horse from a mouse in the laboratory then you just cannot be convinced."[9]

We will reflect upon the nature of scientific evidence in greater detail in later chapters, but now let us turn to the evidence itself. Evolutionary theory is the best explanation of broad patterns of observational phenomena; its hypotheses have been tested and have proven their power *vis à vis* the biological data. Because evolution is confirmed inductively by the method of hypothesis, to appreciate how the evidence supports it we must understand what it is the theory claims. So what are the various hypotheses that make up evolutionary theory?

Evolutionary Hypotheses

As before, we must be careful about terminology. In one generic sense evolution simply means "change over time." It is in this sense that scientists speak, for example, of stellar evolution or cosmic evolution. Stars go through a series of changes from the time that they are formed as huge clouds of dust and gas, contracting under their own gravity until the hot clumps ignite the hydrogen fusion process, and burning in varying degrees of brightness until they die as supernovae. The cosmos itself continues to change since the Big Bang. (As we noted, young-universe creationists contest even these nonbiological forms of evolution; for them the various types of stars we observe are not in different stages of stellar evolution but are completely different kinds of stars, all created uniquely as they are by God.) Biological evolution does include the generic idea of change over time, but it is more specific about the kind of change that is relevant. It would be a mistake to define biological evolution *simply* as "change over time"; you and I are biological organisms and we are changing over time, growing bigger and older and (we hope) wiser, but this is not biological evolution in the relevant sense. Darwin expressed the idea of evolution that he had in mind using the formula "descent with modification," by which he referred not to individual organisms or to modification of just any trait,

but to *lineages of organisms, with common ancestors, descending repro-*
ductively one from another and changing over generations in their herita-
ble traits. This is the basic concept of evolution as it is typically presented
in evolutionary biology textbooks. It is usually what biologists have in
mind when they speak about the fact of evolution. In this book, I will
follow this standard definition but will also sometimes use Darwin's own
formula or speak of the *common descent hypothesis.*

Once we have in mind this notion of common descent—the idea of the
evolutionary "tree of life"—we can move on to ask *how* evolution occurs.
That is, we can inquire into the processes that cause this evolutionary
change. Others before Darwin had speculated *that* evolution occurred; his
revolutionary advance was to discover simple causal processes that could
explain how it happened. Darwin called the mechanism he discovered the
"natural selection" of accidental differences. We will shortly look at it in
detail, but here I will just mention Richard Dawkins's compact formulation
that the Darwinian mechanism involves *the nonrandom survival of ran-*
domly varying replicators. The variations arise both by mutation and by
recombination of the genetic material and they do so "randomly" in the
sense that they might or might not turn out to be ones that are useful to
an organism in its environment.

Darwin discovered the trick that made evolution work, but his was not
the only or even the first attempt to figure out how species changed into
others. Lamarck, who was one of France's greatest biologists (indeed, he
coined the term "biology"), had proposed a mechanism of "transformation" almost sixty years before Darwin published *The Origin of Species.*
According to Lamarck, organisms had an innate power to develop complex
organization and transform sequentially from one species to another. Furthermore, he hypothesized that organisms inherited from their parents
those traits that they had acquired during their lifetimes due to the differential use of their various parts in response to environmental conditions in
which they had lived. On this view, giraffes, for example, came to have
longer necks as a result of constantly stretching them as they reached for
leaves higher and higher on trees, and they passed this acquired trait on to
their offspring. By inheriting these acquired characteristics, each generation of baby giraffes supposedly began life with a slight head start over
the previous generation. Darwin actually thought that some evolution did

occur in a Lamarckian fashion. However, it is important to understand the significant difference between Lamarck's and Darwin's proposed explanations of how evolution occurs: the former says that variations occur that are just those that are needed in response to a given environment, whereas the latter says that variations arise "accidentally" and then survive differentially depending upon environmental conditions. There is still an occasional scientist who will propose a possible Lamarckian mechanism, but so far the evidence has consistently gone against such hypotheses and in favor of Darwinian mechanisms.

Natural selection is the key causal process that sheds light on evolution, especially adaptive evolution (that is, descent with modification in which the modification leads to a functional trait); but there are other causes of evolutionary change such as mutation, recombination, migration, system of mating, and random genetic drift.[10] Some biologists speak of these hypotheses regarding evolutionary processes as "the theory of evolution," distinguishing this from "the fact of evolution" we discussed above. Though common, this locution is misleading in several ways. First, it makes it appear that hypothesis of common descent is not part of evolutionary theory, which is not only historically inaccurate, but also philosophically problematic because it plays an important theoretical role in evolutionary explanations. Second, it makes it sound like theories are to be contrasted with facts, as though theories were never factual. Although there remains disagreement about the relative importance of the various evolutionary mechanisms, there is no doubt any longer that they are factual. Thus it is better to say that the common descent hypothesis and the evolutionary mechanisms are alike part of evolutionary theory—a theory that is factual. The evidence may be stronger or weaker for particular hypotheses, and to say, for example, that the common descent thesis or natural selection is a fact is just to say that the evidence in favor of it has accumulated to the point that it is hard to think that any new evidence could overturn it.

The third major aspect of evolutionary theory involves which pathways evolution took. Evolutionary theory tells us that descent with modification will occur in any system that instantiates the processes of evolution, but it is an additional problem to discover retrospectively which lineages diverged from which, and when, in a particular evolutionary sequence that has occurred. Many evolutionary biologists devote all their research efforts to tracing the branchings of lineages in the evolutionary history of life on earth. Much of this phylogenetic research may be conducted independently of assumptions about the specific evolutionary mechanisms. However, once certain facts about these processes are known, this knowledge can be used to uncover significant additional evidence about the evolutionary sequence. Evidence for evolutionary pathways comes not only from the fossil record, but also from observed patterns of common characteristics of living organisms, including their anatomical features, their embryological development, and their genetic material. Naturally, hypotheses about evolutionary pathways are exceedingly numerous, given the long and complicated history of life on our planet.

Besides such hypotheses about the when and wherefore of specific phylogenetic relationships that explain specific hierarchical patterns of commonalities, pathway hypotheses also deal with more general features of the structure of the tree of life. For example, paleontologists Stephen Jay Gould and Niles Eldredge have argued that evidence from the fossil record does not clearly support the smooth and constant evolution of species as Darwin's tree of life metaphor proposes, but rather suggests that evolution may have occurred in geologically quick pulses, followed by perhaps long periods of relative stasis. If their "punctuated equilibria" model is correct, then perhaps the tree of life is shaped more like a bush, with acutely angled horizontal branches turning comparatively suddenly into long verticals. Disputes about such features of the evolutionary pathways can get pretty exciting and even contentious as scientists test their favorite hypothesis against the evidence and against each other, and there is a tremendous amount that remains unknown about which pathways evolution took. No doubt there are many pathways, especially of very early evolutionary history, that we will never be able to map because insufficient evidence exists to trace them, but we already know a surprising amount about much of the branching structure. We know that humans and bonobos had a fairly recent common ancestor and that a somewhat more remote common ancestor gave rise to all primates. We also know approximately when these branchings occurred. Our date estimates and other knowledge get fuzzier as we continue into the past, but we can still be confident about much of the order of evolution as we follow the lines of common ancestors back to the those from which all mammals arose, back farther to the common ancestor of vertebrates, and on and on to some common ancestor of all life on our planet.

In dividing evolutionary theory into these three main groups of hypotheses—descent with modification, evolutionary mechanisms, and evolution's pathways—I am following a common and useful taxonomy, but of course one could lay out the elements of the theory in greater detail. Working biologists do focus on more specific hypotheses because that is where the action is now, and the fact that they regularly argue, sometimes vociferously, about such points is no sign that evolution is about to come crashing down as creationists often suggest. What we need to keep in mind is that evolutionary theory is a complex of interrelated hypotheses that develops in response to empirical evidence. In the next couple of sections I will lay out just a bit of the evidence for evolution in such a way as to give a rough outline of the history of the development of evolutionary theory. I begin with the evidence that Darwin brought to light and that led him to his revolutionary theory.

Darwin's Own Evolution

Creationists today, even occasionally the intelligent-design creationists, often make remarks to the effect that biologists simply assume evolution without proof or that they believe it only because they are atheists who are hostile to religion. Such claims could not be farther from the truth. Darwin did not simply decide to assume the view wholesale, nor did he discover the phenomena and the laws of evolution all at once. Rather, he uncovered them piece by piece as he slowly became acquainted with the evidence that he was presented by his studies, travel, and research, all the while considering how best to explain what he found. If Darwin assumed anything at first, it was rather the specific creationist hypotheses that were prevalent in his time. It was only gradually that the evidence forced him to reject these in favor of the interconnected hypotheses of the new evolutionary theory he developed.

Darwin's Creationist Beginnings
As an undergraduate entering Christ's College, Cambridge, Darwin had to think seriously about his religious beliefs. Students could not enter Cam-

bridge University unless they could sincerely affirm the beliefs of the Anglican Church. Young Charles read Anglican apologist Rev. John Bird
Sumner's book, *Evidences of Christianity,* and found it a well-argued,
persuasive defense. He wrote of Bird's argument that it disproved skepticism and that it showed that there was "no other way except by [Jesus']
divinity . . . of explaining the series of evidence & probability" of Gospel
events.[11] In his first year at Cambridge, Darwin also read William Paley's
arguments for the Christian Creator and found them utterly convincing.
He committed them to memory. At this time Darwin's interest in the natural world was blossoming (he became an avid beetle collector and a regular
member of field trips led by his botany professor, the Rev. John Henslow,
and research excursions with his geology professor, the Rev. Adam Sedgwick), so he was particularly impressed with Paley's *Natural Theology,*
with its many biological examples and its view of "a happy world" that
teemed "with delighted existence" because each organism had been created
perfectly adapted to its environment. Waxing poetic about the beneficence
revealed by God's Creation, Paley wrote: "In a spring noon, or a summer
evening, on whichever side I turn my eyes, myriads of happy beings crowd
upon my view."[12] Most of Darwin's professors took for granted a progressive creationist view, and their disagreements involved not *if* God
had designed and created organic beings—of course He had, they all
assumed—but whether He had designed each being and its relation to
others to perfection in all specifics (Paley's "teleological" view) or rather
had designed ideal types or general body plans from which extant organisms departed to a greater or lesser degree. The latter "idealist" view was
in the ascendancy, due in large part to the advocacy of England's greatest
comparative anatomist, Richard Owen. His research convinced him that
it simply was not true that organisms were perfect in their adaptive details
as Paley claimed; he identified myriad examples of what appeared to be
obvious adaptive imperfections. Rather, God's design was evident in the
broad biological pattern of likeness-within-diversity—the symmetries and
unity of plan that could be found shared across species and higher taxa.[13]

Attending lectures and special field trips by the best biologists and geologists of his day, Darwin soon was steeped in the state of the art. His professors recognized him as a rising star, and Henslow recommended him for
a position on the survey ship *Beagle* for what was supposed to be a two-

year mapping trip of the coast of South America, but which turned into a five-year voyage and royal collecting expedition for the young naturalist.

Darwin's Evidence

Reflecting upon the expedition years later in his *Autobiography*, Darwin wrote, "I worked on the true Baconian principles, and without any theory collected facts on a wholesale scale."[14] In the most significant sense this is certainly true, in that Darwin filled his collection buckets with no idea in mind that what he found would lead him to a revolutionary new theory. Indeed, as we shall see, had he been looking for evidence for evolution he might not have overlooked much that passed right under his nose. On the other hand, Darwin did not make his collections with his mind a theoretical blank slate but rather took for granted the creationist views—the special creation of species and the detailed hypotheses of catastrophist flood geology—that he had absorbed from his professors at Cambridge. He was also aware of recent challenges to some of these views, such as Charles Lyell's hypothesis of uniformitarian geology, and he could not help but consider which of these various flashlights best illuminated the facts he was to encounter.

For instance, at St. Jago, the *Beagle's* first stop, Darwin observed a horizontal band of compressed shells and corals some thirty feet above sea level, running for miles. Obviously the oyster bed had formed underwater, but surely the Atlantic Ocean could not have dropped its level so far in the brief lifetime of the volcanic island. Somehow the bed must have been raised, but Darwin inspected it closely and found no sign of a catastrophic upheaval. Darwin decided that the evidence of the shell bed supported Charles Lyell's hypothesis of gradual geological change, in this case of a slow rather than a sudden uplift. A few years later on the same trip, Darwin experienced one mechanism of uplift, his first earthquake, and observed in a letter that "The world, the very emblem of all that is solid" felt "like a crust over a fluid."[15] Two weeks later he discovered fresh mussel beds lying above high tide with the shellfish all dead, and he realized immediately what had happened—the earthquake had raised the land a few feet. Now Darwin had seen for himself one step of the gradual processes of geological change as Lyell had seen them previously when an earthquake in Naples Bay had raised up a once-submerged Roman temple. Again and again on the journey, Darwin would encounter phenomena that fit Lyell's hypothesis and went against the catastrophist view. Genesis-inspired geologists held that the whole South American continent had been violently covered with the waters of Noah's Flood, which then receded, leaving terraces and valleys, but west of the Santa Cruz river estuary Darwin observed geographical formations that were incompatible with this picture. There the five- to ten-mile wide river valley was bounded by three-hundred-feet high walls that led up to perfectly horizontal plains on either side. The plateaus were composed of shells and barnacles indicating that they had originally been formed underwater. Darwin concluded that a Lyellian gradual uplift of the plateau and the Andes was the best explanation of the formations, and even Captain Fitzroy of the *Beagle* had to admit that the raised plains "could never have been effected by a forty days' flood."[16]

Although the evidence against the catastrophists' geological view was piling up and Darwin was slowly coming around to Lyell's view, on points of biology he had not departed from the creationist views of his professors. He noted, for example, the paucity of animal life in the hills above Valparaiso and wondered if no new species had been "created" there since the time of the land uprising.[17] But, of course, it was impossible that the geological evidence would have no bearing on biological questions and vice versa. At Port St. Julian, for instance, Darwin had discovered a "Mastodon" buried in a gully in a layer of loam. Back on board the *Beagle* he reasoned through the implications of this find. His professors had hypothesized that either Noah's Flood or major climate changes had caused the extinctions of the ancient megafauna. But Darwin had found no evidence of flooding or catastrophe in the surrounding rocks. Furthermore, since the loam in which it was buried overlay the shelly gravel of the plateau, he reasoned, it must mean that the animal had lived after the shells. And since similar shellfish could still be found in the ocean, the climate could not have changed much since that time. Darwin's biographers, Desmond and Moore, write: "At a stroke Darwin had thrown out the two commonest explanations of extinction. . . . He still believed in Lyell's Creator but had broken with his climatic theory; he was branching out on his own."[18]

This was his mind-set when the *Beagle* reached the Galapagos Islands in September 1835. Darwin was actively weighing the theories he'd been taught against the evidence he'd found; but he was in no way picking out, as many creationists purport, just those facts that would support evolution. He had not yet come to that view and was still very much thinking like a creationist. Indeed, he gathered specimens of the Galapagos finches, which were later to play such a famously important evidential role in his theory, without even noting at the time the critical fact that different species inhabited the different islands, or even properly labeling which island each specimen came from—information that he later had to try to reconstruct or gather independently once he recognized its significance. Remember, on Paley's view, God created organisms perfectly adapted to their environment and so, since there were no significant environmental differences among the islands, Darwin naturally would not have expected that the finches would be different. He had, however, noticed differences among mockingbirds and so collected a greater number of these and did tag them by island. He did collect a variety of other birds, including ones he took to be wrens, "Gross-beaks," and a kind of blackbird. However, he failed to collect any of the Galapagos tortoises since he believed they were not native, though he noted in passing that prisoners taken on board the *Beagle* had mentioned that each island had its own particular kind. At the end of his visit he heard that trees were also peculiar to each island, but by then it was too late to collect more.

On his voyage Darwin also discovered previously unknown fossils of megatherium, glyptodons, and other large mammals. Owen, classifying the fossils Darwin brought back, identified five kinds of giant, archaic armadillos and sloths. Darwin had also collected specimens of modern armadillos he found or that perhaps had been left over from dinner; he had been served roast armadillo by the gauchos he met in Brazil, and in his diary wrote that "[t]he Armadilloes . . . cooked without their cases, taste & look like duck."[19] Darwin thought them quite delicious and at the time had no idea that he was eating species that were probably the evolved descendants of those giant fossil armadillos.

It was only upon his return to England after the five-year trip when he began to sort through and reexamine his collection of over five thousand specimens and his nearly two thousand pages of notes that he started to recognize the clues and patterns therein. He distributed his collection among experts for identification and was surprised, for example, when they reported that specimens he had assumed were the same as European or African species were in fact distinctive South American species. He was also puzzled to learn from John Gould, who was classifying his bird specimens, that he had misidentified many of the birds, including the "Grossbeaks" and "blackbirds," and that what he had brought back was "a series of ground Finches which are so peculiar" that they formed "an entirely new group" of what he eventually recognized as containing thirteen species.[20] Gould also reported that the mockingbirds Darwin brought back from the Galapagos were not varieties, as Darwin had thought, but three distinct species, and that they were close relatives of but not identical to species on the American mainland. As for Darwin's fossils, it was Lyell who drew out the implications of Owen's reports and argued that, though gigantic, they were certainly closely related to the modern species.

I regrettably cannot here lay out all the relevant patterns and show their evidential import in much detail, but I do want to call attention to a few of the more important points that Darwin drew from such information. The first was the geographical pattern of distribution of species. Why did the islands of the Galapagos each have their own different species of finches, mockingbirds, tortoises, and so on, closely related to but still distinct from the mainland species? As Darwin reviewed the information, he began to entertain the idea that the island species were modifications of a single species that had originated on the mainland. The relation between the modern and the fossil forms further supported this idea of transformation. This second, temporal pattern could be seen as a series of snapshots of species changing over time—from giant archaic forms to small contemporary forms, for instance—and helped explain how the biogeographical patterns could have come to be. A third point that struck home with Darwin was the pattern he now saw in the relationship between species and their varieties. Having studied classification at Edinburgh he was already aware of the hierarchical structure of the biological taxonomy, with *species* grouped into *genera* (starting with Linnaeus, all organisms were identified in a binomial system that gives their genus and species designations) and species themselves coming in several *varieties*. What he now realized, upon learning from the experts that he had misidentified species in the field, was that it was not obvious when species belonged to one or another genus, or where varieties ended and species began. Looking just at the names, one might think that they referred to clearly distinct kinds, but looking at the organisms themselves as they were being classified one could see that they graded into one another. What this meant was that while small differences in a group of organisms made for varieties, just somewhat larger modifications would turn them into distinct species. Varieties were incipient species. Here was one of the keys to the problem of the origin of species.

I do not mean to suggest that Darwin figured all this out in just so many straightforward steps. I have tried to indicate that Darwin arrived at his evolutionary theory only in stages in response to the evidence, first in seeing that the explanations he had been given as a student did not account well for what he observed in the field, then by considering rival explanations others had put forward, and finally conceiving his own new alternatives that could better account for the patterns he observed. In telescoping the story, I am necessarily leaving much of it out, but I hope that even this simplified picture will give a sense of the complexity of the issues and the variety and quantity of the evidence Darwin used, and that this will help the reader see just how far off-base is the creationist charge that evolution is simply "assumed."

Though Darwin now had the outline of a theory that could explain the biogeographical and temporal patterns of distribution, and the nested patterns of taxonomy, there remained the most puzzling pattern of all, namely, that organisms were adapted to their environment. In his autobiography, Darwin recalled how he turned his attention to this problem.

After my return to England, it appeared to me that by following the example of Lyell in Geology, and by collecting all facts that bore in any way on the variation of animals and plants under domestication and nature, some light might perhaps be thrown on the whole subject [of the explanation of adaptations].[21]

At this point, Darwin explicitly began to change his method; instead of putting whatever he found into his bucket, he was focusing on trying to answer specific questions. Actually, this was in keeping with the method that Bacon had proposed. Bacon is often criticized as being a naive enumerative inductivist but, as is often the case, this caricature is quite unfair and fails to appreciate his progress towards a more sophisticated methodology. He specifically recommended a focused search procedure that would collect facts about a particular phenomenon (heat, for example) and look for patterns of commonalities and differences that might indicate an underlying cause (motion, as the explanation of heat, for example). This is what Darwin began to do for the puzzle of adaptation, keeping in mind his new transformationist hypotheses. He started to talk to breeders to learn what they knew of adaptations. He became a pigeon fancier to learn how the bizarre varieties of pigeons were shaped. He spent years collecting, dissecting, and classifying barnacles to try to understand their range of variation in nature, and he became the world authority on these creatures. More patterns that supported the evolutionary ideas kept emerging as Darwin trained his newly discovered lights on the facts at his disposal, providing more and more evidence that allowed him to further develop and support his views. Let me now briefly jump forward in time to bring in one more important strand of evidence for the common descent hypothesis, which will then bring us back to the puzzle of adaptation.

One of the main classes of evidence for descent with modification is the hierarchically ordered patterns of common traits that have been found across species. One could easily write a book about this topic, as indeed has already been done,[22] but here I will just mention one pattern that struck Darwin and which all others since him found particularly revealing. In *The Origin of Species*, Darwin wrote:

What can be more curious than that the hand of a man, formed for grasping, that of a mole for digging, the leg of the horse, the paddle of the porpoise, and the wing of the bat, should all be constructed on the same pattern, and should include the same bones, in the same relative positions?[23]

Darwin had learned this information from Owen's discourse *On the Nature of Limbs* (1849). Owen had pointed out this and many other similar patterns that he said favored his view that God had created *archetypes,* perfect patterns, and he further claimed that such similarities were hopelessly inexplicable on the opposing view that God had specially created each particular species for the particular environment in which it would live. By this time, Darwin disagreed with both positions. He recognized that certain such cross-species patterns, known as homologies, were good evidence that the forms had a common historical cause. The general inference from homologies to a common cause is straightforward, but it can become extremely complicated in particular cases. Let me try to explain the idea with a simple analogy.

Suppose two students turn in identical or nearly identical essays. Even though both had had the same task of answering the assigned question, their professor, being familiar with ordinary processes of thinking and writing, would certainly not accept the mere fact of their common goal as a good explanation for the unlikely coincidence in the papers. Being aware as well of the process of copying, the professor will readily recognize that as the better explanation for the match. Furthermore, if it can be ruled out that either student copied from the other then the reasonable explanation is that both copied from some third source; perhaps they started from the same essay in a fraternity term-paper file and made slight modifications to it. The most reasonable explanation of the common pattern is that there was a common cause. It is just this reasoning that supports descent with modification from common ancestors as the best explanation of homologies, for the causal processes of reproduction, whether sexual or asexual, are all a form of slightly less-than-perfect "copying." Thinking of organisms as copying machines also proved to be the key to the puzzle of adaptation.

As the evidence accumulated, Darwin gradually came to accept the idea that species arose by transmutation, but he still could not see exactly how it could happen. His investigations into the breeding of pigeons and other animals and plants led him to believe that selection was somehow the answer, but for a long time he could not see how this could occur in nature.

Then, in 1838, he happened to read Thomas Malthus's *Essay on the Principle of Population*, then in its sixth edition, which argued that populations reproduce exponentially and thus are inevitably driven to compete for limited resources. Clearly, not all organisms could survive the intense competition that resulted when they bred beyond their means. This, Darwin recognized at last, was the key to how selection worked in nature. He wrote later in his autobiography that

... being well prepared to appreciate the struggle for existence which everywhere goes on from long-continued observation of the habits of animals and plants, it at once struck me that under these circumstances favourable variations would tend to be preserved, and unfavourable ones to be destroyed. The result of this would be the formation of new species. Here, then, I had at last got a theory by which to work.[24]

Though he had really been using it unconsciously all along, Darwin now explicitly switched from an enumerative inductive method to the method of hypothesis. The idea of descent with modification was already in hand, and now he had a second hypothesis to test—the new causal mechanism of natural selection. Organisms reproduced with slight variations, and as they competed for what resources their environment afforded them, those whose variations made them slightly fitter than their compatriots would be more likely to survive. Darwin described this population squeeze as being like a thousand wedges that would open up gaps and lead species to gradually split, first into varieties and eventually into new species. This newly found searchlight of natural selection illuminated whole classes of facts and turned out to be far more powerful than Darwin could have imagined. We will return to a more detailed discussion of natural selection later in this chapter and in subsequent chapters, but let me conclude this section with a few brief remarks about a couple of other aspects of Darwin's own evolution.

Darwin's Doubts

I have sketched the broad outlines of Darwin's evidence for his evolutionary hypothesis so far without saying much about the ways in which it simultaneously led him to doubt and eventually reject the creationism of Paley with which he began. One encounters early inklings of his doubt about the teleological view that God had created the world expressly for man even in the very first days of the *Beagle* voyage, which found Darwin wondering why there was so much beauty in the vast ocean where no one lived to be able to admire it. It seemed, he wrote, "created for such little purpose."[25]

Paley and others had not hesitated to make dozens of specific pronouncements about God's beneficent purposes which they thought were revealed in the perfect adaptedness of organisms to their environment, so it is easy to see how one might more and more question the view upon looking more carefully at the details of comparative anatomy. Why do flatfish begin life with eyes set on both sides of the head, given that they live on the sea floor and so one eye has to slowly migrate around the head to the other side so that it can see and not be abraded by the sandy floor? Why do human males have nipples, since only females use them to suckle their babies? The existence of such useless rudimentary or atrophied organs, quite understandable if species arose through transformation of one kind into another, is no testament to perfect original design. Darwin's doubts about design arose cumulatively as he read Owen and as his first-hand research brought him a wider and wider range of examples.

Darwin recognized that the existence of useless rudimentary or atrophied organs is easily explained if species arose through transformations of one kind into another, and he properly took their existence to be further support of his theory. However, though these facts also led him to doubt and eventually abandon the detailed teleology of Paley, he was loathe to draw from them grand conclusions against the possibility of design in a global sense. In a letter in 1860 to Asa Gray, the greatest American botanist of the day, Darwin wrote:

I cannot be contented to view this wonderful universe, and especially the nature of man, and to conclude that everything is the result of brute force. I am inclined to look at everything as resulting from designed laws, with the details, whether good or bad, left to the working out of what we may call chance. Not that this notion satisfies me at all. . . . I feel most deeply that the whole subject is too profound for the human intellect.

Gray promoted Darwinism in the United States, having accepted both the common descent thesis and the mechanism of natural selection, but he tried to "improve" it by suggesting that God periodically intervened to create the variations upon which natural selection worked, much as Newton had thought that God periodically had to nudge the planets to keep them in line. This notion of *directed variation* was adopted also by Lyell, John Herschel, and others, but Darwin rightly objected that not only would it make natural selection theoretically redundant but that it also seemed clearly opposed by the empirical evidence from domestic animal and plant breeding that variations were accidental. John Dewey was later to mock Gray's idea of directed variation as "design on the installment plan."[26]

One of Darwin's rules of research was to always write down not only positive evidence but also any item he encountered that appeared not to fit with his theory so that he would not forget or inadvertently ignore potentially countervailing data. In many cases he was eventually able to account for the problematic items, but he was also diligent in reporting those for which he could not yet find a satisfactory answer. He felt that anyone who looked at *all* the evidence would see that it supported his evolutionary theory. Those who continued to insist upon Paley's happily designed world of delighted existence, were closing their eyes to the evidence. He was certainly thinking back to Paley's words when he wrote: "We behold the face of nature bright with gladness. . . . We do not see, or we forget, that the birds which are idly singing round us mostly live on insects or seeds, and are thus constantly destroying life."[27] Darwin was being characteristically understated in this passage, for he was by then well aware of the constant bloody competitions that went on in the natural world.

In his scientific writings Darwin rarely speculated about the theological implications of his work, and it was only occasionally in his notebooks and correspondence that he would express his thoughts on theological matters. Having looked nature squarely in the face and observed the violence of its life-and-death struggles, he could not reconcile it with Paley's gentle vision of a world created by a benevolent God. On this point he found himself more aligned with an argument against ongoing special creation that had been made by philosopher David Hume in the previous century. In an 1879 letter, replying to a German youth who had written him about religious questions, Darwin wrote:

This very old argument from the existence of suffering against the existence of an intelligent first cause seems to me a strong one; whereas . . . the presence of much suffering agrees well with the view that all organic beings have been developed through variation and natural selection.

We must remember, however, as Sedgwick argued in an 1860 article, that even though Darwin's purely natural evolutionary mechanism might have been "atheistical," this did not mean that Darwin was an atheist.[28] Darwin believed that his mechanisms applied only to the evolution of species after life began. He had no answer to the origin of life itself or to the origin of the world, and he admitted as much in his letter to the young German: "The mystery of the beginning of all things is insoluble by us, and I for one must be content to remain an Agnostic."[29]

Darwin's Deathbed Recantation?

I cannot end this brief review of Darwin's own evolution without mentioning the story that on his deathbed Darwin completely changed his mind, accepted God, and disavowed his theory. One of my students brought up this story in class as an argument that he seemed to think dealt a deathblow to evolution. This is the story of Lady Hope, a noblewoman and Christian temperance crusader who claimed to have visited Darwin as he lay dying and to have heard his confession. She told this story in 1915 at a religious retreat in Massachusetts, continued to elaborate upon it thereafter, and it was publicized in magazines and newspapers, and continues to have wide currency. Darwin's recent biographer, James Moore, wrote a book *The Darwin Legend* (1994) to at last debunk the tale because people continued to ask him about it each time he was interviewed about Darwin. Lady Hope's story was challenged by Darwin's family who tried to get her to stop misrepresenting him. Darwin's son Francis said that he had been present at his father's side during his last days and that he had never expressed any doubts about the theory he had discovered. Indeed, it would seem odd that Darwin would have even thought there was any need to reject evolution to accept God, as Lady Hope and others seemed to believe. In a letter replying to an inquiry about the compatibility of theism and evolution and about whether he believed in God, Darwin had written that it was certainly possible to be "an ardent Theist & an evolutionist," but that because of his own uncertainty about the existence of a personal God he should probably be labeled an agnostic. However, he wrote, he "had never been an atheist in the sense of denying the existence of a God."[30]

In the end, however, whether or not Darwin disavowed his theory on his deathbed is beside the point. Biologists do not accept the truth of evolution on the basis of Darwin's authority but on the basis of the evidence. Evolutionary theory has been out of Darwin's hands from the moment *The Origin of Species* appeared in 1859. Once Darwin published his evolutionary hypotheses and the evidence upon which they were based, these entered the public domain of knowledge, and others took the ball and ran with it. Scientific knowledge is not "owned" by any individual so no individual, even the discoverer, can "take back" a theory. Even if Darwin had disavowed evolution on his deathbed it would have made no difference to its truth. We still speak of "Darwinian evolution" to give Darwin his due as discoverer, but scientists have greatly improved and extended the theory in the century since Darwin's death in 1882. Moreover, not all the changes in our understanding about evolution have been simply additions to the theory; many specific elements of Darwin's own view turned out to be wrong and had to be corrected. On the other hand, many significant problems—data anomalies as well as theoretical weaknesses—that Darwin was unable to resolve in his lifetime have since been entirely accounted for. To even begin to describe the wealth of post-Darwinian evidence goes well beyond the scope of this book (graduate students in evolutionary biology spend several years covering the material), so here I will be able to do no more than to outline a few of the important highlights.

Some Post-Darwinian Developments

From Fossil Jumble to Radioisotope Dating

The fossilized giant armadillos that Darwin discovered in South America helped push him toward his theory of descent with modification, but fossil evidence had otherwise not played much of a role in his theorizing. Fossils were just beginning to come into their own during Darwin's lifetime as a recognized source of information about the past. An early Christian view was that fossils were partially formed creatures that God had started to create from the clay and had left unfinished, though by the eighteenth century this had shifted to the view that they were remnants of creatures that had perished and been buried in a jumble as a result of Noah's Flood. That dominant view began to be undermined toward the end of the century by the discovery of fossils of unknown animals and plants, because it seemed to Christians that God, being all-benevolent, could not have created creatures only to then allow them to become extinct. Did not Scripture say that two of *every* kind was saved on the Ark? Many continued to argue that living representatives of the unique fossil forms would eventually be found in other parts of the world. But the reality of extinction was finally accepted in the early nineteenth century after French biologist Georges Cuvier discovered the first fossilized "proboscidians" (mammoths) and pterodactyls, which everyone acknowledged could not possibly have remained unnoticed in the world. Being able to suggest how new species could originate to replace those that had become extinct was one of the advantages of Darwin's theory. However, it is fair to say that what was known of fossils at the time worked more against Darwin than for him, which just goes to highlight the strength of his other evidence and argu-

recanted his earlier view, giving special attention to the issue of diluvial gravel deposits that he had previously cited as evidence of the Flood.

I think, one great negative conclusion now incontestably established—that the vast masses of diluvial gravel, scattered almost over the surface of the earth, do not belong to one violent and transitory period. It was indeed a most unwarranted conclusion, when we assumed the contemporaneity of all the superficial gravel on the earth. We saw the clearest traces of diluvial action, and we had, in our sacred histories, the record of a general deluge. On this double testimony it was, that we gave a unity to a vast succession of phenomena, not one of which we perfectly comprehended. . . . Our errors were, however, natural, and of the same kind which led many excellent observers of a former century to refer all the secondary formations of geology to the Noachian deluge.[35]

Of course the battle was not won overnight, but like Sedgwick, other Christian scientists who themselves looked at the evidence eventually had to agree that it clearly went against the literal Genesis creationist view. Darwin had a harder row to hoe than Lyell when he put forward his own new theory a few decades later, but in that case as well the evidence convinced people surprisingly quickly, including, as described by historian David Livingstone in his book *Darwin's Forgotten Defenders* (1987), many theologians who are now regarded as the founding fathers of Christian evangelicalism. As we examine modern creationism we thus need to keep in mind that in many ways it is a throwback to an early nineteenth century position. For creationists today to hang on to this view, they must either ignore or try to explain away most of what science has learned in the intervening century and a half as well as much of their own theological history.

The modern revival of creationism in what we now take to be its "classic" form is due in large measure to George McCready Price, who has been called perhaps the greatest of the antievolutionists.[36] Writing in the early part of the twentieth century, Price was a Seventh-Day Adventist who took his inspiration from Adventist founder Ellen G. White, who claimed that God had carried her back to the Creation in one of her visions, and revealed directly that His work had been accomplished over the course of six days in a week that was just like any other week. Price argued that the rock and fossil record was the "backbone" of evolution, but that this record actually refuted that theory. Geologists, he said, reasoned in a circle by dating rocks by the index fossils found in them and then turning around and fixing the age of the fossils by their position in the rocks. Really, he argued, all fossil life-forms had lived contemporaneously, and if there was any order to be found in the rocks and fossils this was simply the result of sorting that occurred during that flood of Noah. In what we might think of as a perverse twist of the notion of natural selection, Price's explanation was that the flood waters, churning up the earth, would first have buried the "smaller and more helpless animals" and shortly thereafter the faster ones who might have managed temporarily to escape the rising waters by scurrying up the slopes. Of course there would be no bird fossils in the lower sedimentary layers, he reasoned, since they would all have flown to the highest altitudes (though one then wonders why the pterosaurs chose to remain on the ground to be buried with their Jurassic and Cretaceous brethren). In later writings Price would simply deny that rocks and fossil layers occurred in a regular sequence.

Scientists mostly ignored Price in the same way that scientists try to ignore creationist views today. The stratigraphic evidence was too strong to doubt seriously. Even Price's disciple Harold W. Grant finally had to accept the reality of the geological column after he visited the oil fields of Texas in 1938 and saw firsthand some of the data upon which geologists based their conclusions. Core samples from deep oil drilling were a rich new source of evidence that confirmed and extended earlier findings. Grant wrote to his former teacher that:

All over the Middle West the rocks lie in great sheets extending over hundreds of miles, in regular order. Thousands of well cores prove this. In East Texas alone are 25,000 deep wells. Probably well over 100,000 wells in the Midwest give data that has been studied and correlated. The science has become a very exact one. . . . The same sequence is found in America, Europe and anywhere that detailed studies had been made.[37]

Wanting still to retain the literalist Adventist view, Grant proposed a way to reconcile "Flood Geology" with the fossil ordering he now felt compelled to accept: suppose the order of fossil life-forms in the column simply reflect the ecological zones in which all those creatures lived contemporaneously, until they were successively buried in the Flood. Grant's expressed hope that Price would not be too disturbed by this revision was not to be fulfilled. Though not many years earlier Price had praised Grant's work to Adventist church leaders as being better in some ways than his own, he now denounced Grant as giving in to the "tobacco-smoking, Sabbath-breaking, God-defying" evolutionary geologists, and all but called him a charlatan.[38]

Of course, Grant's own view was as untenable as Price's. Creationists today who understand the stratigraphic and fossil data agree that the data is not consistent with the hypothesis of a global cataclysmic flood, and as we noted in passing in the first chapter, there are now small factions that reinterpret the Genesis deluge as referring to a local or a tranquil flood. At the Creation Museum, however, one still hears the dominant view derived from Price. In *The Modern Creation Trilogy*, Henry and John Morris continue to explain the ordering of fossils by combining Grant's and Price's earlier flood geology theories (though without crediting them or mentioning the Adventist origin of the views), writing that it was obvious that "a global cataclysm would tend to deposit organisms in ecological burial zones corresponding to the ecological life zones where they were living when caught up by the global flooding . . . [and furthermore that] . . . [a]nimals that are capable of running, swimming, climbing, or flying can escape burial longer than others."[39] Ironically, in their fervor to oppose evolutionary theory, creationists wind up concocting their own version of survival of the fittest (though only temporary survival, to be sure) to explain away the ordering of the fossil record!

Paleontologists can now do even better than establish the relative chronology of the geological column and of fossils, because one of the most important sources of evidence for the evolutionary picture that has been discovered since Darwin, allows rocks to be given an absolute chronology. The discovery of radioactivity and the subsequent investigation of its properties provided a rich source of information that can determine the date of some materials. Willard Libby of the Institute for Nuclear Studies of the University of Chicago won a Nobel Prize in the early 1950s for showing how the analysis of relative levels in organic material of ordinary carbon-12 to radioactive carbon-14 as it decayed to nitrogen-14 could be used to measure the age of organic material, quite precisely, up to about 40,000 years. Subsequently, other radioisotope systems were investigated that, because of their slower rates of decay, could be used to measure other materials of a far older age. Although radioactive dating does not work directly on the sedimentary rocks that contain fossils, these can be dated indirectly by dating igneous rocks that bracket the layers. So as it turns out, fossils do come with dated labels in a way, and this information has allowed paleontologists to give an absolute and not just a relative structure to much of the tree of life and to confirm the age of the earth.

Over the years, creationists have offered many negative arguments to try to undermine this powerful new evidence, but scientists have shown these to be baseless. The evidence for the age of the earth—the precision and strength of radiocarbon dating in combination with independent measures of age and other radioisotope methods—makes the young-earth view untenable, and this is just one of the several reasons for the current struggle we noted in the first chapter between the YECs and the OECs. In their latest opus on creation-science, Henry and John Morris continue to reject the validity of radiometric dating. They suggest an alternative divine "explanation" for the observed radioactive decay elements in the following argument:

[I]t is eminently reasonable that He would create a complete and fully functioning universe right from the start. It follows then, that the mere presence of a so-called daughter element in a mineral would not at all have to mean that it had been derived over long ages by decay from its supposed "parent element." It could just as well—in fact much more likely—have been *created* there to begin with. . . . Thus, the "apparent age" of a given radioactive mineral would have no relation whatever to its true age.[40]

One regularly encounters this sort of argument and probability assignment regarding God's creative design in creationist literature, which we will discuss in detail in later chapters. For now let us just note the Morrises' confident assessment of the strength of their alternative positive Creation hypothesis: "At the very least," they write, "it would be impossible to prove [it] otherwise."[41] They are certainly correct that scientists will be unable to falsify this *Omphalos*-inspired account of radioactivity, so let us here leave it at that. We will return to the subject of radioactivity shortly, but this is a good time to take a moment to crack another creationist chestnut to which this topic bears an interesting connection—the argument from the second law of thermodynamics.

The Second Law
One of the most common arguments that creationists make against evolution is that it supposedly violates the second law of thermodynamics, which

states that closed systems tend towards increased entropy over time. (Entropy is a measure of the energy in the system that is available to do work. "High" entropy refers to "low" energy states, that is, those that are more disordered and thus less capable of doing work.) How can it be that evolution can go against the inexorable trend toward greater entropy identified by the second law? This purportedly unanswerable question has in fact already been answered by scientists, many times; I will be brief since most of the new creationists, to their credit, seem to have recognized that the argument is fallacious and have stopped using it. We now usually hear it only from traditional YECs. At ICR's museum one finds a section devoted to the second law with a prominent plaque which praises it as the most basic universal scientific law, one which is accepted by scientists in all fields, and yet one that directly contradicts evolution. Why is it alleged to contradict evolution? Supposedly because evolution always involves an increase in the ordered complexity of systems, whereas the second law says systems must invariably run towards disorder. This might sound rather devastating to someone unfamiliar with evolution and thermodynamics, but there are several significant misunderstandings in creationists' arguments regarding evolution and the second law which render their point specious.

The first is a misunderstanding of evolution: evolution is *not* always toward increasing complexity. Species can and do become less complex in certain environments. For instance, parasitic species that were once free-living lost complexity as they evolved to become dependent upon their hosts. This might be considered an understandable confusion, since increasing complexity is certainly the more striking feature of evolution, but the mistake is symptomatic of the generally uninformed character of most creationist criticisms.

The second misunderstanding is more significant. When presenting their argument from the second law, many creationists conveniently leave out the part of the definition that limits it to *closed* systems. A system is thermodynamically closed if no energy crosses its boundary, and it is open otherwise. Think of a closed system as a perfectly insulated box that no heat can flow into or out of. Objects within the box might be of different temperatures, and if so they will exchange energy (hotter objects cooling down, cooler objects warming up) until eventually all reach an equilibrium. That is the point at which no further work can be done, because work requires energy differences so that the energy can flow. (You can think of this on the model of a waterwheel set up on a dammed stream; for it to turn, the water must be able to flow from a higher to a lower point—a miller will get no work from a waterwheel set up in a lake.) In an open system, on the other hand, the box is not perfectly insulated, thus allowing the objects within to increase in free energy (decrease in entropy) if energy is flowing into it from outside. If the system of the waterwheel on the stream were closed, then the wheel would eventually stop, after all the water above the dam had flowed by and dissipated its energy. But it is not a closed system. Energy from the sun is constantly pouring in, evaporating the water so that it rises (i.e., so that it goes, thermodynamically and literally, uphill), falls as rain above the dam and keeps the stream flowing and thereby the waterwheel turning. The creationist argument fails to recognize that the second law applies only to closed systems and that the earth is an *open* system. Their misunderstanding goes deeper still, for even if the earth *were* a closed system, evolution would still be possible since, as we noted, some objects in the insulated box may (at least temporarily) decrease in entropy though the system *as a whole* moves towards equilibrium. Thus in neither case is there a contradiction between evolution and the second law.

Creationists have by now heard this explanation many times, so what is their response? The plaque on the wall at the ICR's museum simply adds the claim that increase in entropy applies not only in closed systems but in open systems too. What is one to make of such an argument? Since the second law explicitly applies only to closed systems, it is either a misunderstanding of the second law or else creationists have discovered a remarkable third thermodynamic law. A second response, made by ICR's full-time debater Duane Gish, is that being an open system is not sufficient for evolution and that energy conversion mechanisms are also required but that these do not exist. But no evolutionary theorist ever said that being an open system is *sufficient* for evolution (what is the Darwinian mechanism for, after all); they were simply showing that evolution does not violate the second law, in answer to the creationist challenge that it did. Gish's argument is particularly interesting since it indirectly admits the evolutionists' point but tries to hide this admission by shifting to a new challenge about the need for (purportedly nonexistent) energy conversion mechanisms. But he is wrong on this last point too, for there are plenty of energy conversion mechanisms. The energy stores of organisms do not simply run down until they are depleted, but rather, like the sun-driven rain refills the stream, are regularly replenished. Plants convert solar energy by photosynthesis and animals gain energy by eating plants and each other. These processes might not be optimally efficient but there is more than enough incoming energy to allow considerable waste. Globally, entropy has increased as the second law requires, but locally it has decreased. As the physicist Erwin Schrödinger put it in his classic work *What is Life?* living organisms can "remain aloof" from the second law "by continually drawing from [their] environment negative entropy. . . . What an organism feeds on is negative entropy."[42]

In reality there are no perfectly closed systems (except the universe as a whole), so for us to apply the second law in practice we can only do so as an *approximation* in those smaller systems in which energy exchange with the external environment is negligible. But, as we have seen, even as an approximation the second law is not violated by evolution. Indeed, the second law could turn out to be a driving force in the emergence and evolution of life. Physicist Ilya Prigogine won a Nobel Prize for his work showing how thermodynamical systems far from equilibrium can give rise to order, or what he called "dissipative structures," and this discovery has sparked an active research program into self-organization. Jeffrey Wicken, for example, argues that the second law might be involved in the production of order and information and that it could extend the Darwinian program by "establishing continuities between biotic and prebiotic evolution and in allowing organisms to be understood as elements in ecological patterns of energy flow with *macroscopic trends* operating over and above microscopic particulars."[43] Wicken's own particular view might or might not pan out, but he is just one of a variety of researchers who think that thermodynamic principles might underlie features of the evolutionary processes. In an introduction to a collection of articles from researchers dealing with this topic, philosopher of biology David Hull suggests that a theoretical reformulation might be under way and that "[o]ne of the ambiguities that might well disappear as both evolutionary theory and thermodynamics are reformulated is the sharp distinction between statistical disorder and organizational complexity. One theory might well emerge capable of handling both."[44] Stuart Kauffman has pursued this research program perhaps further than anyone, especially in his *The Origins of Order* (1989) and he argues that "[all] free-living systems are dissipative structures."[45] Such investigations are still in their very early stages, so it is too soon to say what light they will shed, but if these possibilities are borne out and entropy turns out to have a positive causal role in the emergence of life it will be a significant addition to the Darwinian mechanisms. Creationists are certainly wrong to say that evolution violates the second law and it would be particularly ironic if the law turns out to actually play a creative role in evolution.

Thermodynamics, Radioactivity, and the Age of the Earth

Thermodynamical arguments against evolution had been among the first that Darwin had to confront. The theory of thermodynamics, like the theory of evolution, was a mid-nineteenth-century scientific achievement and it did appear to physicists at the time that the second law was problematic for evolution, though initially in a different way. Darwin's gradualist evolutionary theory required that the various geologic ages each had to have been hundreds of millions of years long, but Lord Kelvin's calculations of energy dissipation from the sun and earth using the new thermodynamical methods indicated that the solar system was rather young. Darwin took Kelvin's criticism seriously and it might have been one of the reasons he began to allow the possibility of Lamarckian evolution, since inheritance of acquired traits would not require as long to transform species, but he also argued that not enough was known about the thermodynamics of the earth to accept the calculations as definitive over his biological evidence. In the pecking order of the sciences, biology and the other sciences tend to defer to physics, but this turned out to be one case in which biology was right and physics was wrong. Lord Kelvin's calculation underestimated the age of the earth by one to two orders of magnitude because the physics of his day had not yet discovered radioactivity and nuclear processes, which generate heat. Today, with this new knowledge in hand, the best estimate is that the earth is some 4.5 *billion* years old. Darwin would have been pleased.

Note that YECs, for all their trumpeting of the supposed incontrovertibility of the theory of thermodynamics, rarely breathe a word about Lord

Kelvin's argument that the earth was too young for evolution to have occurred. Why not? Because even though his thermodynamical calculations made no assumptions about radioactivity they still showed that the earth was between 20 million and 400 million years old, which is three to four orders of magnitude older than the ten-thousand year maximum age that their Genesis-based estimate allows. Old-earth creationists have accepted the validity of scientific estimates of the earth's age and no longer find any reason to question radiometric dating, but young-earthers still try to argue that the radioisotope data cannot be trusted, suggesting that rates of decay were not constant but had accelerated at times, thus skewing the data so the earth appears to be much older. They note that on the standard view decay occurs at a relatively constant rate, and thus "if a mineral contains any of the daughter element *at all* it will suggest a very old age," and so even if there are anomalies "whatever 'ages' are suggested will all tend to be very great—and that is why evolutionists favor radiometric dating over all other processes."[46] However, the Institute for Creation Research recently acknowledged that although "numerous young-earth creationists have studied this issue and have published their views and results . . . most publications have been somewhat simplistic and have not addressed the problem thoroughly."[47] Recognizing the fact that "[m]any non-Christians and Christians alike are convinced that radioisotope data justify the long periods of time needed for evolution to occur," ICR joined forces with the Creation Research Society and the Creation Science Foundation and sponsored a conference of ten young-earth creationists from these groups to begin research to confront the issue and hopefully "[turn] the tide . . . away from old-age naturalistic evolution toward the young-earth, God-honoring creation model."[48] After the conference the team concluded that their "principal tentative approach . . . will be to explore accelerated rates of decay of radioisotopes during one or more of the Creation, Fall, and Flood events."[49] The ICR team did not report how it planned to investigate these events scientifically. As we saw, according to them, science can never prove anything about the past. However, this might be the least of their problems. The relationship between radioisotope decay and heat production puts YECs between the proverbial devil and the deep blue sea, for the accelerated decay rate they must postulate to make the data fit their commitment to a young earth would have generated hellish quantities of heat in the past, far greater than current measurements allow, since cooling, of course, must follow the laws of thermodynamics. The creation-scientists at the conference reportedly proposed several theories that might "compensate for this large amount of heat and possibly even result in net volumetric cooling in places." They immediately suggested that this would be evidence of God's creative and sustaining role, because: "Such theories seem to ultimately depend upon supernatural intervention at the time of Creation, Fall, and the Flood."[50]

One of the team, Larry Vardiman, who is also Administrative Vice President at ICR, writes that "[t]his project has the potential to revolutionize scientific thinking on radioisotope dating."[51] If the team finds a way to test for the operation of divine intervention then they will have revolutionized scientific thinking far more significantly than that. But until then, Darwinians need not worry that the supporting evidence gained from twentieth century advances in physics will be overturned.

Before we move on we should note in passing that the intelligent-design creationists argue for just such a theoretical revolution that would allow science to appeal in this way to theistic interventions. It is revealing, however, that they have remained completely silent about this and other similar proposals and typically steer clear of the entire subject of radiometric dating or the second law or the Flood. Phillip Johnson, for one, does not even mention these in any of his books. As mentioned previously, IDCs studiously avoid any topic having to do with the age of the earth because they want to straddle the fence between the YECs and OECs. But this is just the sort of case that we should rightly expect them to come clean on. IDCs, YECs, and OECs all agree on the viability of a science that sanctions miracles. The YECs have here proposed a thermodynamic miracle. How will their theistic science test this hypothesis? Vardiman, at least, is forthright about the proof he takes to be important: "God's intervention is explicitly stated in Scripture (II Peter 3:5,6 and implied elsewhere)."[52]

Mendelian Genetics and the Modern Synthesis

Despite the successes discussed above and others, I do not want to leave the impression that Darwin had gotten the theory exactly right on the first shot and that the subsequent history was an unbroken string of vindications. Much has been learned since Darwin, and as new evidence has accumulated, biologists have been able to correct errors in the theory. Darwin was quite wrong, for example, about the way that inheritance occurs. According to his hypothesis of *pangenesis*, traits are passed on because particles from all parts of the body coalesce in the germ plasm which then, in sexual reproduction, fuses and so combines in the offspring the traits from the mother and father. However, if inheritance worked by this sort of blending mechanism then it would be hard to see how varieties could ever develop and diverge into species, since whatever variations arise would always merge back to the average.

Though it wasn't recognized at the time, the key to the answer to this problem was published just a few years after Darwin's *Origin of Species,* by an Augustinian monk named Gregor Mendel who had been conducting breeding experiments on pea plants. Mendel had discovered that observed traits were transmitted by what he called "factors" and that these were discrete—they could separate and assort independently. The sex cells thus did not each have a full complement but only a random subset of the organisms' genetic traits, so offspring of sexual reproduction were not a blend of the parents' genome after all. Mendel also found that some factors were dominant over others, but that the recessive factors retained their properties even in generations in which they were dominated and could thus be passed on and be expressed in later generations. The laws of inheritance that Mendel formulated were just what Darwin needed, but few read or recognized the importance of Mendel's work when it first appeared, and it was only rediscovered decades later after three biologists independently reached the same conclusions themselves.

There were other problems that Darwinism encountered along the way as well. Reading Erik Nordenskiöld's *History of Biology* (1928), for example, one gets the impression that Darwinism was an interesting hypothesis in its day but that it had been superceded and was no longer taken seriously by biologists. Nordenskiöld's history was published in the 1920s during a period of decline in evolutionary studies that Huxley called "the eclipse of Darwin." It is important to realize, though, that during this period biologists never rejected the general idea of descent with modification or the common descent thesis. In turning away from "Darwinism" they were questioning Darwin's proposed mechanism, particularly natural selection as the force behind the gradual process of evolutionary development. Ernst

Mayr explains that geneticists in this period were largely ignorant of taxonomic literature on species and speciation and so underestimated the role of selection and emphasized evolution by large mutations. Naturalists, on the other hand, were ignorant of the genetic literature and, with mistaken ideas on the nature of inheritance, were appealing to Lamarckian mechanisms to explain evolution. Evolution emerged from its eclipse in the mid-1930s when members of the two groups finally began to collaborate. Field biologists and mathematical geneticists worked together to make evolutionary theory experimental and quantitative, by showing that observed evolutionary phenomena, particularly macroevolutionary processes and speciation, could be explained in terms of known genetic mechanisms. These results are now known as the *modern evolutionary synthesis.*[53]

Field biologists turned the focus of attention from the genetic structure of individuals to that of populations. They demonstrated the high genetic variability of wild populations, turning around the prevailing concept among geneticists that wildtype species were uniform. In examining species as communities of populations, they also developed the *biological species concept*, which identifies species in terms of traits that maintain the group's reproductive isolation (often based on behavioral mechanisms rather than merely on a sterility barrier), thereby keeping each distinct from other species. For their part, mathematical geneticists proved mathematically how small selective advantages have a major evolutionary impact if selection is continued over a long period. Geneticists also demonstrated that what appears to be continuous variation of a trait was actually caused by discontinuous genetic factors that obey Mendelian laws in their mode of inheritance exactly as mutations do. Furthermore, they discovered the process of *pleiotropy*, whereby a single gene can affect several different components of the phenotype, and the process of *polygeny*, whereby a single trait is controlled by several independent genes. These results showed that there is much greater plasticity to the genotype than had been envisioned, which further supports Darwin's view. These and other advances make up the framework for contemporary neo-Darwinism.

One striking aspect of the early contributions of genetics to evolutionary theory in the synthesis is that genetic causal mechanisms were known then only at a fairly high level of abstraction, and little was known about how these were instantiated at a molecular level. The discoveries then yet to be made about the biochemistry underlying genetic mechanisms provided even greater resolution to the picture and are among the most significant post-Darwinian sources of evidence for evolution and evolutionary processes.

Molecular Biology and the Genetic Code

It was not until research was carried out by Oswald Avery and others from 1944 to 1952 that the nucleic acids DNA and RNA were identified and conclusively accepted as the common genetic material for all organisms. This by itself was important new evidence of the relatedness of all living beings. If we had found instead that the genome of every species is carried in a radically different manner, say by "ABC," "XYZ," and so on with no way to combine with or change into one another, it would have immediately falsified the common descent hypothesis. The discovery that all species have the same nucleic acid "alphabet" fit in perfectly with all the other evidence of common descent that biologists had previously uncovered.

Learning that nucleic acids somehow carry the blueprint for the building of proteins was a significant step in another sense as well, for most of the cellular structures that make up living organisms are composed of protein. Proteins also serve as catalysts for almost all chemical reactions in living organisms. Thus, if we understand how proteins are built, we've come a long way toward understanding how all organic life is built, from gross traits that we observe at the macro-scale all the way down to the tiny molecular machines that run our physiological processes at the micro-scale. Protein molecules exhibit such tremendous variety and complexity in their three-dimensional structure that it might make one despair of ever explaining their origin, but it turns out that at a more basic level proteins are fairly simple. All proteins are polymers, that is, molecular chains composed of similar chemical units. The units—the links in the chain—are amino acids, and there are only twenty main types used in biological organisms. The primary structure of a protein molecule is thus a linear sequence of amino acids, and it is simply their order that determines the three-dimensional form of the molecule. The final shape of the molecule is the most important factor in determining its function; but since this shape is a direct product of the primary sequence of amino acids, understanding the origin of proteins is simply a matter of understanding the processes that give rise to differences in these sequences.

Of course, "simply" is a bit of an exaggeration, for there are many pieces to this puzzle. Knowing *that* nucleic acids carried the sequence information for building proteins is one thing, but *how* do these accomplish all that needs to be done? We know that for the Darwinian mechanism to work, the genetic blueprint not only has to be stored and then transmitted for protein formation, but must also be able to be modified to produce new variations, and be able to replicate so that these variations can be inherited reproductively by the next generation. In 1953, James Watson and Francis Crick discovered the now famous double-helical structure of deoxyribonucleic acid and this immediately suggested how these processes could work at the molecular level. Like protein, DNA is a polymer, but rather than being composed of twenty types of links it is made up of four—nucleotides (or "bases") adenine (A), thymine (T), guanine (G), and cytosine (C)—which are attached to a backbone of alternating phosphate and sugar groups. DNA's double chains intertwine together with the bases paired up in a regular fashion so that A always pairs with T, and G always with C. RNA has a similar structure except that uracil (U) replaces thymine. This paired structure means that each strand is a mirror of the other, so when the double-helix "untwines" each can pair up again with spare nucleotides and "re-twine," leaving two double-helices where before there had been only one. We now see how the "copying machine" works. Furthermore, genetic variations are simply variations in the sequence of nucleotides, and new variations can arise if a mutation changes that sequence by inserting or deleting nucleotides or if segments of the strands cross over and recombine in new ways. This gives the basic molecular mechanism for the variation and copying of genetic material. One more key step was figuring out how the genetic material "coded" for proteins.

The story of how the genetic code was cracked is a fascinating case study of the important way that theory and experiment combine in the process of scientific discovery.[54] In 1953, George Gamow, noting the four-base linear structure of DNA and the linear primary structure of proteins, hypothesized that the order of the former completely determined the latter like a template and reasoned that the problem was to figure out how the four-letter "alphabet" of nucleic acid bases could be used to form "words" that would translate into the twenty-letter alphabet of amino acids. If the words were one letter long, then four nucleotides could obviously only code for four amino acids. Two-letter sequences could combine to get only sixteen. A three-letter sequence, which would allow sixty-four combinations, called "codons," is therefore the minimum needed to get all twenty amino acids. In 1961, Marshall Nirenberg and J. H. Matthaei discovered how to add to a test-tube system an RNA message that synthesized proteins, and how to find out what amino acid was synthesized by it, and they showed that the triplet UUU was a codon for the amino-acid pheynlalanine. By the next year Francis Crick's lab had confirmed by genetic studies that the code was indeed a triplet code, and by 1966 all but three of the sixty-four possible triplets had been assigned their corresponding amino acid. The final three codons, UAA, UAG, and UGA, were found to be chain terminators—punctuation that specified the end of an amino acid "sentence."

The code was found to be identical in a wide variety of organisms and cell types: tobacco mosaic virus, yeast, *E. coli* bacteria, plant coat proteins, rat liver, amphibians, and mammals, including human beings. This pervasive pattern led Crick to propose that the genetic code was a "frozen accident" from early evolutionary history. He wrote, "To account for it being the same in all organisms one must assume that all life evolved from a single organism (more strictly, from a single closely interbreeding population)."[55] Subsequent research discovered a few slight deviations from the "universal" code—usually just a change in the assignment of one or two of the sixty-four codons, mostly occurring in mitochondria—and this has led to active research in the evolution of the code itself.[56] The near universality of the code supports the hypothesis that all forms of life on earth are related to one another by common evolution—at some point very early in the evolution of life a single chemistry emerged and all subsequent variation has built upon that structure. Evolutionary theory deals primarily with how life evolves rather than how it originated, and most evolutionary texts usually say no more than that it probably originated on earth just once or a few times. However, Thomas Fox explains that what we now know of the genetic code tends to support the single-origin hypothesis:

There are deviations from the standard code in some genetic systems, and in many (possibly all) organisms, departures from standard coding occur at specific sites. However, all the changes so far discovered are most easily viewed as divergences from a single standard genetic code. None of the evidence indicates that the variations in genetic codes arose as the result of independent origins of life.[57]

There are many other developments, as well as exciting prospects on the horizon, that would be well worth our attention had I enough space to devote to them. If the controversial evidence of "nanobacteria" in a Martian rock is borne out it could lead us to modify the view that life began on earth. Advances in developmental biology, such as the discovery of homeobox genes and homologous developmental pathways, have the potential to extend the evolutionary synthesis in significant ways. It would also be interesting to talk about the many ways that evolutionary theory interacts fruitfully with other disciplines, from economics to psychology to medicine to computer science. However, creationists have yet to turn their attack on these and other developments, so in the concluding sections of this chapter let me return to two more central points of contention: the role of randomness in the Darwinian mechanism and the evolution of human beings.

How Could Beings Evolve "By Chance"?

Chance Caught on the Wing

Jacques Monod described evolution as "chance caught on the wing." This is a beautifully poetic phrase, but for people who do not understand the Darwinian mechanism it is probably more misleading than helpful at first because of the way that it emphasizes chance. Chance is just luck, isn't it? Could it be that the amazingly adapted complexities we find in the biological world—the vertebrate eye, the symbiosis of flowering plants and their pollinators, the molecular machinery of cilia—is just a lucky coincidence? Of course it is not impossible that such things arose fully formed just by chance. The ancient Greek atomists thought that atoms connected at random and that eventually all combinations of atoms actually would occur because of infinite time. But according to current best estimates from physics, evolution did not have the luxury of infinite time, so it would be a happy accident indeed if we had popped fully formed into existence by chance alone; the odds against such an event truly are astronomical. In arguing against evolution, creationists bring up this point more than any other.

To emphasize the improbability of biological life, creationists like to quote astrophysicist Fred Hoyle, who once made a calculation of the likelihood that even a small typical enzyme, say one with just ten amino acid units, could form at random and concluded that it was comparable to the chance that a tornado whipping through a junkyard would stir up the scrap to form a Boeing 747. This is an arresting image, and it becomes more so as we pay attention to the details. We could examine an enzyme at the atomic level and note that for it to function the amino-acid chain must fold into a special three-dimensional molecular structure that will work like a key to lock or unlock other molecules in a catalytic reaction, but this is hard to picture, so let us stick with Hoyle's vivid analogy.

Think of just a wing by itself for a moment. It is no mean trick to get the shape of wings just right so that an object heavier than air is able to take flight and soar in apparent defiance of gravity. Of course birds and bees are not as big as Boeings, so it is much easier for a sparrow or a bumblebee to break the bonds of the earth. Even so, everyone seems to know the story of the aerodynamicist who supposedly proved that bumblebees should not be able to fly. Engineers still have not lived down this embarrassing miscalculation, which dates back to the early 1930s, but the story lives on in common lore and reinforces the apparent miraculousness of the ability to fly. Indeed, insect wings are quite remarkable. The film of which the lifting surface is composed is only 2 to 6 micrometers thick, and it forms a wing that is highly irregular in cross-section, but which nevertheless approximates a conventional airfoil. Engineers have tested the efficiency of the insect airfoil compared to a smooth envelope contour resembling an airplane airfoil and found that the former seems to be superior in both maximum lift and in drag. In an article on the bumblebee myth from which I gathered this information, John McMasters (who, by strange coincidence, is principal engineer for aerodynamics for Boeing Commercial Airplanes), concluded: "The neophyte designer of an airplane wing who approaches the problem from the perspective of aerodynamic optimization might well profit from the study of examples such as this one—provided by mere entomologists."[58] Could it be just fortunate happenstance that such efficient, aerodynamic shapes formed in nature? If evolution is just chance, then how could it ever "catch" a wing, let alone create a fully functional organism?

This seems to be the point that stumps people most often: Doesn't randomization destroy rather than create order? How can mere chance *create* anything functional? The answer to this conundrum is the special secret of the Darwinian mechanism: it is not chance alone that does the work, but random chance together with nonrandom natural selection. It is a supremely simple process, but one whose power is surprisingly difficult for people to appreciate, so let me develop its justification in stages, beginning with the worry about how chance could possibly be a creative force.

Randomness and Creativity
Contrary to what one might expect, the introduction of randomness into a system is one of the most important engines of creativity. Faced with a blank canvas, painters often spur their creative thoughts by splashing a bit of paint at random on the canvas. Jean Arp dropped shapes randomly as the basis for some of his sculptural pieces. The lyrical tunes of George Gershwin's musical play *Porgy and Bess* were written using random elements. People who have investigated the nature of creativity have discovered that this is surprisingly important in creative thinking and is by no means restricted to art.

Edward de Bono was one pioneer in the practical study of creativity. Paul MacCready, who designed the Gossamer Condor, the first significant human-powered airplane, credited de Bono's methods for helping his team come up with creative solutions to design problems. MacCready had taken on the challenge of human-powered flight, a challenge that had inspired but eluded our species since the mythical Icarus strapped feather-coated wax wings to his arms and flapped madly towards the sun, only to plunge immediately back to earth. Among other problems, MacCready's engineering team had to figure out a wing design that could maintain lift for a craft that would only move at the very slow speed that a cyclist turning a propeller with pedal power would be able to sustain. The team succeeded where all before them had failed, and the Gossamer Condor now proudly hangs suspended in perpetual flight within the airspace of the Smithsonian

National Air and Space Museum, together with the Wright Brothers' 1903 Flyer, Lindbergh's Lockheed 8 Sirius, and the Apollo 11 Command Module. I had the pleasure of hearing MacCready speak a couple of years after the first successful flight of the Gossamer Condor, about how his team overcame the difficult design problems they faced, and his recommendation led me to seek out de Bono's work on creativity.

Edward de Bono was primarily interested in creative reasoning—what he termed "lateral thinking"[59]—and in developing practical techniques to improve creative thinking. Reading through his suggestions for stimulating new ideas or creative problem solutions one soon notices that chance appears again and again in a variety of ways. According to de Bono, introducing randomness is a prominent factor in creative processes. Lateral thinking, he writes, is concerned with changing patterns (arrangements of information), and it deliberately seeks out apparently irrelevant information and chance intrusions as a way to generate new patterns. Again and again he mentions the utility of exposure to random stimulation or attending to random inputs. This is the same notion of randomness that appears in evolutionary theory—it is not that mutation has no cause (deterministic or indeterministic) but that the cause is not aimed at producing a particular desirable or advantageous result. De Bono says that the main point is that one is not looking for anything, but is just wandering aimlessly with a blank mind until something just pops out. He often suggests some formal method to generate a random input, such as a routine to select a chance object from the surroundings (e.g., nearest red object) or using the dictionary to provide a random word.

The idea that the introduction of randomness is an important creative force is not idiosyncratic to de Bono; one finds this point reiterated by others who have studied creativity. James Adams, director of the Design Division at Stanford's School of Engineering and member of the design team for Mariner IV, the first Venus spacecraft, writes about creativity in terms of what he calls "conceptual blockbusting" noting that one mental blockage to creative design is having "no appetite for chaos."[60] Koberg and Bagnall describe a procedure they call "morphological forced connections" (whereby one assembles the result of random runs through alternative variations) as being a "foolproof invention-finding scheme."[61] More interesting still is the earlier work of Alex Osborn, originator of the concept of the process of (and coiner of the term) "brainstorming," who was interested in creative imagination.[62] Like the others, he drew no connections to biology but his observations about creative processes have surprising natural counterparts in the biological world. This is particularly striking in his lists of procedures for coming up with new ideas. I'll mention just a sample:

- New ways to use it?
- Other uses if modified?
- New twist?
- Change meaning, color, motion, sound, odor, form, shape?

It is fascinating to read through Osborn's lists and realize just how many of the processes he recommends for generating useful, novel ideas are used regularly by evolution to produce useful new biological structures and functions. For example, he writes of random recombinations, permutations, reversals, multiplications, transpositions, substitutions, and so on. It would be interesting to go through the lists in detail to show how evolutionary processes follow the same patterns, but many of the parallels should be obvious from our earlier brief introduction to the molecular mechanisms of DNA replication.

Can chance create useful novelties? You bet. Random mutations and recombinations are the very springs of creative variation, and as genetic replicators, biological organisms are equipped with both.

From Artificial to Natural Selection

Creationists should be squirming in their seats by now, both irritated and pleased by this talk of randomness and creativity. At the first opportunity they will spring to their feet and argue that using the example of the role of randomness in human creativity, even when applied to the design of plane wings and such, to illuminate how chance could function in biological creativity is all irrelevant since in the former case it is *intelligent human beings* who make the creative design decisions. How does that support evolution? If anything, they would argue, it supports their view that creative design and development requires having an intelligent agent at the helm. I heard intelligent-design creationist William Dembski give a somewhat more precise version of this sort of argument in a talk on what he calls "the design inference." There are, he claimed, only three types of explanation—law, chance and design—and these are mutually exclusive. In trying to account for some phenomenon, one first checks to see whether it can be explained by appeal to laws of nature, and if it cannot, then one checks next to see whether chance is the explanation. If that also fails then the only remaining explanation is intentional design. Dembski calls this procedure his "explanatory filter." He argues that when one runs biological complexity through his explanatory filter, intelligent design turns out to be the only explanation.

Dembski's filter is problematic in several ways, for explanation is far more complex than this simple picture suggests. Here I will mention just one set of problems with the argument that relates to our discussion about the role of chance in evolution. First, the notions of natural law and chance are not mutually exclusive, at least in the scientific sense, since there are statistical as well as deterministic laws. Nor are law and design mutually exclusive; one may properly explain the movements of the hands of a clock by reference to the laws of mechanics that govern their motions, and also explain them in terms of the specific intentions of the clock's designer because we know his purpose was to arrange the components so they would keep accurate time in a conventional system of twelve sixty-minute hours. It is just such a notion that some Christians had in mind when they tried to explain the workings of what Newtonian physics said was a clockwork universe; they held that all the phenomena of the world could be explained in terms of deterministic natural laws and *also* explained in terms of design, God having carefully set the initial conditions and put things in lawful motion during Creation so that things would unfold as He intended. Some theistic evolutionists today hold a similar view, but they add the notion of randomness into the mix by also recognizing indeterministic laws. So, neither are chance and design mutually exclusive. Thus, intentional design always remains a possible explanation even after we have an explanation in terms of deterministic or indeterministic laws, but whether it is the correct explanation is a separate question. There are other problems with Dembski's design inference, but these considerations are sufficient to undermine his simple explanatory filter. We cannot identify design by a two-step process of elimination but rather must have positive evidence for it based, for example, upon knowledge of specific design intentions and the possibility of their having operated through some causal intervention in the situation under consideration.

But let us return to the general point that creationists old and new are trying to make here against evolutionary processes. In one way or another they try to argue that natural processes cannot do the creative job of producing the kinds of order found in the biological world. How could forms so wonderfully complex and functional as eyes and wings ever be sculpted by happenstance? Surely we need a sculptor to explain them. In a way this is correct; it is not chance alone that explains these pervasive amazing adaptations, but chance combined with the sculpting power of natural selection.[63]

Recall that it had been in reading Malthus that Darwin got his insight about how selection could work naturally, but when he published *The Origin of Species* two decades later, Darwin eased his readers into the view by first discussing the artificial selection of domesticated species. In the intervening time he had spent years talking with breeders and learning from their experience as well as taking up pigeon fancying himself to investigate it firsthand. In the *Origin* he first provided evidence to show that there seemed to be no limit on nature's variation—even the oldest cultivated plants, such as wheat, still yielded new varieties—and that heritability of such variation was the rule. He then pointed out the following remarkable feature of domesticated varieties:

[W]e see in them adaptation, not indeed to the animal's or plant's own good, but to man's use or fancy. . . . We cannot suppose that all the breeds were suddenly produced as perfect and as useful as we now see them; indeed, in many cases, we know that this has not been their history. The key is man's power of accumulative selection: nature gives successive variations; man adds them up in certain directions useful to him.[64]

Darwin explained how breeders are quite deliberate about the process of selection, and how they speak of an organism's organization as being "something plastic" that they could shape at will, but he suggested that change of domesticated species over the generations towards useful forms could occur even without deliberate design, and gave examples of what he called "unconscious selection." Mr. Buckley's and Mr. Burgess's sheep flocks, for instance, bred from Mr. Bakewell's original stock fifty years earlier, now looked like two separate varieties even though that had not been anyone's intention. The point is that sculpting a species simply requires some selective force, deliberate or not. The selective force can thus come as easily from constraints imposed upon a species by the unintelligent environment as from those imposed by intelligent human beings. It should come as no surprise, then, that species become adapted to their environment, given that it is that environment that selects over the generations against those individuals that have relatively less functional traits, and thereby allows to reproduce those fortunate individuals whose variations let them fit just well enough within its constraints.

Natural selection thus itself works in a way as a creative force. By favoring lucky fit variants, it increases their representation in subsequent generations, and thereby makes the improbable more probable. It is chance heritable variations together with natural selection that drives evolutionary change and shapes species so that they become adapted to their circumstances.

Darwin's concept of natural selection is often spoken of as the "survival of the fittest" but it was actually not Darwin but Herbert Spencer, one of the first to recognize the power of Darwin's idea and to take on the task of educating others about it, who coined this famous phrase. But perhaps we should say "infamous" phrase, for it has caused no end of confusion and trouble. Darwin himself was not initially happy with Spencer's catchphrase, though he eventually incorporated it into later editions of the *Origin*. One unintentional consequence of describing the mechanism of natural selection with this formula was that many assumed that being *fitter* meant that those who survived the struggle were *better* than those who fell by the wayside. In a narrow technical sense, this is true—to be fit is to be relatively better than one's competitors in being able to survive and reproduce in a particular environment—but when interpreted colloquially it can be easy to slide into thinking that the *better* are also *more deserving* in some moral sense. This pervasive misunderstanding arising from the colloquial uses of the terms led many people to take a particularly unsympathetic view of the downtrodden persons in society, which was not at all warranted by the theory. Society has yet to fully recover from the ills caused by this mistaken view.

The second problem with the formula was that biologists have often defined the technical notion of fitness in a way that seemed to open Darwin's theory to a criticism that has become known as the *tautology objec-tion*. Put simply, a tautology is a statement that is true by definition, such as "A rose is a rose is a rose" or "I am what I am." If Darwin's vaunted theory is tautologous, then it tells us nothing, but rather simply assumes what it purports to explain. Creationists bring up the tautology objection to suggest that evolutionary theory is "dogma incapable of refutation" and thus not scientific. Henry Morris, for one, made this argument in the first edition of *Scientific Creationism*[65] and he reiterated the point in the foreword to the second edition:

[N]atural selection has no predictive value and thus is a mere tautology, stating the obvious fact that organisms that "survive" are thereby decreed to have been the "fittest," but it reveals nothing whatever about how they evolved in the first place.[66]

We find no change in view in the most recent statement of the objection by Henry and John Morris in *The Modern Creation Trilogy*:

For many years, creationists have been pointing out the logical fallacy involved in attributing evolution to natural selection, stressing the inherently tautologous nature of the whole concept. That is, natural selection was supposed to insure [*sic*] "the survival of the fittest," but the only pragmatic way to define "the fittest" is "those who survive."[67]

However, there are many problems with such arguments, some of which arise simply from misunderstandings of evolutionary theory. First, it is not individual organisms that evolve but rather groups or populations of organisms; in the definition of evolution, the key term is "lineage." A lineage evolves when the relative proportion of given heritable traits changes over the generations. This descent with modification might involve only a slight change in the proportion of different *alleles* (that is, different forms of a gene), or it might involve substantial changes in the genome that eventually cause the divergences that form the phylogenetic tree of life. This point should also make it clear why the creationists' focus on just Spencer's phrase is also problematic, for mere survival is not the point. From an evolutionary point of view, it does not matter a whit that one survives to the age of Methuselah if one never has any offspring. As Ira Gershwin put it in *Porgy and Bess*, "what good is livin' if no gal will give in to no man who's 900 years?" Of course there are many other kinds of good in life (which is part of the reason the earlier social Darwinist slide to the moral notion of good was an error), but though different traits can be useful for achieving these, they matter to evolution only to the extent that they contribute directly or indirectly to differential modifications in the lineage. Creationists are also wrong to say that natural selection has no predictive value; and even if that were the case that is not what it means to be a tautology. Another sign that the Morrises do not understand the charge they are making is that they contradict it themselves. Just a few paragraphs after claiming that natural selection involves a logical fallacy because the notion of fitness is a tautology they do an about-face and claim that natural selection is part of God's design:

As a screening device for eliminating the unfit, natural selection is a valid concept, and, in fact, represents the Creator's plan for preventing harmful mutations from affecting and even destroying the entire species. And that is *all* it does![68]

It is true that natural selection eliminates harmful mutations (though it would be better to speak of harmful variations, since these arise not only by mutation but also by recombination), but that is not "*all* it does" since it simultaneously rewards beneficial mutations, and in this way adaptive traits proliferate as natural selection sculpts a species over the generations. But if to say that something is fit is simply to say that it survived, as they argued previously, then to say that something is unfit is simply to say that it didn't survive—so how can natural selection now be lauded as a "device for eliminating the unfit"? Why is it purported to be a logical fallacy when it appears in evolutionary theory, but suddenly deemed valid as part of God's plan?

Creationist writings are rife with such basic misunderstandings and muddled arguments, though the new intelligent-design creationists are usually more careful. Phillip Johnson does not make the point in such a rough and self-contradictory manner, but he too often brandishes the tautology objection, quoting several biologists from the early 1960s and saying "As long as outside critics were not paying attention, the absurdity of the tautology formulation was in no danger of exposure."[69] He even drags out Karl Popper's old mistake on this point, crediting him with making the problem with Darwinism public.

The famous philosopher of science Karl Popper at one time wrote that Darwinism is not really a scientific theory because natural selection is an all-purpose explanation which can account for anything, and which therefore explains nothing. Popper backed away from this position after he was besieged by indignant Darwinist protests, but he had plenty of justification for taking it.[70]

Of course Popper was correct that science should rule out all-purpose "explanations" (this is just one of the reasons, as we will see in chapter 6, that science does not consider the Creation hypothesis), but he was wrong to have thought that Darwinism fell prey to this problem. Johnson tries to make it sound as though Popper was cowed into submission by a Darwinist inquisition, but Sir Karl Popper was never one to "back away" when he thought he was right. So let us set the record straight. First, Popper's original complaint did not apply to evolutionary theory in general, but only to natural selection. He said all along that Darwin's core thesis of common descent was not only testable but "the most successful explanation"[71] of the biological and paleontological data; he took descent with modification to be a "historical fact."[72] Regarding the hypothesis of natural selection, Popper had originally claimed (influenced by the sorts of comments from the same biologists Johnson quotes) that it was "almost tautological" (and even then he continued to defend its utility on other grounds, and to use it in his own theory of knowledge), but once he understood the difference between evolution by natural selection and evolution by random drift he forthrightly corrected his mistake when he had the opportunity to do so in a later article:

The fact that the theory of natural selection is difficult to test has led some people . . . to claim that it is a tautology. . . . I mention this problem because I too belong among the culprits.[73]

The theory of natural selection may be so formulated that it is far from tautological. In this case it is not only testable, but it turns out to be not strictly universally true. There seem to be exceptions . . . and considering the random character of the variations on which natural selection operates, the occurrence of exceptions is not surprising.[74]

But it would not be right to set aside this issue with simply these statements from Popper. We need to see the reason for the confusion, since creationists continue to bring it up despite Popper's disavowal. Let me try to briefly explain the purported problem and its solution with a simple analogy.

Consider the formula: May the best team win. It seems harmless, but the creationist now points out that we determine which team is best by seeing which wins. If that is what it means to be "best," then the expressed wish seems to reduce to "May the team that wins be the team that wins." It is thus vacuous dogma, objects the creationist, to subsequently claim to explain who won in terms of the one team's being "better" than the other. However, we sports fans are not fooled into abandoning the game by such arguments. Of course we do determine which is the best team by looking at its record of wins, and we would certainly explain why it won the trophy by noting its superior record over its rivals. But we understand that this is not the end of the story, and that the reason behind the winning record was the relative strength and skill of the team's players. It was the heft of the forward line, the hand-eye coordination of the quarterback, the speed of the wide receiver, and also the group's properties like cooperative team-work that made one team stand out over the others. Obviously we are not able to measure or even enumerate all the relevant traits with much precision, let alone predict in advance exactly how one set will fare when pitted against another. If we could do this then no actual games would have to be played so we could see the actual record of wins and losses. But even though we do judge on the basis of the record, we do not doubt that it is the physical traits of a team, its superior characteristics and playing ability, that make it better than the others. Understanding this, we also understand that it is possible that the best team might *not* win. Every sports fan knows of a few games in which a clearly better team lost a deserved victory because a bad call from a referee resulted in an unwarranted penalty, or a freak injury sent the star player to the sidelines. This parallels the distinction that biologists make between evolution by natural selection and evolution by random drift, and the mere fact that we recognize such distinctions is by itself sufficient to show that the tautology objection does not hold in either sports or evolutionary theory.

With this analogy in hand, it should also be easier to understand other aspects of the biological concept of fitness and to see more fully why the tautology objection fails. It helps to show, for instance, why it is reasonable to measure fitness in terms of the relative contribution of alleles to succeeding generations. As in the sports example, there is no specific set of phenotypic traits that is better in an absolute sense, but different traits are nevertheless objectively better or worse in a particular environmental situation. The quarterback might have a great throwing arm, but his team could still be at a disadvantage in the rain or snow against a team that has a stronger game on the ground. A similar sort of relativity of fitness applies in nature. Where the limiting food source is large, hard-to-crack seeds, having big strong beaks will be fitter because it helps one feed oneself and thereby live to reproductive age whereas the slender-beaked individuals will starve and die off. Where the limiting food source is tiny seeds sheltered in spaces that require a reach, however, it could be the long slender beak that is the fitter, other things being equal. Of course, in the real world other things rarely are equal; there are a thousand other traits that might also make a difference, positive or negative. As in the case of our football team, it is practically impossible to list and track the whole array of traits in a dynamic environment; the technical notion of fitness offers a way to proceed through this complexity by providing a single common measure by which to compare traits. Other conceptual issues remain, but we need not worry about the measure because, as in sports, we know that there are specific causal relations that determine what it is that makes a specific trait fit in a given situation.

Indeed, scientists have studied a wide variety of such traits both in the lab and in the field, and have even begun to measure such causal relations at work in nature. I'll mention just one of the most well-known and significant of such studies, which was carried out on the very group of islands and species that provided the first clues for Darwin. Biologists Peter and Rosemary Grant have spent more than two decades in the Galapagos islands, with tweezers, sieves, scales, and calipers in hand, measuring the beaks and weighing thousands of finches and then measuring the availability of the various seeds they were feeding upon. In 1977 they observed natural selection in action when a drought killed nearly four-fifths of the finches on the island. Over the course of that year the available mass of seeds had progressively declined, and the average size and hardness of those seeds that remained progressively increased. As the food supply thus shrank, so did the number of finches of the various species in a pattern that reflected their differential abilities to utilize the remaining seeds. Smaller birds—females generally and males of the smallest species—fared the worst. The biggest birds with the deepest beaks were the best off. On average, a difference of half a millimeter in beak size separated finches that could crack enough of the harder seeds to survive from those that could not and thereby died. So natural selection had operated as expected. But would this differential survival translate to inherited differences in the next generation? There were now six male finches to every female that had survived, and the team watched as the females flew from the territory of one male to the next to check out their prospects. Darwin's suggestion that female choice or "sexual selection" played a role in evolution had been perhaps the most controversial aspect of his theory, but here it also was seen at work. The females did not pick their mates at random, but chose the largest of the surviving males, whose plumage was the most mature and who had the deepest beaks. And, sure enough, when their offspring hatched these were on average bigger and had beaks that were 4 or 5 percent deeper than the previous generation. The proportions of other traits had shifted as well, and statistical analysis of the data allowed the researchers to tell which natural selection had favored the most.

Phillip Johnson and other creationists are quick to try to discount the Grants' observations of natural selection and inheritance in Galapagos finches, saying they are "unimpressed" by such minor variations. Much as Frank Lewis Marsh did, they want to see new body plans and new complex organs and not "mere microevolution."[75] But to criticize the finch-beak changes in this way misses the significance of the study, which is that the Grants were able to see the Darwinian mechanism at work, sculpting specific traits. Scientists already had ample evidence that allowed them to infer that natural selection occurred, but the Grants observed the process directly and quantitatively measured its effects. The Darwinian mechanism can be studied in the lab, but this study and others like it are remarkable in demonstrating it in the wild. Johnson's assessment could not be more wrong. The tale of the Grant's research has been told beautifully by Jonathan Weiner in *The Beak of the Finch* (1994), and readers who wish to judge for themselves can find more of the details there.

Darwinian Engineering

Not all evolutionary change is produced by natural selection, because not all variations result in an adaptive difference in a particular environment. Adaptively neutral genetic traits can simply drift at random in one direction rather than another. There might also turn out to be underlying patterns of order that emerge because of the self-organizing properties of non-equilibrium systems, or because of limits imposed on evolutionary change by developmental constraints. However, we can safely say that when biological forms do make a difference, directly or indirectly, to the ability to survive and reproduce in an environment, then the Darwinian mechanism will immediately kick in to produce a more and more satisfactory form. As new heritable variations arise, natural selection will tend to weed out those that are not good enough to do the job and preserve those that do it just a bit better in the competition so that over time the form will tend toward an optimum. Given the basic elements of replication, random heritable variation, and selection, the process works and maintains itself automatically.

Scientists are only now beginning to recognize the full power of the Darwinian mechanism and to see that it can be used in practical design applications. In one early experiment, aerodynamicists showed how one might use the process to improve the design of airplane wings. They built a frame whose shape could be varied, and in a series of steps they introduced small random alterations to the shape and tested how well each configuration performed in a wind tunnel. The configurations that produced less lift were eliminated from the competition, and the winner was used as the "parent" for the next series of random variations. Lo and behold, as the process was repeated over successive "generations," the resulting configurations began to take on the curved shape of the classic airfoil. What is important here is that the aerodynamicists were not making any design *decisions* about the shape as it evolved; indeed, in principle they could have been eliminated from the scene entirely and replaced by a couple of robots. The selections were made simply on the basis of the measurement of the aerodynamic effects, and the shape evolved naturally.

Of course, working with a crude frame model limits the plasticity of the form. Also, the number of replications and repetitions that could be performed was relatively small, especially compared to the number that occurs in evolutionary time. So, though the experiment gave a nice demonstration of the Darwinian mechanism and a hint of its power, such constraints severely limited the practical utility of the procedure. All this has now changed with the advent of powerful computers that can simulate complex structures and that have processors fast enough to churn through a sufficiently large number of replications. A whole field of research is springing up known as *evolutionary computation,* which implements the Darwinian process on a computer and turns it loose on a design problem. The method is essentially the same as described above: Construct a computer model that can represent the problem space and mutate this to produce a large "population" of random variants; then allow a selection function to "weed out" all but the most fit (i.e., keeping those that do somewhat better than others in accomplishing the set task); then take these winners, replicate them with new random mutations, and weed them again; and repeat this generation after generation. One can also make the process more sophisticated by allowing variations to arise through random recombination, for example, by randomly "mating" winners. Evolutionary computation systems of this sort have already been used successfully by industry in the design of complex circuits, manufacturing schedules, control robots, natural-gas transport pipeline networks, and more. The turbine geometry of the engine of Boeing's 777 airplane was designed using evolutionary programming, and Boeing is experimenting with the technique to evolve new wing designs. At Stanford University, evolution of airplane wings in a computer for 150 "generations" came up with a radically novel wing shape.[76] Imagine taking the wings of a plane and folding them about two-thirds of the way along their lengths so they point straight up, and then folding half or more of that length again so the tips point back horizontally toward the fuselage. A plane built with "C-wings" would not need a tail, and its shorter wing-span would allow it to be larger, carrying half again as many more passengers as a 747, while still meeting airport and manufacturing size constraints.

Again, note that this is not a case of the computer scientist surreptitiously "programming in" the design. Though in many cases the programmer does have some knowledge in advance of the problem domain and includes this information to narrow the search space, the program does not "know" in advance what will work. Often the programmer is equally in the dark and is surprised by the design solution a genetic algorithm (one of several kinds of evolutionary computation scheme) comes up with. Such experimental results show something of the power of the combination of chance variation and selection to produce novel and adaptive solutions to complex design problems. Of course, while these systems exemplify the key aspects of the evolutionary process, they do not fully model evolution in nature.

They typically have a single preset fitness function, whereas in nature what it is to be fit itself varies because individuals are in a changing environment and also in competition with one another. In nature there is no fixed "target" except the ability to survive to reproduce and we find a vast range of adaptations that do this in many complex ways. Can evolutionary computation produce *this* sort of complexity? If you fully appreciate the meaning of Darwin's theory and the evidence for it then you should not be surprised to learn that it can, as research in the new field of what is called "artificial life" is beginning to show.

One of the most suggestive demonstrations so far of this is an artificial life environment known as "Tierra,"[77] which was set up by Thomas Ray, a field ecologist at the University of Delaware. Ray programmed a simplified abstract model of a biological system—a block of computer memory that was an electronic "world" in which computer-program "organisms" could move and replicate. He began with a computer language with a very small set of instructions to mimic the size of the genetic code. Recall that in DNA the genetic "program" is a string of Gs, As, Cs, and Ts coded in groups of three, each group representing one amino acid. In Tierra a program is a string of 0s and 1s coded in groups of five, each group representing one of the 32 instructions. A digital organism in this environment is simply a line of instructions that can occupy a space in the memory block. Then Ray programmed a short 80–instruction-long program that did nothing but replicate itself—a digital "organism" he later dubbed the "Ancestor"— and set it running in the virtual environment. If that had been the extent of the system, the ancestor would have done nothing but copy itself until the available memory space was filled. But Ray had also set up the environment to model the two other key Darwinian processes. Chance mutations that simply switched a "letter" in the string (a 0 to a 1 or vice versa) were periodically introduced into the system to mimic mutagenic cosmic rays and replication errors. These were random in just the Darwinian sense, in that they occurred in unpredictable locations without any foresight as to what effect they would have on the program. As in the biological case, most mutations turned out to be neutral or harmful (here in the sense of producing execution errors when the program ran). Tierra also included a "reaper," which mimicked selective death by eliminating the oldest and most defective organisms (ones that erred most often in executing their instructions). Ray set the ancestor running in this environment and watched as it replicated itself and as mutations soon introduced tiny novelties in the sequence of some of the copies. He then left the system to run and went home to bed.

Professor Ray had guessed that there might be some possibility that a program with as few as 76 instructions could evolve, but he said he was "floored" to find when he returned the next morning that Tierra had evolved an organism only 22 instructions long that could replicate six times faster than the ancestor. More astounding still was that a veritable menagerie of other unexpected digital organisms had evolved, which exhibited novel interactions and surprising functional diversity. Some large organisms arose, including one with 23,000 instructions, but these could not compete against the smaller and faster ones and became extinct. Some programs could not replicate on their own but could do so by parasitically making use of the code of a host. Hosts then evolved that were "immune" to the parasite, and later new parasites arose that overcame that acquired resistance. "Hyperparasites" evolved with an innovation that allowed them to steal compute-time (the amount of processing time allotted to them) from the normal parasites. Moreover, after driving the normal parasites to extinction the hyperparasites formed mutualistic groups with each other that allowed them to cooperate in copying each other—but then a "cheater" evolved that could invade their groups. One organism evolved a way to execute three instructions in a row instead of the standard one. Ray didn't understand what was going on in this case, but computer scientists recognized it at once as a programming trick called "unrolling the loop" that increases efficiency.

Remember, Ray defined no explicit fitness function for the properties that emerged—there were no preset "targets" in the system. Programs running in the environment would simply compete with each other in the sense that they do better or worse at acquiring compute-time (Tierra's analogue of energy) and computer memory (its analogue of territory). Which specific sequences were "fit" at any given time depended upon the surrounding conditions in the environment, and these changed (as they do in a biological ecology) depending on what the other organisms were doing. The novel properties that arose had done so without any prior design or any directive interventions by a human operator.

Evolutionary computation and artificial life systems open up an important new source of evidence about evolutionary processes. As the first system of its kind, Tierra gives only a glimpse of the possibilities. An abstract environment like Tierra with its digital organisms is especially compelling as an illustration of the creative power of an evolutionary system; but one can also choose to simulate known biological organisms, as other researchers are now attempting to model, for example, fish and insects and simple ecologies. In either case, researchers can run these systems repeatedly, performing experimental manipulations to study the effects of different initial conditions, mutation rates, types of recombination, kind and strength of selective forces, and so on.

If I have succeeded in getting you to think like a philosopher you will understand the key point in all this: Although it is an analogy to speak of Ray's Ancestor as an "organism," and although digital "fish" are only mathematical models of real ones, the evolutionary processes that drive these artificial environments are not simulations but the real thing. The Darwinian mechanism is completely general; it is instantiated in reproducing biological organisms, but it can also be instantiated in other systems. Computers are terrific at repetitive tasks and so they proved a natural setting, but any system with the requisite properties (minimally: random variation, replication and selection) can evolve. Biochemical engineers, who typically design molecules for industrial uses by hand, are just in the last few years beginning to put the Darwinian process to use to come up with, for example, new and more efficient enzymes that will be useful in food processing, detergents, and as catalysts in organic solvents for industrial and pharmaceutical applications. Darwinian engineering can push beyond the limits of the designer. As Nobel Laureate Manfred Eigen, who now is chief scientist at a new company with the self-explanatory name *Evotech*, has said: "That's why these evolutionary technologies are a great advantage—you can solve [design] problems in ways you never would think of."[78]

In speaking of bumblebee aerodynamics, McMasters had suggested that engineers could learn a few things about wing design from entomologists. It seems that they have now also learned a few things from evolutionists. Birds and bees have evolved efficient airfoils just by doing what birds and bees do—making other birds and bees and letting natural selection take care of the rest. That is the Darwinian evolutionary process—not chance alone, but chance caught on the wing—chance captured and channeled by natural selection.

"Is Man an Ape or an Angel?"

Darwin's achievement was to discover a powerful set of truths and to support them with clear evidence. I began this chapter by pointing out how obvious even to the untrained eye are the similarities of musculature between human beings and bonobos. Such patterns of similarity have a simple explanation—our species are close evolutionary cousins, both having descended with modification from relatively recent common ancestors. With perhaps a couple of exceptions such as Michael Behe, creationists uniformly reject this evolutionary thesis that we are descended from apelike ancestors. This disagreement concerns the pathways of evolution, in particular the pathway that led to the origin of human beings. Although many evolutionary pathways are still unknown or controversial, much of the broad outline is known, as well as some of the details. Nothing that scientists have learned would suggest that human evolution from earlier primate forms is in doubt. In this last section let us look briefly at this issue.

In his *Descent of Man*, Darwin laid out the original case for human evolution. He identified as part of the argument a wide range of traits that we share with apes and other primates. Not only do we have the same general body shape but we also share much of the detailed structure of our bones, nerves, blood-vessels, internal viscera, and even similar fissures and foldings of the brain. Embryological development is also very similar. Darwin noted a wide variety of indirect connections as well. For instance, the two-way transmission of diseases between man and animal evidenced the close similarity of tissues and blood. He noted that monkeys suffer from apoplexy, inflammation of bowels, and cataracts as we do, and that medicines often produce the same effect on monkeys as on us. The abundant evidence of this sort that Darwin provided was itself strong support for the thesis that human beings and other primates that exist today share a common ancestor, and much more evidence has been gathered since.

In Darwin's time scientists were only beginning to study fossils, so Darwin had no examples of possible ancestral forms of human beings, though his theory predicted that such should exist. Today the fossil record of intermediates prior to humans is surprisingly detailed and getting more so. As new fossils of hominid forms continue to be discovered, these add branches to the human family tree and sometimes force scientists to revise their maps of the paths of our recent evolutionary history. (This can occur in much the same way that people trying to trace their family genealogies at times have to redraw their tree when they run across the name of a previously unknown relative who at first appears to be a great-great-grand-father but then, upon uncovering some of his letters, turns out to be a distant uncle, say, or a cousin several times removed instead.) One hotly debated puzzle within physical anthropology involves the evolutionary relationship of modern human beings to Neanderthals, the brawny homi-nids whose bones were first discovered in a cave in Germany's Neander valley and from which we get the popular "caveman" stereotype. The bone structure of the classic Neanderthal skulls—lower jaw sloping back without a chin, the nose, upper jaw, and teeth thrust forward, eyes set in deep sockets, eyebrows on prominent bony ridges, forehead low and slop-ing back, but with a 10 percent *larger* brain capacity than ours—is very distinctive, as are their teeth and limb bones. At Neanderthal sites, archeol-ogists find only crude stone tools, but regular evidence of the use of fire. Although there is no mistaking Neanderthals for the modern human spe-cies, even their primitive use of technology is sufficient to show that they were more intelligent than apes.

From the evidence of their bones (found in locations ranging through southern European Russia and into Central Asia almost as far as Afghani-stan), we know that after having existed for some 300,000 years, Neander-thals disappeared about 30,000 years ago. The question is whether they disappeared because they gradually evolved into modern humans, or whether they were a branch species that became extinct, losing out to a separate evolutionary line from which all modern humans are descended that emerged out of Africa only a few dozen millennia ago. Evidence estab-lishes that both *Homo sapiens* and Neanderthals existed simultaneously in Europe for a period, which is consistent with both hypotheses. As late as the 1980s, anthropologists trying to solve this and other pathway puzzles could only search patiently for more bones, waiting for new finds that would provide additional information to test the hypotheses. However, with the revolutionary advance of techniques in molecular biology in the 1990s, especially the polymerase chain-reaction (PCR) that now allows biologists to amplify even tiny bits of DNA to the point where they can be sequenced, previously inaccessible evidence is coming to light that can help answer such questions. In June of 1997, Svanta Paabo's lab in Germany announced that it had successfully extracted and sequenced DNA from a Neanderthal skeleton. Comparing differences between the Neanderthal sequence and that of modern humans, Paabo could calculate an estimate of when the Neanderthal line diverged from the line that eventually became *Homo sapiens*. The DNA evidence showed that the split probably occurred about 600,000 years ago, thereby supporting the "out of Africa" hypothesis. Even partisans of the opposing hypothesis expressed admiration for the work's quality and excitement about its implications. So, is it now conclusive that the Neanderthal is not our direct ancestor, not a direct link to more distant ancestors, but is instead a recent cousin? No, not conclusive, for further evidence could conceivably force a reevaluation—that is the nature of inductive evidence, after all—but this test certainly weighs heavily against the direct path hypothesis.

No doubt some creationists are now hoping that this particular result will be overturned, for they previously argued that Neanderthals were not a "missing link" because they were fully human. Neanderthal skeletons had been systematically misinterpreted, they said, and were not a stocky earlier species at all but simply human beings who had suffered from rickets. Of course, other creationists will now happily relegate the Neanderthal to the category of being neither human nor a missing link. As they see it hominid fossil finds are either fully human or fully inhuman and never the twain shall meet. Both views exhibit the same all-or-nothing mind-set, which in this case wants to rule out any possibility of a connection between human beings and apes.

However, the important fact about Neanderthals and the various other early hominids we know of from fossils is that they exhibit both human and ape-like features. They are clear intermediates. When tracing our personal family histories we might be more interested in finding out first about our great-great-grandparents than our great-great-grand-uncles and aunts and their children (our cousins many times removed) but they are all part of our family tree. Hominid intermediates allow paleontologists to draw more of the map of nearby evolutionary pathways, and as evidence of human evolution it matters rather little that some of these hominid species are direct ancestors whereas others like the Neanderthals might be off on a bit of a side branch.

From measurements of the sequences of people from around the world, we know that modern human beings differ by an average of eight variations in the sequence of mitochondrial DNA that Paabo's lab examined. Paabo found that the Neanderthal specimen had twenty-seven differences. By way of comparison, chimpanzees differ from people by fifty-five variations. This kind of new detailed evidence from molecular biology allows us to calculate that the line of descent that led to us and the line that led to the chimps diverged from a common ancestor about seven million years ago, give or take a few million years. The branch that led to contemporary gorillas had already split off a few million years earlier. This may seem a pretty wide margin of error until one acquires an appreciation of evolutionary time; to find a common ancestor with monkeys takes us back an order of magnitude further, some thirty million years ago.

In 1864, the British statesman and novelist Benjamin Disraeli expressed not only disagreement with Darwin's recently published evolutionary theory but moral outrage for it as well:

Is man an ape or an angel? I, my lord, I am on the side of the angels. I repudiate with indignation and abhorrence those newfangled theories.[79]

Disraeli's dilemma expresses what is probably the main concern of creationists, namely, that humans cannot be descended from ape-like animals. I must admit to finding this a rather silly worry. Given that we have so many traits in common with the great apes, why should it matter that we also share a common ancestor with them? Of course the commonalities include traits of which we might not be very proud. Humans and primates alike are susceptible to the vices of tobacco and alcohol, for example. Darwin himself had witnessed monkeys smoke tobacco with pleasure, and he knew of reports that primates suffered the effects of intoxication.

Brehm asserts that the natives of north-eastern Africa catch the wild baboons by exposing vessels with strong beer, by which they are made drunk. He has seen some of these animals, which he kept in confinement, in this state; and he gives a laughable account of their behavior and strange grimaces. On the following morning they were very cross and dismal; they held their aching heads with both hands, and wore a most pitiable expression when beer or wine was offered them, they turned away with disgust, but relished the juice of lemons.[80]

Could it be because of such ignoble similarities that creationists object to our kinship? Certainly the wife of the Bishop of Worcester seemed to have such worries in mind, as do creationists today who argue that when students are taught that they are descended from animals it should be no surprise when they start acting like animals.

However, from a moral point of view, where we came from is far less important than where we are going—where we set our sights. Back when Darwin was still just beginning to toy with the idea that transmutation could answer the question posed by John Herschel, England's most renowned scientist in his day, of how new species arose to fill the gaps left by extinction, he had written: "If all men were dead then monkeys make men.—Men make angels."[81] The fact of the matter is that we have our vices and our moral weaknesses and we just have to deal with them; we are what we are now, however it is that we got here. School children (and more than a few adults) have been acting "like animals" since well before the discovery of evolution, so it is rather unfair to blame such behavior on Darwin.

Indeed we are also being quite unfair to animals by focusing only upon their aggressive side, as we do with expressions like "brutish," "beastly" and "animalistic." Although it is true that nonhuman animals sometimes exhibit violent behaviors that we would rightly find morally repugnant in human beings, they also exhibit a wide range of behaviors that we think of as virtuous in ourselves. In 1997, people around the world saw evidence of this when a young boy visiting a zoo fell into the gorilla habitat. Spectators watched in horror as a large gorilla approached his unconscious body. They were all expecting the worst, but what the gorilla did was examine and then gently lift the little boy in her arms. She then carefully carried his limp body over to the door of the enclosure and waited until the keeper came to get him. All was well. So much for our prejudices about what it means to act "like an animal."

Such behavior is especially evident in those species that are closely related to us. We differ in our DNA from the common and pygmy chimpanzees by only about 1.6 percent, so it is not surprising that we find in them exceedingly familiar traits that go beyond the physical. For example, chimps may greet one another with a hug or a kiss, and Jane Goodall noted that "if they have been separated for a week or more, they are likely to fling their arms around each other with grunts or little screams of excitement."[82] They reassure each other with an extended hand and a gentle touch. They share food with those expressing need. They soothe those in distress. They also care for their young and for each other in ways that humans can easily empathize with. Goodall observed many such friendly and altruistic behaviors among the chimpanzees of the Gombe. I will mention just one example she gives that concludes a chapter full of such cases. On several separate occasions she observed a young female, Little Bee, bringing fruit down from the trees to her mother, who had become too weak to feed herself.

[Little Bee] climbed down to the old female with her mouth and one hand full of palm nuts and laid the fruits from her hand beside her mother. Quite clearly she had some understanding of the needs of the old female. This ability to empathize, so highly developed in our own species, prompts much altruism in humans. If we know that another, especially a close relative or friend, is suffering, then we ourselves become emotionally disturbed, sometimes to the point of anguish. Only by helping (or trying to help) can we hope to alleviate our own distress. Was Little Bee, I wonder, motivated by a similar kind of emotion? Whatever the answer, it is evident that chimpanzees have made considerable progress along the road to humanlike love and compassion.[83]

The variety of new evidence that supports Goodall's assessment is impressive but it would have come as no surprise to Darwin, who spent years investigating animals' emotions and eventually wrote a book on the subject entitled *Expression of Emotion in Man and Animals* (1872). One of the main objections that people had made against his thesis of human evolution was that humans had mental traits that were totally different in kind from those of lower animals and so could not have evolved from them. Darwin responded with a systematic survey of the purportedly unique human qualities, from language to shame, showing that they could be found in rudimentary forms elsewhere in the animal kingdom.

The difference in mind between man and the higher animals, great as it is, certainly is one of degree and not of kind. We have seen that the senses and intuitions, the various emotions and faculties, such as love, memory, attention, curiosity, imitation, reason, &c., of which man boasts, may be found in an incipient, or even sometimes in a well-developed condition, in the lower animals.[84]

Darwin also confronted head-on the issue of morality. He agreed that only human beings could now properly be called moral beings, but he argued that one could find a precursor to the moral sentiments in the social instincts. As we have seen, fellow feeling may clearly be observed in animals, especially primates. Sympathy is the keystone of morality, Darwin argued, and as intelligence and foresight increased these would together combine to produce the moral sense:

[T]he social instincts . . . with the aid of active intellectual powers and the effects of habit, naturally lead to the golden rule, "As ye would that men should do to you, do ye to them likewise"; and this lies at the foundation of morality.[85]

Darwin even proposed a biological variation of John Stuart Mill's Utilitarianism—the ethical theory which holds that one should always act to produce the greatest good for the greatest number—that he thought could lend support to human evolution by linking his theory to the best philosophical work on ethics of his day. In this final bold move Darwin went beyond the evidence—one cannot simply "biologize" a moral theory without trespassing beyond the limits of science—but he was certainly correct to point out that there was nothing about our *capacity* for morality that was different in kind from traits found in other animals.[86] The commonalities we find with other primates is persuasive evidence that the intellectual and emotional characteristics that allow moral action in human beings are the result of our evolutionary development.

We cannot help but recognize ourselves when we observe friendly and altruistic behaviors in our primate cousins, and it is equally reasonable to say that humans exhibit "chimp" qualities as to say that the chimps exhibit "human" qualities; really they are simply traits that we have in common. Of course we may wish that our species expressed love, compassion, and the other moral virtues more consistently than it does, but we are not hindered in our striving for such an ideal by knowing that we have evolved from ancestors that had these qualities in a smaller measure. Indeed, knowing that moral improvement can and has occurred might just as easily help and inspire us in our efforts.

We thus should not be bothered by Disraeli's dilemma, for we are neither apes nor angels but simply human beings, somewhere in between. Our species has evolved biologically from "ape" ancestors and this in no way prevents our evolving morally toward an "angelic" ideal. Evolution in all its complexities is no longer a "newfangled" theory but a set of truths that is well established by the evidence.

3

The Tower of Babel

If we possessed a perfect pedigree of mankind, a genealogical arrangement of the races of man would afford the best classification of the various languages now spoken throughout the world; and if all extinct languages, and all intermediate and slowly changing dialects, were to be included, such an arrangement would be the only possible one.

—Charles Darwin

Therfor was callid the name of it Babel, for there was confoundid the lippe of all the erthe.

—Genesis 11:9, translated by John Wyclif (1382)

Origins

For creationists, all science, at least all "True Science," has a biblical basis. Creationists ignore or discount the weight of scientific evidence for evolution because they believe it is necessarily inferior to the evidence of the Bible, which alone contains God's revealed literal truth. They therefore begin by looking to Scripture to learn how the world came into being, how life was formed, or how humans were created, and they then interpret the empirical evidence so it will fit that picture. These sorts of questions are grouped together by creationists as issues of "origins." The terms "origins" or "origins science" are common, almost identifying shibboleths, in the writings of creationists. In part, this terminology is a way to avoid directly mentioning Genesis in secular contexts, but it also harkens to the creationist's central idea that by knowing a thing's origin one can know its God-given purpose. A correct understanding of origins is, therefore, a prerequisite to understanding God's plan for us, and the Bible is the only true authority for such questions. The exhibits in ICR's Museum of Creation and Earth History include references to biblical passages so that one can check the "evidences" that support their answers. According to creationists, good Christians put their faith in these divinely revealed truths, interpreted literally or robustly, and should not be swayed by the secular theories of the sciences, at least when they conflict with the biblical accounts.

Creationists are demanding that their beliefs on these matters be introduced as a respectable alternative view in geology, biology, and other science classrooms. When we think of the "creationism controversy," it is primarily the conflict with biology and secondarily the conflict with geology that come to mind. However, the stories of the Creation of the universe, the earth, the animals, and of Adam and Eve, are not the only accounts of origins to be found in the Bible. The diligent reader of Genesis also discovers in it the divine origins, for instance, of our human feelings of guilt and shame, of human mortality, of the optical effect of rainbows, of our dislike of snakes, of the pain of childbirth, and of the difficulties of agricultural production. The sweep of creationism is broad, indeed, but so far we have not seen publicized attacks from creationists on these other fronts in the same way that we have seen their united assault upon evolutionary theory. But these should all be considered equivalent from the creationist point of view for they are all purportedly revealed truths; so the consistent creationist has to endorse and promote them in the same way. Will psychology departments soon hear creationists' demands to include the story of Eve's temptation by the serpent to eat the fruit of the Tree of Knowledge of Good and Evil and to give it balanced treatment with the studies by Piaget, Kohlberg, and Gilligan of moral development in children? Will medical schools soon be told that their obstetrics classes should teach that God made childbirth painful as punishment for Eve's disobedience? Will agronomy professors and Future Farmers of America clubs in public schools soon be required to discuss God's curse upon the land when He expelled Adam and Eve from Eden as the origin of the low fertility of the soil and the reason that farmers must sweat to make it bear fruit?

Psychologists, obstetricians, and agronomists probably will find it farfetched to think that creationists would seriously challenge their sciences in this way. Such challenges seem ridiculous, too ridiculous to be taken seriously by legislators. But biologists find the attacks upon their science to be equally ridiculous—yet that does not seem to negate the seriousness of the threat. Indeed, when creationists make headway it is often because scientists do not take them seriously soon enough. Then an antievolution law is passed or a creationist candidate wins election to a school board, and scientists suddenly wake to find the damage done and must scramble to repair it. Of course, the biological case touches the prejudices and excites the passions of the general public in a way that the other issues do not, so it is less likely that creationists would be able to find enough support to press their agenda on these other issues. Most people do not get worked up about agriculture and psychology in the same way they do about evolution. Until they study and appreciate the beauty of the evolutionary picture, many of the lay public seem to find it unappealing or even a little disgusting to think that they are descended from other primates. They often prefer the idea that God intentionally formed man in His own image in a unique act of special creation, and creationists are able to exploit this preference. On the other hand, creationists are politically astute enough to realize that people are less sympathetic to some of the other biblical origin stories, like the view that God continues to punish us for Adam and Eve's original sin with the pain of labor at birth and with toil in the fields, so they do not mention these in the public debate. Still, the creationist viewpoint must regard these and other such cases in the same way it regards evolution. So, if creationists succeed in their attack upon biology, then the other sciences fall without resistance because the creationist arguments apply in the same way to them.

Scientists who have made the effort to debate creationists and to show the mistakes in their "scientific" case against evolution typically express frustration that creationists continue to make the same criticisms and arguments even decades after they have been rebutted. Even more frustrating to them is that audiences for these debates, usually comprised of Christians who are not scientists, continue to be taken in by these bad arguments. One reason audiences side with the creationists is that they do not have sufficient background knowledge in science to understand and adequately evaluate the arguments; but another reason could be that many people are predisposed to reject the distasteful idea that we humans evolved from ape-like ancestors. My thought is that it might help people see the weaknesses of creationist arguments if they could examine them in a context in which they did not already have a preference for the creationist conclusions. We could do this with any of the cases mentioned above, but instead let us look at another example in which the equivalence to the biological case is especially clear—the question of the origin of languages.

Everyone is familiar with the great variety of human languages—there are over five thousand spoken in the world today—and most people have studied English, Spanish, French, German, or Latin in high school, and perhaps gone on later to learn Portuguese, Hindi, Chinese, or Arabic. How did these and the many other languages arise? Linguistic theory holds that languages evolve from one another. Paleolinguists trace our modern languages back to their earlier forms, they find evidence of when languages branched off from one another, they examine the remains of languages that are now extinct, and they chart the lineages of their gradual development. We will look into linguistic evolution and the evidence for it in some detail shortly, but first let us ask how creationists should view this evolutionary picture of the origin of languages. The answer is straightforward: they should reject it in the same way and for the same reasons they reject biological evolution.

The creationist opposition to this scientific picture of the evolution of languages has the same source as their opposition to the evolution of species, for in the book of Genesis we find not only the story of the creation of animals and humans, but also the story of Yahweh's creation of the different languages of the world—the story of the Tower of Babel.

Creationist Linguistics

As in the other origin stories, the creation of languages allegedly took place at a specific place and time. According to the Genesis account, Yahweh created today's languages at Babel in the land of Shinar (Babylonia). The story begins following the great Flood. After a year in the Ark, Noah finds that the waters that had covered the whole earth and submerged the highest mountains have finally receded. God bids Noah to disembark with his family and all the animals, and He promises them that He will never again curse and destroy the earth because of the evil that contrives in the heart of man from infancy (Gen. 8:21). Setting out the rainbow as a sign of this Covenant, God sends the survivors forth to be lords of the earth, and, as in Eden, to once again be fruitful and multiply. Noah's three sons, Shem, Ham and Japheth, and their wives (whose names are not mentioned) take God's commandment seriously, as they must since the entire world will be repopulated by their offspring (Gen. 9:1–19). They also begin again to till the land, sweating now to make it bear fruit. Noah grows grapes and becomes inebriated drinking wine. Ham has the misfortune to see his father drunk and naked, and for this Noah curses him and his children to be slaves of his brothers' families (Gen. 9:20–27). However, the work of repopulating the world goes on and in the three centuries following the flood the brothers have many children and grandchildren, who disperse to form the nations of the earth "according to their tribes and languages" (Gen. 10:5, 20, 31).

Presumably, Noah's family originally all spoke the same language (Augustine and nearly all the early church Fathers were certain it was Hebrew, but later Christians disputed this), so how did they come to be divided in this way? How did these linguistic differences arise? The next chapter of Genesis provides the explanation.

At first, we learn, everyone does speak "the same language, with the same vocabulary" (Gen. 11:1–2). This linguistic unity makes the descendants of Noah powerful and ambitious, and, settling in the plain of Shinar, they begin to build a town marked by a tower, the top of which would reach heaven itself. Yahweh notes this activity and decides to stop it once and for all by removing the source of their power.

They are all a single people with a single language! This is but the start of their undertakings! There will be nothing too hard for them to do. Come, let us go down and confuse their language on the spot so that they can no longer understand one another. (Gen. 11:6–8)

With this, Yahweh creates the multiplicity of languages, and the people, now unable to communicate in a common tongue, are forced to stop building their town. They abandon their tower and scatter in confusion across the face of the earth. This place was called Babel, the story goes, because it was there and then that God created the confused babble of differentiated languages.

This biblical story of the genesis of the diversity of human languages is especially significant in our inquiry not just because it is another example of a creationist tale of origin, but because the linguistic case is such a close analogy to the biological one. As it holds for the origin of life forms, the creationist account also holds that languages were specially created by God. This is in direct opposition to linguistic theory, which states, like its biological counterpart, that languages have evolved naturally from earlier forms. The thesis of descent with modification that Darwin proposed for the origin of species applies in virtually the same manner to the origin of languages. Furthermore, the mechanisms of that evolution are much the same, as are the kinds of evidence that support both biological and linguistic evolution. Given this similarity, it is not surprising that criticisms that creationists make against the evolution of species apply equally against the evolution of languages. Therefore, if one thinks creationist arguments undermined evolutionary theory then one must accept that they also undermine linguistic theory. My sense is that it will be easier to recognize the absurdity of the creationist arguments against the evolution of language, and that this will serve to reveal their weakness on the biological side as well.

However, before I continue with this argument I want to head off a possible objection. Certainly the creationism controversy is just about biological evolution, isn't it? What upsets creationists is the Darwinian idea that organisms, including humans, evolved from one another rather than being separately created. This is the story that creationists are lobbying to have taught in biology classrooms. So, the objection concludes, discussing "creationist linguistics" is irrelevant speculation; asking what creationists would say about the origin of languages is a purely hypothetical question.

I must admit that when I first thought of the idea of criticizing creationist arguments against biological evolution by way of an analogy with linguistic evolution I did think of it as a hypothetical scenario. After all, neither the Arkansas "Balanced Treatment" Act nor the more recent proposed laws and curricular changes said anything about rejecting linguistic evolution and teaching the special creation of languages or any of the other Genesis accounts of origins. Nevertheless, I thought that the analogy would be illuminating and was prepared to defend the hypothetical approach as being fair and reasonable. The parallels between the two cases from the scientific point of view, though not perfect, are clear. Furthermore, from the scriptural point of view, the cases are identical. To be consistent, creationists would have to agree that their acceptance of the biblical view applies equally to all accounts of origins, and that the arguments they give regarding biological evolution apply in the same way to linguistic evolution. Besides, analysis of hypothetical questions lies at the very heart of philosophical method. Since well before Socrates, philosophers have profitably made use of even wildly implausible hypotheticals—imagining, for example, that the world we observe is to reality as shadows on a cave wall are to the objects that cast them—and as a philosopher I was ready to defend the utility of hypothetical questions in general and of the biology/language analogy for creationism in particular. As it turns out, however, I will not have to rely on these points, because creationist linguistics is not a hypothetical scenario.

No fictional creationist linguist is needed, because at least some creationists do explicitly reject linguistic evolution and hold to the special creation by God of different languages at the Tower of Babel. That is the reason for a beautiful display of the Tower of Babel one finds in ICR's Museum of Creation. Indeed, many creationists hold that it is also at Babel that we find the origin of the separation of the "one people" into different nations. We rarely hear creationists express these and other such views in public debate, but I soon discovered them when I began to read their books and pamphlets. (I also found that almost every book in my university library that dealt with linguistic evolution was defaced by comments in the margins that criticized any mention of evolution, or by marks at places where an author mentioned some unsolved problem or highlighted the differences between language abilities of human and nonhuman animals.) Here I quote from Henry Morris, the most well-known creationist. In *Scientific Creationism*, he tells us that:

There really seems no way to explain the different languages except in terms of the special creative purpose of the Creator. Evolution has no explanation either for language in general or the languages in particular. Exactly when or how the Creator transformed the primeval language of the original human population into distinctive languages of different tribes and nations . . . can perhaps be determined by a close study of the records of prehistory. But this is not a problem susceptible to scientific evaluation.[1]

In *The Biblical Basis of Modern Science,* Morris is more specific about what those "records of prehistory" tell us. Not only is supernatural intervention required to explain the origin of language itself, but:

As far as the great proliferation of *different* languages among men is concerned, the biblical account is likewise the only satisfactory explanation. If all men came from one ancestral population, as most evolutionary anthropologists believe today, they originally all spoke the same language. As long as they lived together, or continued to communicate with one another, it would have been impossible for the wide differences in human languages to have evolved.[2]

Morris goes on to say that anthropologists postulated the idea of different races to try to save the evolutionary story, but he argues that "the miraculous confusion of tongues at Babel does provide the only meaningful explanation for the phenomena of human languages."[3]

The original real divisions within the single human race, Morris claims, were the linguistic divisions imposed by God, which led the families at Babel to disperse. Strangely, he then appeals to the evolutionary Founder Principle (whereby a small breakaway group of individuals founds a new population that can lead to a new species), and postulates that these small inbreeding family groups could "quickly develop new physical attributes —even so-called 'racial' characteristics—that would characterize their respective descendants."[4] His main point, however, is that:

. . . there is good reason to accept the biblical record of the confusion of tongues at Babel as the true account of the origin of the different major language groups of the world. Evolutionists certainly have no better answer, and the only reason why modern scientists tend to reject it is because it was miraculous. To say that it would have been impossible, however, is not only to deny God's omnipotence but also to assert that scientists know much more about the nature of language than they do.[5]

So as not to leave the impression that only a few contemporary creationists hold this view of the supernatural creation of languages, let me point out that there is a long history of Jewish and Christian scholarship devoted to various aspects of the issue. One strand involved an ongoing search to rediscover that original language which, created miraculously by God as a gift to Adam, was assumed to be perfect. As Adam had fallen from his supposed perfection when he ate the apple, so also had language in a sense fallen after God's curse at the Tower of Babel created the multiplicity of linguistic forms. Some scholars believed that the original perfect language was a language of things themselves—a Language of the World—but that it could not be read without knowing the symbolic key. With the key in hand, however, one would also be able to understand the hidden meanings in Scripture that lay beneath references to earthly things like plants, stones and animals. Others thought it was a language of images and they looked to Egyptian hieroglyphics as a source of the ancient wisdom. Many believed that, whatever its form, the language of Adam was the language of Creation and so would be magical, conferring special powers to one who might learn it. Linguist Umberto Eco traces these and other lines of this Genesis-inspired research in his fascinating study *The Search For The Perfect Language* (1995), describing them as a series of failures founded on a dream. Creationists, of course, have a different set of concerns, and try to make the phenomenon of human language work in one way or other to support their view about divine Creation. Intelligent-design creationists John Oller Jr. and John Omdahl, for instance, focus on human language ability as part of a specious argument that we were supernaturally created in God's image.[6] Their argument, as well as creation-science's rejection of language evolution, may be seen as a resurrection of the ancient Bible-based belief in the supernatural origin of languages.

Creationists have not devoted the energy to attacking linguistics to the extent that they have attacked biology and geology, so we do not find their array of specific arguments directed specifically against linguistic evolution. Nevertheless, because the analogy between the case of biological and linguistic evolution is so close, many of the arguments they give against the former apply equivalently to the latter. In what follows I will show how this parallel works.

The Evolution of Languages and Species

Darwin himself recognized the similarity between biological and linguistic evolution, noting that "the formation of different languages and of distinct species, and the proofs that both have been developed through a gradual process, are curiously the same."[7] As we shall see, the parallels between the two theories are striking. Let me emphasize that most of what I will be discussing is the evolution of languages, not the evolution of language. The former involves the transformation of languages one into another, while the latter involves the origin and evolution of our linguistic ability and of language itself. Linguist Noam Chomsky, famous for his theory of a universal grammar, writes that:

The study of language falls naturally within human biology. The language faculty, which somehow evolved in human prehistory, makes possible the amazing feat of language learning, while inevitably setting limits on the kinds of language that can be acquired in the normal way.[8]

Language has often been taken as a unique, distinguishing characteristic that separated humans from other animals and thus as a reason to think that humans could not have evolved from them. Darwin discussed this point in the *Descent of Man* and adduced evidence that other animals do possess rudimentary forms of language. Though the study of animal language remains controversial, it is fair to say that recent evidence has further supported Darwin's argument that the differences are of degree rather than of kind. Despite the interest of this issue I will not pursue it here for a couple of reasons. First, much that has been written on the subject of glottogenesis, the origin of language, has been rather speculative, and it is only recently that good evidence has begun to accumulate.[9] In this sense the subject is similar to the study of the origin of life, which is also still in its infancy. The main reason, however, is that my purpose here is to draw a parallel with biological evolution, and the latter is typically taken to include not the origin of life but rather the development of new life-forms from earlier ones. Creationists do focus on the ultimate origins question, but we will have to deal with that later. Here we begin with the basic thesis that evolution of languages occurs.

Sprung from Some Common Source

In 1860, the German biologist Ernst Haeckel started hounding his friend August Schleicher to take a look at the newly translated book *On the Origin of Species*, thinking that, as an amateur botanist, Schleicher would enjoy Darwin's work. It took a couple of years for Schleicher to get around to it, but when he did it made a profound impression. Schleicher responded that he read and reread "this incontestably remarkable work" and that he was able to observe all the principles of Darwin's theory—variation, inheritance, and selection—at work in his garden, where the "struggle for life," he noted impishly, is more commonly known as "weeding."[10] But

Schleicher was not just an enthusiastic gardener, he was also one of the most influential comparative linguists of the century, and he found that Darwin's theory of the gradual development of biological species resonated immediately with his own study of languages. He quickly published an open letter describing how "The rules now, which Darwin lays down with regard to the species of animals and plants, are equally applicable to the organisms of languages . . . as far as the main features are concerned."[11] Let us review a few of these main features in turn.

The first is the fact of variation. The large-scale variation among languages is obvious—English is different from German and both are different from Hopi—but languages also vary at smaller scales. English spoken by the British is noticeably different from that spoken by Americans, which led George Bernard Shaw to observe wryly that England and America were two nations separated by a common language. Variations of language can be found all the way down to differences between individual speakers. Schleicher noted that such variation in language is the counterpart of variation among organisms of a biological species: "It is well known that the individuals of one and the same species are never altogether and absolutely identical; it is the same with the individual of speech; 'native accent' is always more or less strongly developed."[12]

People most commonly observe variations of native accent when traveling from one region of the country to another, as I did when I moved to Texas. I cannot say that I have acquired more than a touch of the Texas accent in the six years I have lived here, but I have come to love its leisurely, confident, and expansive character. Texans still sometimes greet friends with a "howdy!" (though not "howdy, pardner!" as in the movies). Linguist E. Bagby Atwood has studied and mapped usage of a long list of Texas expressions such as the unique Christmas-morning greeting "Christmas gift!" as well as its characteristic regional vocabulary. For example, rather than saying that milk has soured, Texans would say that it has gone "blinky." Many characteristic regionalisms are dying out, but one still regularly hears people drawl the warm greeting "Welcome y'all!"[13]

I was particularly struck by this Southern contraction of "you all" since I had moved from Pittsburgh, Pennsylvania, which covered this pronoun form—second-person plural, for those who recall their grammar-school grammar—with its own distinctive local term "yuns," a single-syllable variation of "you-uns," which itself is probably a contraction of "you ones." Native Pittsburghers pronounce their word "yuunz." One hears other variants like "you folk," "you guys," even "yous" and "yous guys" elsewhere. What is interesting about this profusion of variations is that the term "you" by itself is already a second-person plural form. Of course, it is also now used as the second person *singular* pronoun, though that role used to be filled by the pronouns "thou" and "thee." The plural "you" also used to be distinguished from these familiar singular forms as an honorific form to be used when speaking to a social superior, but linguistic historians note that this began to change in the seventeenth century when Quakers rejected the honorific use of the term as "a denial of the equality of all men" and argued "for employing the two forms solely on the basis of number."[14] Eventually "you" came to be used for both singular and plural second-person pronouns. Now we find variation arising anew with "y'all" and the profusion of other second-person forms in different regions of the country.

The second feature common to biology and linguistics is that there are patterns to the variation. Variations in speech patterns often are clustered geographically, and a regional variant of a language may be sufficiently distinctive that we recognize it as a dialect. The Texas form of speech does not qualify by itself as a unique dialect—though there is a fairly well defined linguistic boundary line that separates it from the speech patterns in Louisiana on the east, it extends through an indeterminate portion of the other surrounding states. Atwood places Texas within the division he calls *Southwestern* of the *General Southern* dialect. Other easily recognized American dialects are found in the northeastern states, from the quizzical, laconic dialect of eastern New England to the in-your-face machine-gun-fire "tawk" of the Big Apple, New York City. We still classify all these as forms of English but there is no denying their distinctive differences. This recognition reveals a second sort of pattern in the variation. Classifying these dialects as different *variants* of English is the same as recognizing different varieties of a biological species. (Think of the many dialects of English as being linguistic equivalents of the many varieties of dogs—the British dialect of Cockney is a terrier; Texan Southwestern is the "hound-dawg"; and in the Bronx they speak Pitbull.) Similarly, just as we can look closer to see finer differences we can also take a wider view and lump together languages under broader family groupings as species are grouped under genera. This recursive pattern—kinds of kinds within kinds—leads taxonomists in both biology and linguistics to classify their respective entities in a branching hierarchical structure.

Schleicher explained this analogy in detail. He warned that the terminology could be easily confused, but that in general one could say that:

The species of a genus are what we call the languages of a family, the races of a species are with us the dialects of a language; the subdialects or patois correspond with the varieties of the species, and that which is characteristic of a person's mode of speaking corresponds with the individual.[15]

In taxonomizing languages and dialects, linguists face many of the same problems that biological taxonomists do. How, for instance, should one distinguish a dialect/variety from a language/species? In some cases the differences between forms of speech are sufficiently great that it is easy to say they represent different languages, and in other cases the differences are sufficiently small so that it is easy to say they are merely dialects. However—and this is the key point—there is no precise place at which linguists must draw the division between separate dialects of one language and separate languages. From the "sufficiently great" to the "sufficiently small" is a fuzzy area of in-between. Recall how Darwin failed to carefully label all the finches he collected in the different Galapagos islands because he had taken them to be merely varieties of the same species instead of each being a different species. We will return to this issue shortly.

The next step is to recognize that languages vary not only across regions but also through time. It is hard to see more than hints of this change in one's own lifetime, though even the young can glimpse it on occasion when listening to their grandparents. From my childhood I remember that my father would revert to using the pronoun "thee" when talking with his mother or father since my grandparents still spoke the old Quaker "plain speech" at home. To get more of a sense of how language has changed one must look to old written records. One could pick up any ancient text and immediately notice differences. Most obvious at first glance are differences in the font, or typestyle. The University of Texas at Austin has one of the original Gutenberg Bibles on display, and when one first looks at the beautifully illuminated pages it is hard to even recognize many of the letters. Looking through old Bibles is an excellent way to see language evolution over time (a "proof" that should be especially persuasive to creationists), for Bibles provide one of the best preserved literary sequences. Examining the form of speech in a contemporary Bible and digging back through earlier editions is the linguistic equivalent of the geologist's observations of a series of fossilized clams in progressively deeper layers of sedimentary rock. One finds not just differences in the shapes of the letters, but in the spelling of words, then in the words used, and finally in the grammatical structures of sentences. Let us look at just one well-documented sequence, the first few lines of the Lord's Prayer. With the exception of the contemporary version, the others are taken from a collection in the textbook *In Forme of Speeche Is Chaunge: Readings in the History of the English Language*.[16]

Our Father, who is in heaven, may your name be kept holy. May your kingdom come into being. May your will be followed on Earth, just as it is in heaven. (Contemporary)

Our father which art in heauen, hallowed be thy name. Thy kingdome come. Thy will be done, in earth, as it is in heauen. (King James Bible, 1611)

Oure fadir that art in heuenes, halewid be thi name; thi kyngdoom come to; be thi wille don in erthe as in heuene. (Wycliffite Bible, c. 1395)

Fader ure þu þe ert on heofene. sye þin name gehalged. to-become þin rice. Gewurð e þin gewille. on eorð an swa swa on heofenan. (Late West Saxon, after 1150)

Fæder ure þu þe eart on heofonum; Si þin nama gehalgod. to-become þin rice. gewurþe oin willa on eorð an swa swa on heofonum. (Anglo-Saxon Koiné, before 1000)

Pater noster qui es in caelis, sanctificetur nomen tuum: adueniat regnum tuum: fiat uoluntas tua sicut in caelo et in terra. (Latin, c. 4th century)

Much can be learned by analysis of such patterns of variations. Creationists are often dissatisfied with contemporary translations of the Bible and cleave instead to the King James Version, but if one were to check a current King James Version one would notice even slight differences between it and the original, quoted above, from 1611. As one looks at earlier and earlier versions, the differences become progressively greater. As far back as the Wycliffite Bible it is still pretty easy for a contemporary English speaker to make things out, but if one were to be presented directly with the Late West Saxon or Anglo-Saxon Koiné version the meaning would not be obvious and many would not recognize it even as English and conclude it was a foreign language. Where does one language leave off and the next begin? Kinship terms are among the words most highly resistant to major change, as we see above in the variations in the word "father." The transition from the Latin reflects a regular transformation of the "p" sound to "f," and the "t" sound to the softer "d" and then "th." If we were to compare the same line in German, French, and other languages both through and at a time, the patterns of variation become even more apparent.

Such gradual changes can be found in a wide variety of language traits. Written script changes slowly over time in the shape of its characters. More pronounced changes can be found in the pronunciations of terms and in the rhythm and accent of sentences. Even grammatical form gradually shifts. For example, some languages are structured primarily by word order and others primarily by inflection (endings), but all languages use a little of both, so one type can change into another. Linguists point out that "have" seems to be in the process of becoming an inflection in spoken English as more people are saying "coulduv," "shoulduv," and "woulduv." Add up enough of these small differences of pronunciation, vocabulary and so on over time and they yield distinctive dialects. Over longer periods of time dialects might become distinct languages, just as biological varieties can diversify into distinct species.

The gradual transformation of languages and the significance of the time required for such evolution had begun to be appreciated even before Darwin drew the same conclusion for species, and here we see another reason that creationists have to deny linguistic evolution in order to retain the literal biblical view. John Herschel, for one, realized that much longer periods are needed to account for the transformation of languages than is allowed by the 6,000-year age of the earth then accepted based on Scripture:

Words are to the Anthropologist what rolled pebbles are to the Geologist—Battered relics of past ages often containing within them indelible records capable of intelligible interpretation—and when we see what [little] amount of change 2000 years has been able to produce in the languages of Greece & Italy or 1000 in those of Germany, France & Spain we naturally begin to ask how long a period must have lapsed since the Chinese, the Hebrew, the Delaware . . . had a point in common with the German & Italian & each other.—Time! Time! Time!—we must not impugn the Scripture Chronology, but we *must* interpret it in accordance with *whatever* shall appear on fair enquiry to be the *truth* for there cannot be two truths.[17]

These comments appeared in an open letter Herschel had written in 1837 about what he took to be the "miracle of miracles"—the origin of new species in place of extinct ones. (It was this letter that had led Darwin to quip in his notebook about apes changing to men and men to angels.) Herschel had no idea by what laws this "miracle" occurred but he expected that these would eventually be discovered. Little did he know that it would be just the following year that Darwin would discover (though not reveal) them, and that they would be closely analogous to the processes that produced languages.

We now have enough pieces of evidence to begin to put them together to get the first major evolutionary conclusion for language. We observe the current variations of forms of speech and their patterns of regional distribution. We note the hierarchical, branched structure of subdialects within dialects within languages within language families. We further see variations over time—tiny changes of accent and vocabulary arising in our own lifetime and bigger differences emerging as we look at the available records of the near and more distant past. The conclusion to be drawn is clear. Even though the record of transitional sequences is quite incomplete, the evidence taken together clearly supports the thesis that the different linguistic kinds we now observe arose through gradual transformation from earlier, ancestral forms.

Thus, exactly the same sort of observations and reasoning allows both biologists and linguists to draw the basic evolutionary conclusion. Here we have seen only a tiny fraction of the specific empirical evidence that linguists refer to and an even smaller fraction of the total observational evidence available to biologists, but even in this bare outline the case for common descent is powerful. When Schleicher read *The Origin of Species,* he immediately recognized that Darwin's argument for descent with modification from common ancestors established for biological kinds what he had already accepted for linguistic kinds.

What Darwin now maintains with regard to the variation of the species in the course of time, through which—when it does not reveal itself in all individuals in like manner and to the same extent—one form grows into several distinct other forms by a process of continual repetition, that has been long and generally recognised in its application to the organisms of speech. Such languages as we would call, in the terminology of the botanist or zoologist, the species of a genus, are for us the daughters of one stock-language whence they proceeded by gradual variation.[18]

Before continuing on to describe parallels between the mechanisms of biological and linguistic evolution we should take a moment to head off a possible worry that might seem to lessen the significance of the similarities we find. By focusing on the work of Schleicher, our discussion could seem to suggest that he was an exception, or even that we find similarities between the biological and linguistic theories only because of some direct influence; it would not be very telling to find in his work discussions of evolutionary development from common ancestors if he simply had picked these ideas up from what he had read in Darwin. However, such an impression would be erroneous. Though Schleicher is the most important advocate of the linguistic evolution thesis in the nineteenth century, he was in no way a radical. Indeed, the idea of the common descent of languages is certainly not attributable to Darwin's influence, for it had already been proposed in the previous century. Although there are even earlier precursors that could be cited, historians of linguistics give the credit for the hypothesis of common descent of languages to the eminent Sir William Jones.

Founder of the Asiatic Society, which for a period rivaled the Royal Society in influence in the sciences, in his day Jones had the reputation of being the greatest language scholar of all time and one of the greatest translators. Jones was the first Westerner to study the ancient language of Sanskrit, adding it to the list of some twenty-eight languages he learned over his life, and he noted surprising similarities in key vocabulary items and grammatical features between it and several other languages. Jones reasoned that the striking common patterns he discovered could not have been a coincidence, and they led him to formulate what was to become one of the most significant theories of modern science. Jones stated the hypothesis in his annual Presidential Discourse at the February 2, 1786 meeting of the Asiatic Society:

The *Sanscrit* language, whatever by its antiquity, is of a wonderful structure; more perfect than the *Greek*, more copious than the *Latin*, and more exquisitely refined than either, yet bearing to both of them a stronger affinity, both in the roots of verbs and in the forms of grammar, than could possibly have been produced by accident; so strong indeed, that no philologer could examine them all three, without believing them to have sprung from some common source, which, perhaps, no longer exists. . . .[19]

Jones actually went further and suggested that both Gothic and Celtic and perhaps even the old Persian language might be included in the same "family," which he was to call "Indo-European," and that later linguists would discover included yet more languages. Of this hypothesis, Jones's biographer Garland Cannon writes:

Its total repetitions have made it one of the most quoted formulations among all scholarly formulations in all disciplines. In the history of ideas it stands as one of the great conceptions that attempt to explain human beings and their intellectual accomplishments and that discriminate Truth from mythology and speculation. In its ultimate implications it may be comparable to the discoveries of Galileo's and Copernicus's breakthrough into scientific astronomy . . . and to Darwin's integrated explanation of change in organisms, where the expansive truth replaced the constricting false. It belongs in this elite set of conceptions of world knowledge.[20]

We have in Jones's formulation the core and germ of the modern science of comparative linguistics. The theoretical framework he proposed included, first of all, the fact that languages formed identifiable groups. This was important in itself, but even more significant was the fact that it was proper to think of these as families. Most things are similar in one way or another and can be classified together on the basis of some shared property (Abraham Lincoln, redwood trees, and the Statue of Liberty are all tall, for example), but to say that a group is a *family* is to attribute a special characteristic to it, namely that they share features *because* they are related to one another as relatives are. This point ties in with the second major theoretical breakthrough of Jones's formulation, the *explanatory* thesis that languages spring "from some common source," that they are descendants of a common ancestor. In this thesis we actually have two revolutionary ideas. The first is the introduction of time. Instead of regarding languages as static entities, Jones now envisions the possibility of their changing over time. We now so take for granted the fact of language evolution that it is hard to realize what a radical break this was from the traditional, Christian view that languages, divinely created, were immutable in the same way and for the same reason that biological species were taken to be immutable. The second is that regarding languages as evolving diachronically in this way provides the explanation for the observed patterns of family resemblances and differences in just the same way that the biological theory of evolutionary descent explains patterns of relatedness among species of organisms. The final major theoretical element of Jones's formulation is contained in the comment that the ancestral language forms—in this case, Indo-European—might "no longer exist." To suggest that ancient, ancestral languages could be extinct also broke from the tradition of immutability, and Jones was careful to hedge his pronouncement on this point with the "perhaps," leaving open the possibility that Indo-European was still spoken somewhere. (The same theological worry about extinctions would hold on the biological side as well. Christian theologians took it that God considered His Creation to be good and could not imagine that He would allow entire kinds of creatures to die out. As we noted, the fact of biological species extinctions was not accepted until Cuvier's discovery of fossil mastodons in the Paris basin at the turn of the century.) Beyond these key theoretical points was, of course, Jones's specific discovery of the outline of the Indo-European family.

Jones was revolutionary in his vision of a science of linguistics, but circumstances did not permit him to see his full vision realized. He had founded the Asiatic Society on the model of an ideal scientific community that Bacon had described in the *New Atlantis,* but he found that his fellow countrymen were not eager to include Indian pundits and scholars in their intellectual circle. Jones's reputation and his profession of belief in God protected him from theological criticism, but we should not be surprised to learn that his view of the evolution of languages encountered some initial prejudicial resistance of a familiar sort. Cannon notes that "There was a continuing influence of the ideas of supposed immutability [of language] and of God's dispersion of languages after the destruction of the Tower of Babel. Some scholars were trying to locate the Garden of Eden and to verify Newton's dating of Jason's search for the Golden Fleece. . . ."[21] Jones continued to develop and refine his hypothesis in subsequent Discourses and other writings until his death at age forty-eight, but the detailed comparative work that cemented the evolutionary thesis was carried out in the next century by linguists such as Rasmus Rask, Franz

Bopp, Jacob Grimm (better known to most of us today as one half of the Brothers Grimm), Friedrich von Schlegel, and August Schleicher, to whom we now return.

Schleicher discussed not only the fact of linguistic evolution—the thesis of descent with modification from common ancestors—but also added to the work begun by Jones and others of reconstructing the pathways of descent. He drew a tree diagram of the then best assessment of the branching genealogy of Indo-European and compared it to Darwin's proposed phylogeny for plant and animal species. As fascinating as it would be to look closely at these and more recent maps of such pathways and the evidence that supports them, such an excursion would take us farther than we need go for our current purposes. Many of the genealogical relationships, in both linguistics and biology, are known with considerable precision, but others remain puzzling. In addition to the Indo-European group of languages, linguists have traced other language phyla, such as Altaic, Afro-Asiatic, Dravidian, Uralic, Eskimo-Aleut and so on. After Jones's work, the taxonomic order in the complex profusion of languages could be explained and further investigated within the theoretical framework of language evolution. How different was this picture from that of James Parsons who, writing before Jones's formulation, explained European languages "with reference to the biblical myth of Noah's sons; the dispersing tribes of Shem, Ham and Japhet [being] held to account for the distinction between Semitic (Jewish, Arabic), Hamitic (Egyptian, Cushite) and Japhetic (the remaining) languages."[22] Paleolinguists today have dug with considerable confidence back to Proto-Indo-European, Proto-Altaic, and Proto-Uralic languages and so on, and some, such as Joseph Greenberg and Merritt Ruhlen, are trying to push back to a proto-proto language they call "Nostratic." As one would expect, it becomes more difficult to be sure of specific relationships the further back in time one delves, so the work of Greenberg and colleagues remains controversial; but there is no reason to think that the *general* pattern of evolutionary transformation from common linguistic ancestors would have been any different in kind in the past than what we observe now.

The Mechanisms of Linguistic Evolution

In the previous section we looked at the linguistic equivalent of Darwin's basic evolutionary thesis of descent with modification. Schleicher and other linguists had already recognized the evidence for the truth of this view before Darwin published the *Origin*. In this sense, linguists were evolutionists well ahead of Darwin. But this by itself is not enough to conclude that Schleicher was a *Darwinian* evolutionist. After all, the idea of evolutionary development had been proposed long before Darwin for biological kinds as well. What Darwin had done was to provide the hypothesis with a broad base of evidential support and, especially, a coherent mechanism by which it could work. The strong parallels between linguistic and biological evolution with regard to the common descent hypothesis are clear, but do the parallels also hold for mechanisms?

As we have seen, one key element of the Darwinian mechanism is inheritance. The traits that help certain individuals survive and reproduce in an environment better than other individuals in their population are more likely to get passed on to the next generation than are the traits of the individuals who do not live long enough to reproduce. If offspring bore no resemblance to their progenitors, if children's traits in each new generation were unrelated to their parents' traits, then species would never evolve adaptations to their environments. It is the iterative process of natural selection, inheritance with minor variations, and again, selection that drives Darwinian evolution. Inheritance of traits is crucial, and we now know how this works in considerable detail in the replication of the genetic material. Although general features of linguistic ability are coded in DNA, however, variations of language traits are not. We might be born with our mother's chin or lips, but we have to learn our mother tongue. So how can we say that language meets the inheritance requirement?

Here I must ask you again to think like a philosopher and try to imagine the concept of inheritance at a higher degree of abstraction. Put aside the concrete instantiation of biological inheritance. Forget about cell division and DNA molecules. Think not of the specifics of how the genetic material gets replicated, but focus just on the concept of replication itself. Abstractly, the molecular process is simply a mechanism for making copies. Receiving one's genetic inheritance is not like inheriting a family heirloom, in which case one gets the object itself. Rather, to say that a child has inherited her parents' genes is to say that she received a copy thereof. Once it sinks in that the essence of inheritance in the Darwinian process is simply copying it is easy to take the next step to seeing Darwinian evolution at work in other contexts.

Think back now to the Gutenberg Bible. The many copies of the Gutenberg Bible are the multiple "offspring" of the "parent" print blocks. If we go back even earlier we can distinguish iterative copying from printing. Prior to the printing press, copies of the Bible were made by those devoted monks who literally copied them by hand, often without even being able to read what they were copying. So we have mechanical printing and transcribing by hand as two ways of making copies of written language. Moving back still further to a time before the advent of writing, we know that important religious stories were passed on through an oral tradition in which the tales would be repeated and memorized, as far as possible, word for word. To see the process of inheritance at work in spoken language, we need only take one final step in abstraction and recognize that learning a language is also a matter of copying, in this case by imitation. We begin to acquire language by imitating the speech of our parents or early caregivers. My sister always chuckles when she or I use some odd word that we obviously acquired from our mother, whose endearing malapropisms and unintended neologisms continue to delight us. We all inherit many of our speech patterns—not just the language itself, but also idiosyncratic vocabulary, accent, and cadence—from our parents. But since this inheritance comes about by imitation rather than through genes, our individual way of speaking is not limited to what we get from our parents and we continue to add to our rich linguistic inheritance whenever we learn, consciously or unconsciously, the language traits of other speakers (and eventually writers) we encounter.

We have already established the *occurrence* of variation in both languages and species, but how similar are the two with regard to the mechanisms for the production of that variation? Again, we must be prepared to think of the mechanisms abstractly to see the connections, since of course the specific physical processes that generate variations in genes will not apply directly to memes, to return to Dawkins's useful concept. With this in mind the applicability of many of the sources of variation is clear.

Think first of mutation. The monks who spent their days prayerfully writing out the Bible had tight quality control, but if one happened to make a minor error, that mistake could be carried on when the next illiterate monk made his own copy from that original, and in this way a "mutation" would be inherited by subsequent generations in that lineage of copies.

Such a mistranscription or mistranslation might have been what led to an amusing passage in the Bible saying that it is easier for a camel to pass through the eye of a needle than for a rich man to enter the kingdom of heaven. Although the surreal image of trying to thread a camel through a needle does quite dramatize the difficulty, I have always found the proverb a little odd. The metaphor made more sense after I heard that the term was probably meant to be "rope," which differed from "camel" in the original language by just one letter.

Linguists also recognize that dialects can combine in a manner that is analogous to biological hybridization. Often this is tiny and seemingly inconsequential in the short run. For example, the Texan "Howdy" is sometimes hybridized with the northern synonym "Hi" and one hears "Hidy." Occasionally, however, the process happens wholesale, as we find in the contemporary adoption of hundreds of English words into Japanese—a result of the required national training in English of all Japanese pupils beginning in middle school. Even in such cases, however, new variations arise in numerous tiny ways. Not only do the Japanese pronounce the adopted English words differently than native speakers (sometimes to the point of nonrecognition), they also use them with interesting new shades of meaning and sometimes in combined grammatical forms. One of my favorite examples of this last sort of transformation is the clever verb "Makku-doh-naru" which means "to eat at McDonald's." Though hybridization occurs in the biological world, it is never on this massive scale which leads to significant differences between how languages and biological entities can evolve.

Jones had introduced the idea of language families and their common descent, but it was Schleicher who pushed the idea that they formed a family *tree*, with languages forming by *diverging* through time, like the limbs of a tree. But we now know that languages can evolve by merging; English, for example, began as a branching from a Germanic stem but subsequently blended significantly with French. Schleicher was perhaps a bit over-eager to match linguistic evolution to the Darwinian form as far as he could, so much so that he failed to appreciate that the phenomenon of language blending was a significant difference from the biological mechanism.[23] (A rival "wave" model, proposed by Johannes Schmidt later in the century, allowed for both divergence and convergence, but it had its own weaknesses.) Because Schleicher's one-way only mechanism is wrong, it is sometimes said that the "evolutionary" model of language development is wrong, but this is misleading in several ways. At worst it simply means that *Schleicher's* evolutionary model taken as a whole is wrong. In fact, many other evolutionary models were being discussed even at that time. Better to say that one or another element of his model is wrong or that his model is incomplete. In this sense, Schleicher's view is very much on a par with Darwin's own view, which of course has been greatly supplemented and refined on the basis of new evidence since he proposed it. Much of what Darwin hypothesized about the causes of variations, for instance, we now know to be false. As it turns out, one of the things Darwin was wrong about was that evolutionary change always happens by divergence. Genetic technology now allows us to incorporate genes from one species into another. Though this form of gene transfer is still in its infancy and may yet be curtailed by regulation, we have a proof of the existence of genetic merging. Of course this is an "artificial" process, but we also know of at least one important case of natural merging. Biologist Lynn Margulis has shown that mitochondria, the power-generating organelles in cells, were originally independent bodies but somehow fused in some cells, maybe at first as parasites, at an early stage of evolution. Margulis has suggested similar mergings for other cellular organelles, though these other hypotheses remain controversial. The basic point for us here is that the mechanisms of "borrowing" and "merging" can occur in biological evolution as well as linguistic evolution, and that the difference is of degree. The image of the branching tree is a useful model in both sciences, but it clearly is a closer approximation for the evolution of species than for the evolution of languages.

Does the third element of the Darwinian mechanism—selection—also apply in the case of the evolution of languages? It seems so, at least in some cases. Here I will just briefly mention vocabulary evolution, since that is the most obvious in people's experience and because we can document some of the environmental changes that cause their natural selection. One reason that the term "blinky" has been dying out is that in an era of refrigeration it is simply not as common that milk turns, and the term "sour" is taking over since it survives because of its broader applicability. Similarly, many of the characteristic Texan farm terms are disappearing simply because increasing urbanization has decimated rural life and culture. Sadly, real Texas cowhands are a dying breed, though one still encounters a few of them among the faux urban cowboys at the Broken Spoke, one of the last true Texas dance halls. Its owner, James White, remains an unapologetic Texas "good ol' boy" and each night reassures the crowd that they will continue to find the best chicken-fried steaks in town there and that "thar ain't never gonna be no hangin' ferns or none o' that Pierre water here . . . jus' good Whiskey an' real country music." The crowd loves to hear White make his nightly pledge as he introduces the band, but the authentic cadence and color of his speech is heard with decreasing frequency among the patrons themselves as the older generation of native Texans gives way to a younger generation ignorant of the farm or ranch environment. Discussing the evolution of Texas vocabulary, linguist Bagby Atwood notes that "Decreasing familiarity with rural life and with pre-mechanized days is most clearly evident in the waning knowledge of the horse and his uses. This is reflected in the obsolescence of the terms *nicker, near horse, off horse, . . . wiffletree. . . .* Growing unfamiliarity with cows is also apparent."[24] Economic change and increasing urbanization is selecting out a form of language that in an earlier day had been highly adaptive but which does not fit as well in the new environment.

Common Patterns of Evidence

Having seen some of the commonalities between linguistic and biological evolution, we should not be surprised to find that linguists and biologists also share inferential methods and kinds of evidence. For instance, probably the most important data that linguists appeal to in reconstructing the historical evolution of languages are the common patterns found in extant languages. Sound-to-meaning associations, distinctive spellings, odd exceptions to grammatical forms and such are for linguists like the homologous structures and vestigial organs that biologists find, a parallel which testifies in both fields to common ancestors and histories. Darwin himself had recognized this, noting that "We find in distinct languages striking homologies due to community of descent, and analogies due to a similar process of formation."[25] Also, just as patterns of biogeographical distribution of species provided important evidence for Darwin, so too does linguistic geography function for linguists. By charting a dialect map of the United States, for example, linguists are able to garner evidence of origin, migration, contact, and so on of language groups. Linguists also examine remnants of extinct languages, ancient scripts carved in stone that might be called linguistic fossils. Spoken language, of course, until very recently, could leave no traces. Schleicher noted that, were it not for written records, linguists would have had had a far more difficult time discovering evolution than zoologists and botanists, but that, "As it is, we are better off for materials of observation than the other naturalists, and therefore we have forestalled you in the idea of the non-creation of the species."[26]

Using written records and other evidence, then, linguists can reconstruct the gradual development of the lineages of languages, finding when they branched off from one another in a similar way that biologists look at gene flow in populations and chart the lines of biological evolutionary branching. Each field independently provides evidence of the power of Darwinian processes and of the explanatory power of evolutionary theory. There is much more evidence that we could examine, but let me close this section with one recent striking case in which the lines of evidence in both linguistics and biology are coming together.

In a series of publications beginning in the early 1960s, geneticist Luigi Luca Cavalli-Sforza and his colleagues showed how statistical analysis of current genetic data from human populations can be used to help trace our recent evolutionary history. Cavalli-Sforza had been a student of Sir Ronald Fisher, the Cambridge mathematician who played an important role in mathematizing evolutionary theory in the modern synthesis, and he had been inspired by Fisher's statistical analysis of human blood groups. By analyzing data involving twenty genetic variations from fifteen native populations around the world, Cavalli-Sforza was able to draw an evolutionary tree showing how these populations branched off from one another. Over the next two decades he continued to collect further information, and in 1984 he and his colleagues published a revised and expanded tree for forty-two populations based on a much larger data set of one hundred and ten genes. For our purposes, the important point here is that the history of evolutionary branchings (in this case all within our species) was inferred simply from information about genetic distances—

information culled from current human populations and with the same kinds of mathematical techniques that are used to draw the larger evolutionary tree of life (in that case across different species) from genetic data. The reconstructed tree of recent branching of human populations provides a rough estimate (within a twenty percent statistical error since the data set is still relatively small) of when major population migrations occurred and their relative order. Cavalli-Sforza checked the figures from his tree against what was known independently from archaeological data and found a reasonable match. They fit, for example, the view that human beings originated in Africa and spread from there some 100,000 years ago in stages across the world, colonizing first Southeast Asia and Australia some 55,000–60,000 years ago, then Asia and Europe some 35,000–40,000 years ago, and eventually emigrating to Northeast Asia and America.[27]

Then, in a groundbreaking 1988 paper, Cavalli-Sforza and his colleagues began to look into whether they could find a similar connection between genetic evolution and linguistic evolution.[28] They took the most recent taxonomy of the world's languages by linguist Merritt Ruhlen and compared it with their genetic tree. Ruhlen's families of languages matched branches of the tree that were close together with just a few exceptions that could be explained on other grounds, such as native language replacements by conquerors who imposed their own languages. Moreover, Ruhlen's groupings of families of languages into superfamilies, indicating the most ancient postulated linguistic branchings, matched the major branches of the genetic evolutionary tree.

Cavalli-Sforza recounted his research in a recent book and reviewed the explanation of the similar structure of the two evolutionary trees:

The explanation is simple. During modern humanity's expansion, breakaway groups settled in new locations and occupied new continents; from these, other groups broke away and traveled to more distant regions. These schisms and shifts took humanity to very remote areas where contact with the original areas and peoples became difficult or impossible. The isolation of numerous groups had two inevitable consequences: the formation of genetic differences and the formation of linguistic differences. Both take their own path and have their own rules, but the sequence of divisions that caused diversification is common to both. Their history, whether reconstructed using language or genes, is that of their migrations and fissions and is therefore inevitably the same.[29]

There remain controversies and questions about aspects of this research that will have to be resolved once linguistic and genetic data improve, but the general results appear to be sound; two statistical methods agree independently that the probability that the similarities between the genetic tree and the linguistic classification are due to mere chance is negligible. Thus, it looks like Darwin's prediction about the relation of the pathways of linguistic and human evolution, made a hundred years earlier, has proven to be correct.

Designed Languages
The parallels between the evolution of species and of languages are impressive, but I do not want to suggest that the analogy is perfect. Though obviously struck by the numerous similarities, Schleicher correctly noted that "no more than the *principles* of Darwinism could be applied to the languages. The realm of speech is too widely different from both the animal and vegetable kingdoms to make the science of language a test of all Darwin's inductions and their details."[30] We have already seen, for example, that the evolution of languages differs from that of species by virtue of the ability of languages to blend and thus converge in ways that are only recently possible for species. There are several other differences that are worth noting.

It is possible that there is more than one root of language trees; linguists occasionally discuss the "monophyletic" versus "polyphyletic" hypotheses, but the origin of language itself, like the origin of life in biology is an exceedingly difficult problem to investigate. Nothing in evolutionary theory *requires* a single origin, and there is not yet much evidence one way or another, though Darwin did think there probably was only a single root of the tree of life (here on earth). Most linguists accept the single source thesis as a working hypothesis, if only for the sake of simplicity. Creationists are more definite than linguists in favor of the thesis of a single original language, because Genesis says that at first everyone spoke the same tongue—the language of Adam. In any case, creationists reject that current languages are evolved descendants of that common source as they reject the common descent hypothesis in biology. Thus, that there is a possible disanalogy regarding the numbers of roots is of minor significance to the creationism debate.

A more significant disanalogy involves the possibility of intentional language modification. While some new words arise by chance, such as my mother's unintended neologisms, many new words are coined not at random, but with some intended purpose in mind. However, intentional neologisms alone do not a language make. Moreover, the occasional attempts to control the development of language, most notably by the French, have met with little success.

Over the years there have been a few dozen languages such as Volapük and Monling that were intentionally designed. One of the first to propose and try this was the philosopher René Descartes (French, to be sure) back in the early seventeenth century (his idea was to use numbers to represent words and concepts). However, with the exception of Esperanto, which claimed several million speakers scattered around the world at the mid-twentieth century, designed languages have rarely been adopted beyond linguistic hobbyists. Interestingly, a common argument given against them is that even if the world were to adopt such a language, different peoples would tend to pronounce it with their own accents so it eventually would again evolve into a new series of distinct languages. Of course, designed languages typically were not created from scratch but were stitched together using common Indo-European word roots and simplified grammatical structures. It is not surprising that the first lines of the Lord's Prayer rendered in Idiom Neutral looks rather familiar:

Nostr patr kel es in sieli, ke votr nom es sanktifiked; ke votr regnia veni; ke votr volu es fasied, kuale in siel, tale et su ter.[31]

Despite their derivative nature, these designed languages are notably different from naturally occurring ones. Immediately apparent is that the syntactical rules of artificial languages are completely regular—designed that way for ease of use—in stark contrast with the irregular verb forms, inconsistent rules for forming plurals, haphazard assignment of nominal gender, and so on that are the bane of every student of natural languages. In Esperanto, for example, which is written with a simplified Roman alphabet, there is only one sound per letter, and word accent is always on the next to last syllable. All nouns end with -*o* and adjectives with -*a*, with the suffix -*j* to indicate a plural form. Verbs in the present tense end in -*as*, in the past tense -*is*, future tense -*os*, and in -*u* for the imperative. Natural languages, on the other hand, despite the commonalities of form and the Chomskyan expectation of an underlying universal grammar, are full of unexpected twists and bizarre exceptions to every rule. Such variability is not much of a conceptual stretch given that natural languages are "organic." Schleicher, writing before the attempts to design a language really got going in the late nineteenth century, used the same terminology when he observed that "Languages are organisms of nature; they have never been directed by the will of man. . . ."[32] Like plants, languages grow and develop in strange and uncontrollable ways, continually breaking free from every prescriptive grammarian's attempt to constrain them.

In this contrast we see one of the weaknesses of the creationist's design argument. The intuitive notion we have of what something would look like were it intentionally designed does not correspond to the way the natural order appears. We expect that languages that are designed and specially created would be structured in a simple and regular manner, as is Esperanto, rather than be the jerry-built jumble that is natural language. Natural languages are not formal constructions imposed from on high but, as Walt Whitman noted, have their bases broad and low, close to the ground. Poets seem to understand this best of all. These languages were not created whole and immutable, as creationists contend, to be scattered from the Tower of Babel, but rather developed into and from one another over time, through piecemeal constructions and unplanned transformations. Every language is, in Emerson's words, "a city to the building of which every human being brought a stone." Though not perfectly Darwinian, this ongoing transformation is clearly an evolutionary process and a close analogy to biological evolution in its most significant aspects.

Creationist Arguments

When the evolutionary geneticist Theodosius Dobzhansky reflected on his exchange of correspondence with Frank Marsh, who had broken with his more conservative fellow creationists in the mid-1940s by accepting small-scale evolutionary change but who could not bring himself to draw the inference to large-scale evolution, he drew the lesson that "no evidence is powerful enough to force acceptance of a conclusion that is emotionally distasteful."[33] I have taken the time to spell out some of the structural parallels between linguistic and biological evolutionary theory in the hope that it can provide a therapeutic way through the emotional block for those who find it unpleasant or troubling to think of themselves as being descended from prior animal forms. Given that people typically do not seem to have the same prejudice against linguistic evolution, it should be easier to judge the evidence with a bit more detachment and to recognize its strength.

If one is to follow the creationist line in opposing biological evolution in favor of the creationist's interpretation of the Genesis story of special creation, then consistency demands that one also deny linguistic evolution and accept the story of God's special creation of languages at the Tower of Babel. Similarly, if one acknowledges that the theory of linguistic evolution is well confirmed, then one should do the same for the biological theory. The evidence that supports the evolution of species is of the same kind and is as incontrovertible as that which supports the evolution of languages, so accepting creationist biology is as absurd as accepting creationist linguistics.

In the remaining sections of this chapter I will continue to draw out the parallels between linguistics and biology, looking at several of the most common objections that creationists make against evolution. If these work to undermine biological evolution they should work in the same or a similar way against linguistic evolution. Again, my hope is that with less of a prior bias against the latter it will be easier to see the weaknesses of the criticisms generally and thus easier to understand why they fail to undermine biological evolutionary theory.

"Evolution Is Never Observed"

One creationist argument that people find especially convincing is the following: If evolution is true then we should be able to observe it, but we never see one species turn into another, so it must be false. The argument appears in many different guises. If humans evolved from monkeys in the past why don't we see monkeys turning into humans any more? There are many kinds of dogs, a creationist will admit, but you never see a dog turn into a cat, do you? Or, more boldly, creationists may simply assert that no scientist has ever seen any evolutionary change occur whatsoever. Isn't science supposed to deal only with what can be observed? How can evolution be scientific if it is not or cannot be observed? On the face of it, this apparently straightforward criticism might look pretty strong, but there are so many problems with it that it is hard to sort them out.

First, many such criticisms are the result of simple ignorance, faulting evolution for things it does not claim. Criticizing biological evolution because we do not now see monkeys changing into humans is like criticizing linguistic evolution because we don't see Spanish changing into Gujarati. These are current species of language, descended from others, and as they continue to modify we would expect them to develop into new languages, not one into the another. It is a common but erroneous view to think that humans descended from monkeys, which is perhaps why some creationists think that monkeys should still be changing into humans, but the true evolutionary picture is that humans and monkeys are cousins, each species having branched off from a common ancestor. The same holds of dogs and cats, though in this case the common ancestor is even more distant. Convergent evolution, in which similar sorts of traits arise in separate lineages, is possible, and we know that it has occurred in a limited degree in the past under conditions with strong selective pressures from the environment, but convergence of dogs into things that look like cats, say, is so highly unlikely that it is not a serious possibility. It does not make sense to object that we do not observe something that evolutionary theory says we should not expect to occur.

Leaving that aside, what about the claim that evolution cannot be true because scientists never observe any evolutionary development? Let me reply to the challenge in stages. To begin to see what is wrong with this argument against biological evolution let us first look at it how it applies to language evolution. If the argument is good it should apply even more strongly to the evolution of languages, for not only is it true that we have not observed the natural origin of a completely new language, but also those that we have seen originate have all been ones like Esperanto that were purposefully designed by human intelligence. Doesn't this mean that we should conclude that all languages are intentionally designed and that they did not evolve from one another? Certainly not.

Has a scientist ever observed one language evolve into another? In one simple sense the answer to that question is clearly "No." Even though languages evolve much faster than biological species it is still too slow a process to see a new language emerge in a person's lifetime. However, this does not mean that we cannot know that they have been formed from others by descent with modification. We can know this even without observing it directly because we can observe the evidence for it. Jones did not observe Sanskrit and the other Indo-European languages evolve from a common source, but he did observe the linguistic homologies that were persuasive evidence of their evolution from a common source, and linguists today extend this evidential comparison using sophisticated statistical methods that bring a precision to the task that he could only have dreamed of. Furthermore, as we saw above, linguists can observe and chart the accumulation of linguistic "mutations" and the other processes of linguistic change and when we put these together with the other evidence we may thereby infer that "language speciation" has occurred and is occurring still. In *this* sense we observe the evolutionary development of languages all the time. Exactly the same is true for biological evolution.

At this point, creationists are bound to object that this notion of observational inference is not what they mean by "observing evolution." Many creationists seem to hail from the most literalist possible county in Missouri: When they ask biologists to "Show me" evolution, they seem to expect nothing less than to be presented with one species transforming into another, with intermediate forms of each characteristic feature. In his "Back to Genesis" talk, John Morris shows a painting of an armadillo-like creature covered partly with scales but with some of these half-changing into feathers that stick out in odd places as well. If reptiles evolved into birds why do we not observe creatures halfway between the two like this? More often, as we noted, creationists will say that no one (except God) was around to observe when life began or when species originated—and so these topics cannot be scientific.

This is a serious misunderstanding of how science works and of the nature of observation. First of all, as I discussed in the previous chapter, science is much more than a collection of direct observations but also relies on the inference to the best explanation. So, for example, although we may not have direct observations of half-scale/half-feathers, we nonetheless have good evidence of their evolutionary relationship, because we know of mutations in chickens that cause the scales that normally cover their legs to be converted into feathers. Second, there is no clear break between observation and theoretical inference. What we might think of as "direct observation" is rarely so simple. When we observe a microorganism through a microscope, for example, we are not just seeing it through a series of glass lenses but also through the conceptual lens of optical theory. The importance of theory is even more apparent for observations with electron microscopes, which work on completely different theoretical principles than light microscopes. In each case the theory has in a sense been built into the microscope. We do not see with our eyes alone; we see with our brains.

Though we often do not recognize it, such inferential processing is involved in even ordinary situations that are comparable to cases of evolutionary development. Consider what it would take to say that we had observed any sort of gradual development, say of a tree. In my yard is a mighty live oak that is estimated to be well over a hundred years old. In the spring I can observe the tiny green shoots of the new twigs (though hard as I try I can never actually see them grow), and every couple of years I notice that I have to prune a limb that has drooped low enough to scrape the eaves. In the literal "show-me" sense I have never observed that tree grow. In that sense, perhaps, all I have seen is a series of mental snap shots taken at different times. In the inferential sense, however, it is quite reasonable to say that I have seen the tree grow for I can mentally "connect the dots," as it were, of those snap shots. I was also not around to see that tree when it was a seedling newly sprouting from an acorn (probably no one observed that acorn fall and sprout), but given what I know of the process of growth, and having seen other trees in other stages of growth I can conclude with a high degree of confidence that that was what happened. The growth of Schleicher's tree of languages took centuries longer and the growth of Darwin's tree of life took longer still by many orders of magnitude so these are harder for us to see in our mind's eye. Nevertheless, by taking the time to review the evidence we can connect the dots of the linguistic and biological observational data and here too observe evolutionary change continuing to occur around us. To close one's eyes to this as creationists do and insist upon the special creation of each language and each species of plant and animal is as absurd as insisting upon the special creation of each individual tree and twig because we did not observe its origin directly.

Once we are attuned to the nature of observation and understand what evolutionary theory actually claims, it becomes clear that scientists have observed evolution in as clear a manner as one could desire both in the lab and in nature. Aside from observations from many lab experiments and field studies like the Grants' that I described in the previous chapter, scientists have also observed, for example, how insects have evolved resistance to various pesticides and how disease-causing bacteria have evolved resistance to penicillin and other antibiotics. (This last example involves a serious issue in public health, and understanding the evolutionary processes that lead to resistant strains has recently helped the medical profession begin to change the way it prescribes antibiotics so as to not further accelerate their evolution. Interestingly, pharmaceutical companies are beginning to use a form of Darwinian engineering to help develop more powerful drugs.) Biologists even know of cases of speciation in the wild within the last half of the twentieth century. In the plant genus *Tragopogon*, for example, two new species (*T. mirus* and *T. miscellus*) have evolved by a process known as allopolyploidy. Most *Tragopogon* species are diploid, which means that they have two sets of chromosomes. The new species were formed when one species accidentally fertilized a different one and produced an offspring with four sets of chromosomes. This mutation resulted in an interfertile tetraploid that could not fertilize or be fertilized by either of its two parent species types and thus qualified as a different species by being reproductively isolated.[34] In plants, allopolyploids have often evolved into distinct phyletic lines. It is even possible to induce allopolyploid speciation in lab settings, in some cases, by exposing plants to the chemical colchicene.

"Where Are the Missing Links?"

Recently, creationists have modified their view that evolution does not occur. In the face of the many examples in which evolution has been directly observed, they now agree that evolution can be demonstrated, but they now say that this is merely "microevolution." Intelligent-design creationist John Ankerberg, for example, writes:

Microevolution or strictly limited change within species can be demonstrated but this has nothing to do with evolution as commonly understood.[35]

Such statements, and they now appear commonly in creationist literature, are astounding to biologists who, in the first place, do commonly understand evolution to include microevolution and, in the second place, do not define the latter in terms of "strictly limited change within species." We will return to the scientific view of the matter shortly, but first let us look a little closer at the objection.

What Do the Fossils Say?
When creationists now say that the evolutionary change observed today has nothing to do with evolution "as commonly understood," how could they be understanding the term? Obviously they do not mean what biologists mean by the notion of evolution. Again, most commonly what creationists seem to have in mind are things like a transformation from monkey to man, which is the popular stereotype of evolution. Species boundaries cannot be crossed, they insist. If species could transmute as Darwinians say then why do we not see thousands of intermediate forms everywhere we look? This is probably the second most common criticism that creationists make: Where are the missing links?

Rasmus Rask, one of the linguists who followed Jones, is paraphrased as noting about linguistic evolution that "The more languages and dialects you take into the comparison, the more gaps you are able to fill by intermediate forms."[36] Even so, the "fossil record" of languages, especially early languages, is very incomplete in much the same way as is the biological record. In both fields there are several cases of clear transitional series, but in most cases one must infer the evolutionary changes based on other evidence. Speaking of the families of languages, Schleicher noted that linguists were fortunate in a few cases to have fairly good written records from a comparatively early time to examine, and he pointed out that what was learned from these "may be otherwise supposed in respect of other families of languages, which do not possess those exponents of their earlier forms. We therefore know positively from the observation of collected facts that languages change as long as they live. . . ."[37]

Creationists are rarely satisfied with such an answer. Recalling the metaphors for kinds of induction discussed in the last chapter, we can say that creationists simply place fossils in buckets rather than examining them in the light of theory, and because of this they continue to argue that the missing links are devastating for evolution. ICR's Duane Gish argues that the fossils say "No!" to evolution, and his argument basically comes down to pointing out gaps in the fossil record.[38]

But the so-called problem of the missing links is no news to evolutionists, and it poses no greater a threat in biology than in linguistics. After all, absence of evidence is not necessarily evidence of absence, especially when it comes to the fossil record. Paleontologists always have to contend with the possibility of preservational artifacts—absences of fossils that are due not to absences of organisms but absence of the conditions necessary for fossilization. Look around today and you can see for yourself that most of the organisms you come across are not making it into the fossil record. It takes rather special combinations of physical factors—usually those of swamps or estuaries where remains can be buried in sediment, be compacted and, if lucky, remain undisturbed for millions of years—for the bones or imprints of an organism to achieve a measure of immortality in stone. To then become part of the scientific body of evidence, they have to erode in such a way as not to be destroyed, and then found by someone who recognizes their importance. Furthermore, from what we know of evolutionary mechanisms, speciation events are likely to occur in isolated populations, and competition will quickly eliminate the less fit of closely similar forms. Both processes make it even more unlikely that there will be a smooth, continuous fossil record of intermediates. Thus, it is not at all surprising that there are "missing links" in the fossil record and this is not good evidence against evolutionary transmutation; on the contrary, given what we know, it is what we expect to see. Given the difficulties involved in fossilization in the first place the record of intermediate forms is remarkably good and continually getting better. One of the most striking new fossil discoveries involves the transition to the aquatic whale from a land mammal; surely a whale with feet counts as a transitional form. Hardly a month goes by without a major announcement of new fossil evidence that reveals more and more about the branchings of the tree of life. Even more exciting is the whole new category of fossil information that advances in genetic technology are just beginning to open up.

Faced with the slow but steady uncovering of such new fossil evidence, Johnson and other new creationists typically move quickly to what has been long considered to be the most spectacularly large gap in the fossil record, what has been called the "Cambrian explosion." Darwin himself knew of the apparently sudden appearance of animal fossils at the beginning of the Cambrian period and pointed out in *The Origin of Species* that this was problematic for his thesis of gradual evolution. At the time, however, the best he could do was hypothesize that such fossils would eventually be found. Given that the investigation of fossils was just getting off the ground in his day, this suggestion was not unreasonable. Today we have substantially more fossil information and our picture of the pathways of descent is slowly becoming clearer, but it is true that fossil evidence remains scarce for the transition into the Cambrian period, and evolutionary biology can still say rather little about those early evolutionary trails. But how big of a problem is this? Was there really not enough time, as creationists suggest, for evolution of animal phyla to occur? The Cambrian may reckon as a sudden explosion from the point of view of geologic time, but we are still talking about millions of years and this could have been quite sufficient for evolution. Much of the talk about a sudden "explosion" during the Cambrian period might be exaggerated, for example, given that recently several independent lines of new evidence suggest that animal phyla began to diverge well before the Cambrian period, during the mid-Proterozoic period about a billion years ago.[39] Moreover, Pre-Cambrian organisms were relatively small and lacked hard parts, so their fossil record is even more sparse than usual. Certainly, the Cambrian explosion is a problem, but it is far too soon for us to pronounce it unsolvable.

Micro- and Macroevolution

Creationists' arguments about missing links in the fossil record and their claim that species are divided by unbridgeable gaps reflect a particularly unfocused form of species essentialism, the view that each species is set apart from every other by an immutable "essence." As they see it, God forms animals "according to their kinds" (Genesis 1:25–26) and they believe this implies fixity of species. It is fine, most now say, to allow limited evolution to occur within a species. However, they insist that microevolutionary changes are "strictly limited within species" and cannot extend beyond to account for macroevolutionary change from one species to another.

Some creationists today will even say that this explicit acceptance of small evolutionary changes is not a modification of their view at all, given that they have always held that the different human "nations" or "races" that went their separate ways at Babel were all evolutionarily diversified descendants of Noah's family and, originally, of Adam and Eve. However, they forget their own history and the split between the members of the antievolution Religion and Science Association (RSA), founded in 1935 by Price and Dudley Joseph Whitney, over the possibility of limited speciation within created types. Whitney, Marsh, and other more liberal members came to accept this possibility, which Price and others opposed. It should come as no surprise that the rhetoric of that debate is familiar; compromising by allowing speciation with types, said RSA conservative Byron C. Nelson would "[open] the door of evolution so wide that I, for one, don't see a place to shut it."[40] Of course, Nelson was right, for recognizing the efficacy of evolutionary mechanisms for evolutionary change within a type is tantamount to accepting it generally.

There is no essential difference in kind between microevolution and macroevolution; the difference is simply a matter of degree. The glossary of a typical evolutionary textbook will define each as "vague" terms referring, respectively, to "small" evolutionary changes within species, or to evolutionary changes "great enough" to classify species within different genera or higher taxa.[41] How much of a difference is "enough" varies widely from case to case. Darwin demolished the old Aristotelian notion that species are immutable in the same way that Galileo had destroyed the Aristotelian conception of the two-sphere universe. According to Aristotelian cosmology, which the Church adopted and modified to fit with its theology, the earth was at the center of the universe, and there was an essential difference in kind between its perishable terrestrial realm below the sphere of the moon and the eternal celestial realm above. Change and disintegration occurred in the sublunary region (the "microcosm"), but this had nothing to do with events in the superlunary region (the "macrocosm"). Although it was Copernicus who first seriously challenged the earth-centered view and Kepler who worked out the planetary laws for a sun-centered view, it was really Galileo who brought down the old system by providing new observations and new physical mechanisms that undermined the terrestrial and celestial essentialism. His telescopic observations that discovered the moons of Jupiter and their motions showed "terrestrial" patterns in the celestial region, and his new physics showed that supposedly distinctive "celestial" patterns of motion could be explained in terms of the same forces and laws that determined terrestrial motions. Similarly, Aristotle had held that all species were characterized by some defining essential characteristic that differentiated them from other species, and Darwin's discoveries overturned this view forever. How ironic that long after Christians come to terms with Galileo's science, which broke up the cosmological essentialism that they had taken to be a necessary part of their theology, some are still tripping over the same issue of essentialism, this time in the linguistic and biological realms.

Of course, to say that there is no essential difference between micro- and macroevolution is not to say that there are not interesting and perhaps significant differences at some higher taxonomic levels. For example, spurred in part by the punctuated equilibria model of evolutionary change, proposed by Stephen Jay Gould and Niles Eldredge, many biologists expect that there are still discoveries to be made of biological constraints that might have been relevant in the development and stability of the approximately thirty-five body plans of the major phyla. The developmental biologist Rudolf Raff has written about some of the research that is beginning to bloom in this area.[42] It is likely that additional causal factors will be confirmed as research in developmental biology broadens the modern evolutionary synthesis. However, this will do nothing to change the basic facts of evolution that are already well confirmed. Nor will it negate the reasonable inference from small changes observed in what is taken at one time to be one species, to accumulations and divergences of these changes to such a degree that we should classify later populations as distinct species. It simply will not do for creationists to admit that microevolution is true within species and then turn around and claim that this has nothing to do with the macroevolution of one species into others. If creationists think each species has some essentially distinctive, immutable features that prevent speciation then let us hear what they are. Of course, they have nothing of the kind to offer.

Incredibly, Phillip Johnson thinks that he has dismissed the inference from observed small-scale changes to large ones with a single argument. He says that Darwinists need to supply "a scientific theory of how macroevolution can occur," and that for them to infer that "small changes add up to big ones" by reference to a principle of uniformity is simply to appeal to "an arbitrary philosophical principle."[43] But there is nothing arbitrary about a principle that regards observed causal mechanisms as operating uniformly, and, unlike creationism, evolutionary theory has identified a clear set of relevant mechanisms. Consider the chromosomes and sequences of DNA not only within species but across genera, phyla, and other higher order taxa. Whether we are looking at ducks or daisies, nematodes or human beings, we find the same genetic material, the same or similar genes, and so on. Start with the genome of one species and with only a fraction of a percent change—some additions, deletions and rearrangements of those Ts, As, Gs, and Cs—you have the genome of a different species. Furthermore, the processes that transform and select these sequences are common across the board. Thus the theory, the observational evidence, and the uniformity principle work hand in hand here, as they always do in science. The burden of proof is on the critic to show that there is some barrier—a difference in essential kind rather than degree.

There is no fixed number of differences in the sequences that distinguishes change within a species from change across species, and it is hard to imagine how one could even attempt to draw such a line. Do creationists define acceptable "microevolution" as anything below a hundred changes? A million? Pick any number; somewhere along the way there will be a sufficient difference to distinguish species; why would creationists think then that it is reasonable to have just so many changes but no more? A hundred but not a hundred and one? A million but not a million one hundred? How can they accept that small changes accumulate up to that number, and then turn around and say that scientists are appealing to an arbitrary principle when they infer that those same small changes just added up a bit more?

It is a pity that Johnson fails to practice what he preaches. In demanding that Darwinists supply a theory of how macroevolution can occur (and blithely dismissing the well confirmed theory they have), he himself is completely silent about how it is supposed to happen according his own creationist alternative. Indeed, though "intelligent design" is proffered as the "better theory," Johnson and other IDC "theorists" consistently refuse to reveal anything positive about their theory. It is easy to point out missing links in any scientific theory for science lays its commitments and its evidence on the table and identifies in every research report the gaps still to be filled. On the other hand, it is impossible to find empirical holes in Johnson's creationism, for his notion makes no empirical contact with the world at all; the "theory" of intelligent design begins and ends with God's creative power. It is not just links that are missing here; what is missing is any semblance of a theory at all.

"Science Cannot Explain X"

Explanatory Gaps in Linguistics
Though it has interestingly distinctive features, the previous argument regarding "missing links" can be thought of as one of what is certainly the largest class of creationist arguments. This is the general argument that begins by pointing to some fact that evolutionary science supposedly cannot explain and concludes thereby that evolution must be false and creationism true. For example, the previous challenge might have been rendered: "Evolutionists cannot explain the gaps in the fossil record; therefore Darwinism is wrong and organisms must have been specially created by an intelligent designer." Here let us examine the general argument, looking at it first in a couple of linguistic cases.

As is true in biology, linguistics has no shortage of as yet unexplained facts. We have already mentioned the problems involved in learning about the earliest language forms. But there are also puzzles remaining even with contemporary languages, some of which are so distinct that they currently defy attempts to discover a clear connection to other languages. Japanese is but one example. Though its writing system incorporates Chinese characters, this is the result of an artificial adoption of that foreign script by Buddhist scholars at a time when Japanese did not have its own; the languages themselves are as different as English and Hawaiian. Some linguists argue that Japanese might be related to Korean or to Finnish, themselves problematic cases, but the links are far from clear. It turns out, however, that Japanese culture provides a theological explanation for this uniqueness. According to the Japanese Shinto religion, their island nation was specially created by the gods. We might imagine a Shinto creationist arguing that their origin story of special, intentional design is proved because linguists cannot trace a pathway connecting their language to an earlier form. But does it make sense to accept this "alternative theory" just because linguists do not have a specific explanation for the origin of Japanese? Certainly not. The overwhelming evidence for language evolution and for the general applicability of its mechanisms certainly allows the linguist to infer that Japanese as well should have a common ancestor with other languages from which they evolved. At this level we have a perfectly acceptable explanation even though the particular explanation (i.e., the specific pathway of descent) eludes and might continue to elude us.

The Origin of Language and the Origin of Life

Of course, if we were able to trace ancestral languages all the way back we would expect to find a single original language (or perhaps a few), and at that point we could not explain it in terms of evolution from a previous one. Here we would not be dealing with the origin of particular languages but with the origin of language itself. This is the counterpart in linguistics to the difference in biology between the origin of species and the origin of life.

I previously mentioned Noam Chomsky's comment on the evolution of the language faculty, but he is often cited by creationists to support their view against linguistic evolution. As they do in the biological case, creationists posit a supernatural cause for both the origin of language itself and for the origin of the different languages. Henry Morris refers directly to Chomsky's work in explaining God's special creation of the different language forms at Babel:

[T]he "phonological component" of speech (or its surface form) is the corpus of sounds associated with various meanings, through which people of a particular tribe actually communicate with each other. Each phonology is different from the phonology of another tribe, so that one group cannot understand the other group. Nevertheless, at the "semantic" level, the deep structure, the "universal grammar" (the inner man!), both groups have fundamentally the same kinds of thoughts that need to be expressed in words. It was the phonologies, or surface forms of languages, that were supernaturally confused at Babel. . . .[44]

Characteristically, intelligent-design creationists say nothing of Babel, but they too cite Chomsky in their attack on the possibility of an evolutionary account of the origin of human language capacity. For instance, Oller and

Omdahl quote Chomsky's pessimistic appraisal of Karl Popper's suggestion that higher language could have evolved from a lower stage of vocal
gestures; Chomsky had written that the "stages" Popper mentions are not
analogous and that there was "no reason to suppose that the gaps are
bridgeable."[45] They do take care to acknowledge that Chomsky does not
himself draw an inference to a "Designer," but suggest that one would
have expected him to do so, noting his cryptic comment that "what is
important for the behavior of an organism is its 'special design.' "[46] Actually, some psychologists had already alleged that Chomsky was a "crypto-
creationist," but Chomsky's MIT colleague Steven Pinker, a distinguished
linguist in his own right, dismisses this charge, explaining that Chomsky
thinks that human language capacity arises from physical laws. Pinker
himself has argued persuasively over the years that evolution by natural
selection is indeed the best explanation for "the language instinct" and
that the purported gaps are not insurmountable.[47] However, gaps in our
knowledge do remain, and we should admit that for both the origin of
language and the origin of life science can as yet tell us very little. Because
these are the areas in which scientists have the least evidence and few
confirmed hypotheses, they are the favorite spots for creationists to press
their attacks.

On the biological front, just as Johnson heads straight for the problem
of the Cambrian explosion, so do other intelligent-design creationists such
as Walter Bradley and Michael Behe[48] swoop directly in on the problem
of the origin of life. The creationist textbook *Of Pandas and People*, put
out by the Foundation for Thought and Ethics,[49] begins with the question
of which of "the two theories"—evolution or intelligent design—better
explains the origin of life. Creationists regularly try to make this question
the test of Darwinism, even though the origin of life has never figured as
a significant part of evolutionary theory. Behe's book *Darwin's Black Box*,
for instance, presents a definition of biological evolution aimed to shift
attention from the origin of species to the origin of life.

Evolution is a flexible word. It can be used by one person to mean something as
simple as change over time, or by another person to mean the descent of all life
forms from a common ancestor, leaving the mechanism of change unspecified. In
its full-throated, biological sense, however, *evolution* means a process whereby
life arose from non-living matter and subsequently developed entirely by natural

means. This is the sense that Darwin gave to the word, and the meaning that it holds in the scientific community, and that is the sense in which I use the word *evolution* throughout this book.[50]

But this is certainly not the sense that Darwin had in mind, as Behe should know full well given that he quotes Darwin's dismissal of the problem of the origin of life, in the same breath as the problem of the ultimate origin of the photoreceptors, as being not part of his theory: "How a nerve comes to be sensitive to light hardly concerns us more than how life itself originated."[51] Neither is it the standard meaning of the term today; in a 1997 survey of the definitions of biological evolution given in twenty-three major textbooks one finds not a single definition that includes the idea of life arising from nonliving matter. Nor is the limitation to development "entirely by natural means" part of any of these definitions.[52]

However, let us not quibble about Behe's definitional sleight-of-hand. Although it is not currently a part of evolutionary theory itself, the origin of life is of course a topic of research in biochemistry, and it would be a natural move to extend Darwinian explanations to the molecular level, given the identification of molecular mechanisms that instantiate the abstract notions of variation, inheritance and natural selection. And although there is no definitional limitation to development only by natural means it is certainly true that science will look for natural processes (though not necessarily Darwinian ones) to explain the origin of organic molecules and the original life-forms. Research into this topic has started only relatively recently.

In the 1920s, biochemists A. I. Oparin and J. B. S. Haldane had each speculated about possible scenarios for the spontaneous generation of organic molecules in a "prebiotic soup" early in earth's history, but it was not until the 1950s that Stanley Miller and his professor, Nobel Prize winner Harold Urey, at the University of Chicago, performed the first experiments to see whether Oparin's and Haldane's hypothesized atmospheric conditions could indeed produce organic molecules. The positive results of these experiments began to inspire further research, and in 1972 the International Society for the Study of the Origin of Life (ISSOL) was founded with Oparin as its first president. Today ISSOL includes almost 400 scientists from around the world whose work touches in some way on the topic. Creationists usually cite the Miller-Urey experiments and then quickly dismiss both the assumptions and the results to date of all such research as insignificant. Intelligent-design creationist Percival Davis and Wayne Frair, for example, write:

Such experiments have produced interesting and often surprising data, but it has not been possible to synthesize anything even closely resembling a self-reproducing organism from simple substances, although . . . amino acids, . . . polypeptides, sugars, nucleotides, and adenosine triphosphate (ATP) have been synthesized. But if we compare these with a living creature (for example a single-celled protozoan, the amoeba) the difference is comparable to comparing a few nuts and bolts with an automobile.[53]

Of course, they are right to suggest that the origin of life remains largely a mystery, but they do not seem to be interested in contributing to the ongoing laboratory work—some of which is described by de Duve in *Vital Dust* (1995)—that is slowly uncovering initial pieces of the puzzle and leading to further understanding. In *Of Pandas and People*, Davis and co-author Dean Kenyon, an erstwhile origin of life researcher who became discouraged about the whole field and subsequently lent his name in support of YEC activism, dismiss the possibility of future discoveries in origin of life research as nothing more than "promissory materialism."[54] We will look further at the import of the reference to materialism in the next chapter, but here let us just note what the preferable creationist alternative is. According to Davis and Frair:

We accept by faith the revealed fact that God created living things. We believe God simultaneously created those crucial substances (nucleic acids, proteins, etc.) that are so intricately interdependent in all of life's processes, and that He created them already functioning in living cells.[55]

In other words, what they are offering is to replace the tough empirical labor required by "promissory materialism" with the satisfied complacency of pious spiritualism. Such faith may be laudatory in religious contexts (though even here it might be wise to remember whom it is said that God helps), but it is hardly a recipe for scientific progress. As one might expect, however, Davis keeps quiet about God and faith in *Pandas*, but instead points out "the amazing correlation between the structure of informational molecules (DNA, protein) and our universal experience that such sequences are the result of intelligent causes. This parallel strongly suggests that life itself owes its origin to a master intellect."[56]

We will look at the purported inexplicability of informational molecules later, but here let us remain on track with the general argument. By now the creationists' program should be more than clear. It is a familiar one and it is not confined to biology. Considering the history of ideas of the origin of language, Giorgio Fano writes,

The first solution that comes to mind when searching for the origin of a thing or of an institution which seems mysterious and difficult to explain is that of a supernatural cause. . . . In relation to the problem [of the origin of language], explanations of this kind are encountered with great frequency from the dawn of speculation through to the thinkers of the nineteenth century.[57]

In linguistics, such explanatory short-cuts have been long discarded just as they have been in biology. Creationists want to bring them back in both fields and beyond.

God of the Gaps

Take a moment now to recall the creationist arguments we have briefly surveyed—the gaps in the fossil record, the Cambrian explosion, and the origin of life and language—and you will notice that they all have the same form. Each challenges scientists with some specific puzzle, claims that it cannot be explained scientifically, and concludes that we need God to explain it. Creationists have long lists of such challenging questions at the ready. Walter Brown posts one representative list of "questions to ask evolutionists" on his web page. Many are old stand-bys, some going back as far as Paley: "Where are the billions of transitional fossils that should be there if your theory is right? . . . How could organs as complicated as the eye or the ear or the brain of even a tiny bird ever come about by chance or natural processes?" Many challenge astrophysics: "Why do we have comets if the solar system is billions of years old? . . . Where did all the helium go?" There are souped-up variations of the classic chicken and egg problem: "Which came first, DNA or the proteins needed by DNA—which can only be produced by DNA?" One even finds ultimate philosophical questions: "Where did matter come from? What about space, time, energy, and even the laws of physics?" One of the big ones these days (we just saw Davis and Kenyon mention it above) concerns the origin of life and of information: "If it takes intelligence to make an arrowhead, why doesn't it take vastly more intelligence to create a human? Do you really believe that hydrogen will turn into people if you wait long enough?" These questions are among the most common, but one could cite many more; creationist books and talks are composed almost entirely of such challenges.

I will not now continue to address the challenges one by one but will instead step back and look at the general argument form itself. When the creationist brings up one or another such puzzle in a debate and demands "Explain that!" how should we take these challenges?

First, note that it takes far more time and care to answer one of these questions than to pose one. This is especially true of scientific questions whose answers may be complex and difficult to explain in nontechnical terms. In a debate, a scientist cannot hope to have time to address even a small fraction of the challenges a creationist can make. Given that it takes a full semester-long undergraduate course to present a basic introduction to evolutionary theory alone, to try to "defend evolution" against all these challenges in an hour or so is sheer folly. One is forced to quickly give superficial simplified answers, leaving the false impression that that is the level of evidence upon which evolutionary theory is based. For this reason, I recommend that scientists politely decline to participate in the standard sort of general evolution versus creationism debate, which is inherently biased. Many of the challenges creationists bring up have been answered repeatedly by scientists, though this fact may seem obscured when creationists continue in debate after debate to noisily shoot the same blanks.[58]

But what about the questions we cannot answer? We should admit up front that science does not have solutions to many of the harder questions creationists pose. Some deal with genuine puzzles. Of course, this is not surprising, given that creationists often get the puzzles by quoting scientists who are themselves always pointing out unanswered questions that deserve investigation. Those researching the origin of life express irritation that creationists taunt them with puzzles that they themselves brought up as interesting research questions. Still, what are we to make of such unexplained gaps? Don't creationists have an argument here? Indeed they do, and it is so common that it has a name: it is the God of the gaps argument.

We have seen how it goes. Point to some hole in a scientific theory and conclude that, since science cannot explain it, the creationist alternative (divine intervention) must be right. However, claims that "science cannot explain *X*" are ambiguous; they slide over an important logical distinction that we must not ignore. Consider the difference between "X is *unexplained* by science" and "X is *unexplainable* by science." The former means that *as of yet,* X remains to be explained, while the latter expresses the stronger notion that science cannot explain X *in principle*; that is, now or ever in this or any possible circumstance. There is also an intermediate category of things that might be unexplainable for some *practical* reason.

If we were confronted with some empirical fact of the second sort for which scientific explanation was impossible—what we may call an "uncrossable" gap—then we would have to give serious consideration to the creationist alternative. This would not be just a puzzle, but a profound puzzle. This is not to say that such a profound explanatory gap would automatically (that is, deductively) prove the existence of God, for there are many other nonscientific options that would have to be considered. Certainly, for example, other supernatural powers besides God's could be put forward to bridge the gap and we would need additional (nonscientific) arguments to adjudicate among them. There are also other philosophical options that would have to be canvassed, such as whether the X is something that could stand without explanation; maybe there are some particular gaps that we do not really need to get across. Still, identifying an uncrossable gap would be a real step forward for the creationist because they, unlike the scientist, have the ability to cross such an abyss by a supernatural leap of faith. One might say that for creation-science and theistic science, this is just part of the power of the cross. In a later chapter we will look in more detail at other purportedly uncrossable gaps proposed by the new intelligent-design creationists, but it is fair to say of the standard puzzles posed by creationists that the only ones that come close to being uncrossable are the ones having to do with ultimate or "limit" questions. If, for instance, we ask for the explanation of why there is something rather than nothing, we have moved beyond the limits of empirical science and into the realm of philosophical metaphysics. One does occasionally hear a scientist pronounce on such questions and claim, for example, that science has shown that there is no ultimate purpose to the universe, and in such cases creationists have a legitimate complaint, for such claims overextend scientific method. The best one can say in defense of scientists who make such claims is that they restrict them to more popular venues where it is understood that they may mix in more personal reflections. Even so, it would be better if they would explicitly note when they move from scientific to philosophical reflections. The proper scientific attitude toward such questions is principled agnosticism, since scientific method can have nothing to say about profound puzzles such as these.

That admitted, we should now note that, except for ultimate questions, none of the creationists' puzzles look to be uncrossable in the sense we have defined. It is true that there are many questions that are unexplained by science and others that might turn out to be unexplainable because of practical limitations, but these are not the same as being unexplainable in principle. These are crossable gaps. Of course, crossable gaps can still look pretty wide and deep, especially to novices, and creationists are masters at rhetorically enlarging the gaps to make them appear to be uncrossable. Lehigh University biochemist Michael Behe, certainly the most important national figure among the intelligent-design creationists after Phillip Johnson, is particularly skilled in this art. Let us look at a couple of examples to see how he cleverly deploys metaphors to magnify explanatory gaps.

Behe is a curious case. Unlike many intelligent-design theorists, he is forthright in stating that he believes the physicist's standard account of the age of the earth as billions of years old. Moreover, though his endorsement seems a bit strained, he states for the record that he accepts Darwin's thesis of descent with modification: "I find the idea of common descent (that all organisms share a common ancestor) fairly convincing, and have no particular reason to doubt it." He also agrees, again somewhat vaguely, that Darwin's mechanism—natural selection operating on variation—can explain "many things." But there is one thing he says it can't explain: "I do not believe it explains molecular life."[59] In *Darwin's Black Box,* Behe uses a variety of metaphors to persuade the reader that Darwinian evolutionary theory does not have the power to explain the gaps—what he calls the "canyons" and "unbridgeable chasms"[60]—that purportedly exist at the molecular level. His first metaphor begins by comparing evolution to a person jumping over a ditch.

Suppose a 4-foot-wide ditch in your backyard, running to the horizon in both directions, separates your property from that of your neighbor's. If one day you met him in your yard and asked him how he got there, you would have no reason to doubt the answer, "I jumped over the ditch." If the ditch were 8 feet wide and he gave the same answer, you would be impressed with his athletic ability. . . . If the "ditch" were actually a canyon 100 feet wide, however, you would not entertain for a moment the bald assertion that he jumped across.

But suppose your neighbor—a clever man—qualifies his claim. He did not come across in one jump. Rather, he says, in the canyon there were a number of buttes, no more than 10 feet apart from one another; he jumped from one narrowly spaced butte to another to reach your side. Glancing toward the canyon, you tell your neighbor that you see no buttes, just a wide chasm separating your yard from his. He agrees, but explains that it took him years and years to come over. During that time buttes occasionally arose in the chasm, and he progressed as they popped up. After he left a butte it usually eroded pretty quickly and crumbled back into the canyon. Very dubious, but with no easy way to prove him wrong, you change the subject to baseball.[61]

In this story Behe is slyly identifying the clever neighbor with the eminent paleontologist, Stephen Jay Gould, who is almost as well known for his love of baseball as he is for his brilliant essays explaining evolution. One has to admire Behe's use of language here and throughout the book. He is especially good when writing within his own area of biochemistry and giving the reader a sense of the forms of organic molecules beginning with everyday images, such as his descriptions of how tubulin molecules automatically combine one with another, stacking like cans of tuna with the beveled bottom of one fitting snugly into the straight-lipped top of the next, to form the microtubules that make up cilia. Behe also describes the blood-clotting system using an analogy of the Rube Goldberg machine and a scene from a famous cartoon in which the rooster Foghorn Leghorn gets clobbered with a telephone pole as the last step of such a machine, and he compares the intercellular transport system with the steps an automated space probe would have to take to recycle old batteries.

It does not take a scientist long, however, to realize that there is little substance to Behe's argument and that he is relying on vivid, memorable metaphors like the ones above to make his case. Evolutionary biologists see no problem with the idea of slowly rising and eroding "buttes," because they can observe the slow process of genetic change in the lab and in the wild. But the reader without the scientific training needed to evaluate the biological examples themselves might too easily rely on Behe's analogies to interpret the lesson from them. In every case, the message Behe urges the reader to take home is that the biological world is chock full of not just uncrossed but uncrossable gaps.

In a chapter entitled "Road Kill," Behe replays the story of unbridgeable chasms, this time with a tale of a groundhog trying to cross lanes of traffic, which purportedly illustrates a problem for evolution. He begins with a description of the automotive dangers groundhogs face even on a quiet rural road.

Usually you're driving along . . . when all of a sudden a small, round shape waddles out of the darkness into your lane. At that point all you can do is grit your teeth and wait for the bump. . . . The next morning all that's left is a little stain on the road, other cars have obliterated the carcass. Nature red in tooth, claw, and tarmac.

In Behe's next image the road has turned into the Schuylkill Expressway which is "eight or ten lanes wide in certain stretches" with thousands of times the volume of traffic. One can predict the next extension of the metaphor.

Suppose you were a groundhog sitting by the side of a road several hundred times wider than the Schuylkill Expressway. There are a thousand lanes going east and a thousand lanes going west, each filled with trucks, sports cars, and minivans doing the speed limit. Your groundhog sweetheart is on the other side, inviting you to come over. You notice that the remains of your rivals in love are mostly in lane one, with some in lane two, and a few dotted out to lanes three and four; there are none beyond that. Furthermore, the romantic rule is that you must keep your eyes closed during the journey. . . . You see the chubby brown face of your sweetie smiling, the little whiskers wiggling, the soft eyes beckoning. You hear the eighteen-wheelers screaming. And all you can do is close your eyes and pray.[62]

This supposedly illustrates a basic problem for gradualistic evolution, which would maintain that the highway was not crossed all at once but one lane at a time. Behe says he has a better explanation—God's intelligent design. Better? Let us put it in terms of Behe's story to see how the intelligent-design "theorist" must imagine how the groundhog crossed this uncrossable highway. According to IDCs, God's design is necessarily for a purpose, so we must suppose that the groundhog and his sweetie must literally have been a match made in heaven. Taking Behe's metaphor to its logical conclusion, what his alternative "explanation" comes to is just this: God must have sent down Cupid to fly the lovesick little fellow over to his sweetie. Even if we were to agree that the odds were greatly stacked against the groundhog's crossing the highway on his own, surely this is still a more reasonable working hypothesis than to jump to the conclusion that he got across by some divine airlift.

Creationists often accuse Darwinians, sometimes fairly, of giving little more than imaginative "Just-So Stories" when reconstructing evolutionary pathways. Behe has told his own tales, but I do not mean to criticize them simply on the ground that they are analogies, because analogical arguments can be extremely valuable. However, one must be very careful when making arguments from analogy to see to it that the analogy is a fair one—the points of analogy really must be "just so" if the analogy is to succeed in doing its work. For this reason, we must always be wary when evaluating such arguments lest we be misled. Up to this point, I have accepted Behe's analogies and criticized them as presented, but I now want to suggest that they are not just misleading but also betray a fundamental misunderstanding of Darwinian mechanisms. Behe has made a terrible blunder in both of these two critical analogies.

Remember, his analogies are intended to function as criticisms of gradualistic Darwinian evolution. In both stories Behe describes a single organism who has either just purportedly crossed or is about to try to cross a seemingly impossible evolutionary gap—the neighbor by jumping from one temporary butte to another across a canyon, the groundhog by setting out to meet his sweetie. However, according to evolutionary theory it is not individual organisms but *populations* of organisms that evolve. As we have seen previously, it is this mistake that makes some people think that evolution is wrong because we never see dogs changing into cats. We cannot think that Behe's groundhog is supposed to stand in the analogy for a population, for in the story we see others from his population, his sweetie waiting on the other side, and the carcasses of his dead rivals that litter the first few lanes of the 2,000 lane highway he must cross to meet her. One might forgive Behe this minor infidelity, but he compounds it by inexplicably leaving out of the analogy all of the very elements that do the explanatory work for Darwinian gradualism. Keeping in mind that it is a population that evolves, recall how the Darwinian processes operate: on the average those individuals in the population who are even slightly more fit to their environment will have a better chance than others to survive, reproduce, and thus pass on those fit characteristics to their offspring, who will then repeat the process, followed by their slightly fitter offspring, and so on. So how should Behe have told the story to make it a fair analogy?

Instead of having our groundhog prayerfully inching out where angels fear to tread, toward his sweetie, and past the dead bodies of his unsuccessful rivals strewn about the first few lanes of the superhighway, to represent the Darwinian picture correctly Behe should have had Mr. and Mrs. Groundhog and the whole great population of groundhogs striking out *en masse*. Behe is right that most would not survive even the first lane and if they continued straight on then fewer and fewer would be left after each lane. But wait . . . gradualistic evolution does not claim that a population just heads across a gap in this way. Rather it observes that Mr. and Mrs. Groundhog and those of their fellows who have successfully made it past the first lane (perhaps because they stepped just a little quicker than those who failed to make it) stop to have a bunch of kids. With the population now more or less returned to its former numbers, Ma and Pa then retire and leave the second generation to tackle lane two. The casualties still will be legion, but this time the whole group starts off being on average a bit fleeter of foot than the previous. Again, those whose slightly fitter characteristics allow them to survive the second lane and reproduce yield the race across lane three to the third generation. With each generation, new variations arise, and though in many cases these will hinder rather than help in the race, those few with useful new traits (not just increased swiftness but perhaps also sneakiness, better hearing, larger litters, and so on) will likely carry them forward to their offspring and in this way each generation—naturally selected by the traffic—will turn out to be better adapted to their dangerous environment. Mr. and Mrs. Groundhog never themselves cross the entire superhighway; it is their distant descendants, now quite modified, who will be found on the other side. If these descendants were to look back after their journey at the descendants of other groundhogs from the original population who never moved out into the highway environment, many would no doubt find it hard to believe that they are related as cousins to those slow and dim-witted creatures. However, one of them might, if he could, write a daring book like Behe's and argue that they are in fact descended from a common ancestor, but that their journey across was literally, and not just metaphorically, miraculous.

Now that we have corrected Behe's blunder so that his analogy better models Darwinian gradualism does the gap still appear uncrossable?

Hardly! Of course, we should still take seriously the problem posed by the 2,000 lanes of traffic, but a bit of thought and a little more investigation shows that the situation is not so grim after all. Behe actually inadvertently suggests a related answer to the problem in the second step of his story where he mentions that the Schuylkill Expressway is eight or ten lanes wide "in certain stretches." If we are confronted with what appears to be a large explanatory gap in one place, whether they be "organs of extreme perfection" or the explosion of forms at the boundary of the Cambrian, a reasonable explanation is that the crossing was made somewhere else, at a different stretch of the highway, perhaps back on that rural road and then through quiet back alleys, where the lanes were fewer and narrower. This possibility identifies a standard form of evolutionary gradualism that may be made concrete in particular cases. The reader who is interested in learning about this in more detail should read Richard Dawkins's book *Climbing Mount Improbable* (1996), which shows how the gradualist approach finds ways again and again around apparently uncrossable gaps.

Of course, we are in no position to bridge every chasm; there remain any number of gaps that have yet to be explained, allowing creationists to bring up different puzzles each year and quietly drop old ones that have been solved. However, the history of evolutionary biology is a history of apparent explanatory gaps that scientists have closed one by one. Recall how the discovery of Mendelian genetics provided the solution to the problem of blending inheritance. When I was an undergraduate biology major, my professors noted that no one could explain the mechanism by which the spindle fibers formed and segregated chromosomes during cell division; fifteen years later the newly discovered answer would fill in that gap in the textbooks. It would be easy to multiply such examples, for evolution has been and remains a progressive, successful research program.

When we consider Behe's criticisms in light of this consistent pattern, his book becomes particularly significant in that it indicates just how far creationists have had to retreat to find significant explanatory gaps in evolutionary theory. Behe is certainly correct that molecular biology has identified a host of new and heretofore unappreciated puzzles for evolutionary biologists. That they have yet to solve, or in many cases even to begin to address these puzzles, however, is mostly the unsurprising result of the fact that molecular biology is still a very new subdiscipline. Many of the most significant molecular techniques that are now allowing biologists to look inside the black box of the cell were developed just in the last decade or two. The opening of this final box has indeed revealed a new level of complexity that has yet to be explained. But what should we conclude from our ignorance about such matters? Should we applaud and encourage the new generation of graduate students in molecular biology who are now eagerly turning their attention to investigating, and perhaps discovering the solutions of these puzzles? Or should we, as Behe and other creationists suggest, judge that these explanatory gaps are uncrossable by evolutionary or any natural theory and conclude that intervention by a divine intelligent designer is the only possible explanation?

We will return to a further examination of supernatural explanation and the creationists' proposed inference to intelligent design in chapter 5, but here I hope the conclusion to draw is obvious. All the creationists' challenges that "Science cannot explain X" are nothing but what philosophers call "arguments from ignorance." To point out that we are ignorant of the scientific explanation of X is hardly good reason to conclude that God is the explanation. To call upon God to plug our explanatory gaps in such cases is tantamount to intellectual indolence. No doubt God could plug these or any purported gap, but surely God has better things to do. We should judge Behe's and others' arguments in the same way that Fano suggests we take the theological explanations that have been regularly proposed over the centuries for the origin of languages: "From a scientific point of view they represent an *ignava ratio* which pretends to offer an explanation while declaring the problem insoluble."[63]

"Evolution Is Just a Theory"

When creationists press their agenda at school board meetings and on the editorial pages of newspapers they often argue that evolution is "just a theory." They try to introduce legislation and lobby for changes of wording in science curricula to emphasize this view. One notorious example occurred in 1996 in Tennessee. Creationists had successfully put such language into Senate Bill 3229 (sec. 49–6–1012), a proposed law that said:

No teacher or administrator in a local education agency shall teach the theory of evolution except as a scientific theory.

Biology teachers reading through to just this line found it a bit funny and rather perplexing, since of course evolution is a scientific theory and it would seem bizarre to legislate this. They did not laugh at the punch line that followed, however, which said:

Any teacher or administrator teaching such theory as fact commits insubordination . . . and shall be dismissed or suspended.

It was as though Tennessee creationists were bucking for a replay of the Monkey Trial seventy-one years later, and this time Scopes would not get away with just a guilty verdict and a fine. Fortunately, reason prevailed in this instance and the measure was defeated 20 to 13 after a five-hour floor debate, saving Tennessee from the embarrassment of being seen as a laughingstock. But as we have already noted, other states have passed less draconian but equally absurd laws mandating that in classrooms evolution should be taught "as theory, not fact."

The contrast that creationists want to draw is clear. When they say that evolution is only a theory they put the emphasis on "only." In the logically slippery way they portray the situation, facts are true, theories are not facts, evolution is a theory, and thus evolution is not a fact and so is not true. These claims have been repeated often enough by creationist activists that it is now one of the most common arguments that one hears from people when they oppose evolution in a public forum.

Evolution is only a theory. It should not be taught as fact in our public schools. If the evolutionist theory is taught, then it is only fair that Creationism be presented. It takes more faith to believe in evolution than it does to believe in divine creation.[64]

They also draw the conclusion that creationists want:

Creation and Evolution are just theories. "Theory" is just a fancy word for "guess."[65]

Along the same lines is the claim that evolution is just an "assumption." We find this kind of charge reiterated throughout creationist literature. In *A Case for Creation*, Wayne Frair and Percival Davis write,

Every age has possessed certain unquestioned presuppositions that served as foundations for its most popular philosophies. Such a presupposition in our day is the theory of evolution.[66]

My guide at ICR's Museum of Creation took pains to emphasize this point as well. Phillip Johnson sometimes makes the same charge that evolution is an assumption, although elsewhere he is more subtle, arguing that it is scientific naturalism that is an "arbitrary assumption" and that evolutionary theory depends upon it. We will look in detail at this last more sophisticated argument in the next chapter, but here let us look at the more common complaint.

It is certainly true that evolutionary theory is a scientific theory, but it is equally certain that it is not a "guess." So what are we to make of the creationists' challenge?

In part, this is simply a problem of terminology; as we saw before, the way terms are used in everyday contexts often are significantly different from how they are used in the context of a technical discipline. What in ordinary speech we might call "just a theory" corresponds more closely to what scientists would call an "untested hypothesis." When a scientist puts forward a new hypothesis about how the world is (or, put another way, proposes a model of some aspect of the world) it is typically an *educated* guess, but it does not have to be even that and may have been thought of at random or in a dream. The real scientific work consists in taking this hypothesis and checking to see whether it can survive the rigors of logical and empirical testing. Hypotheses are tested within an explanatory framework and if they are confirmed they may be added to that framework. The whole structure of hypotheses and models and their explanatory interrelationships constitutes a theory.

Look back now at the theory of language evolution and you will see the outline of such a structure. Jones's common source hypothesis was originally proposed and tested in the context of a relatively small number of languages that were recognized as Indo-European but then became part of a general theoretical framework that included the notion of language families, mechanisms of linguistic variation, transmission and transformation, rules for identification of homologous structures that indicate pathways of descent, and so on. All these explanatory elements and more are what make sense of the changes of spelling, symbols, words, and grammar, the patterns of spatial and temporal distribution of languages, and the myriad other patterns of commonalities and differences linguists find among languages. This whole interconnected structure of confirmed hypotheses is the theory of the evolution of languages. The theory of evolution of biological organisms, as we have seen, has a similar complex structure. To confuse this scientific, technical notion of theory with the colloquial notion of a theory as just a guess or an assumption is simply a mistake that arises out of ignorance of science.[67]

Once we are clear about what a theory is as the term is used in science, we should recognize that it does not stand in contrast to the notion of a fact. Quite the contrary. Jones's original hypothesis of common descent is not "*just* a theory." It is not an "assumption." It is not a "guess." Languages do descend one from one another. That is a fact. And that fact is part of a strong fabric of interwoven facts that are all part of the theory of the evolution of languages. Theories may include statements of direct observations, models of mechanisms and processes, conclusions about unobservables expressed in a technical vocabulary, and so on, and all these may properly be considered factual. At this point, the theory of the evolution of languages has such overwhelming evidence supporting it that one finds it hard to conceive how it could possibly be overturned, though of course it might be refined, modified, or extended. The same can be said of other scientific theories such as that of the cellular organization of living organisms (cell theory), that dealing with the attraction of masses to one another (theory of gravitation), and so on. Atomic theory, the special and general theories of relativity, quantum theory, and, yes, the theory of biological evolution are all similarly well established.

This is not to say that all elements of an accepted theoretical framework are equal. Certainly there are some hypotheses that will be better supported than others. The facts that form the core of the theory are those that are most well confirmed, and they are taken for granted and used in testing other hypotheses. In other areas of the theory that are under development, new hypotheses that are still being tested and integrated into the theory might be quite tentative and not yet sufficiently supported to be accepted as factual. Philosopher of science Imre Lakatos discussed how scientific theories can typically be divided into a "hard core" set of hypotheses surrounded by a "belt" of less critical ones.[68] In both the theories of linguistic and biological evolution there are still many open questions about pathways of descent, for example, and hypotheses about specific relationships might be revised on the basis of new evidence and analysis. In some areas of development, the debates can become quite animated as scientists hammer out the merits of competing hypotheses. This is all part of the process by which a hypothesis proves its mettle.

To call something a scientific fact is simply to say that it has been so well confirmed evidentially that it is hard to imagine how it could be overturned, though we always have to keep an open mind that new evidence could do so, if only in principle. This seemingly paradoxical feature of scientific facts—high confidence together with potential fallibility—is probably what is most difficult for the layperson to fathom. If something could still possibly turn out to be false, then we shouldn't call it a fact, some would say. This is a very common attitude among creationists, who (in keeping with their religious beliefs) desire nothing less than absolutely certain truth. But really the scientific attitude is not so far removed from our ordinary notions of truth and fact. I expect that most people taking the stand in a trial would state their age for the record and not hesitate to swear to it as a fact. Most of us base this on the same evidence—the testimony of our parents, our physical appearance relative to others, the date on our birth certificate. If a prosecuting attorney were to rudely intimate that your parents were concealing the truth about your birthday or that you had misread your birth certificate you would have good reason to feel offended and would be able to give reasons that your statement was true, but you would have to admit that you could not be absolutely certain, given that more than one person has discovered as an adult that they were a few months older than they had been led to believe. However, it would be outrageous for you to have to say that your assumed age is not a fact just because of the off-chance that new information in the future could force you to admit that you had been wrong. This is essentially the status of scientific facts, except that scientists are trained to be more cautious before admitting a hypothesis into the realm of accepted fact in the first place and are more explicit about keeping open the possibility that new evidence might require revision.

More than other sciences, biology has had to defend itself from attacks from outside the scientific community by those who are unfamiliar with scientific use of terms and who misinterpret the notion of "theory" as meaning "not a fact." As we noted, biologists who have responded to these attacks using ordinary terminology have sometimes inadvertently confused the issue by mixing ordinary with scientific terms, calling the history of evolutionary change "the fact of evolution" and calling evolutionary mechanisms "the theory of evolution." Here let us put things clearly. Biologists take Darwin's thesis of the history of descent with modification from common ancestors to be a fact. The key evolutionary mechanisms of variation by mutation and recombination, genetic inheritance, natural selection, random drift, and so on are also known to be factual. Many broad features of the evolutionary pathways are also accepted as facts. All these core conclusions are based on such overwhelming observational and experimental evidence, both indirect and direct, that it is highly unlikely that they could ever be overturned. These are all parts of evolutionary theory and they are also all facts. There are other evolutionary hypotheses that have not yet garnered sufficient evidence and whose "facthood" is still in question, especially ones having to do with particular pathways of descent or with the relative importance of natural selection versus drift, for example, as the cause of some particular biological feature. It is also accepted that the theory of evolutionary processes is incomplete, that many details of the mechanisms have yet to be worked out, and that there could be as yet unknown processes working in tandem with the known mechanisms that are important in generating the patterns of order and disorder that characterize the biological world. As research uncovers more about these processes, we can expect that new findings will supplement and refine evolutionary theory but not undermine the factual elements that the evidence has already established.

Is It All Simply a Matter of Faith?

To wrap up this chapter, there is one further important creationist argument that we must discuss, namely, that whether one believes evolution or not is simply a matter of faith. Phillip Johnson writes, "[S]cientific materialists have faith that they will eventually find a materialistic theory to explain the origin of life even though the experimental evidence may be pretty discouraging for now. Because they have faith in their theory, Darwinists believe that common ancestors for the animal phyla once lived on the earth, even though those ancestors can't be found."[69] One of my students expressed it this way: "It is nothing but a matter of faith on which somebody's views are based. A person with a faith in religion would be more inclined toward creationism. A person with a faith in science would be more attracted to evolutionism." Many creationists express this attitude, though many make the charge while in the same breath contending that the creationist hypothesis is not faith but is based upon scientific evidence. Percival Davis, one of the primary authors of the IDC *Pandas* textbook, was more forthright in his earlier book, stating explicitly that the creationist hypothesis is based on faith. He and co-author Frair paraphrase a line from Hebrews (11:3), that "By faith we understand that the worlds were prepared by the word of God," and they emphasize the importance of *faith* in contrast with evidence by italicizing that word and inserting after it the phrase "not scientific data."[70] The basic contention in these and similar statements is that belief in scientific or religious propositions is all "just a matter of faith" and that there should be no question which faith to choose.

However, although it is true that one can have both scientific and religious beliefs, all beliefs are not created equal. Beliefs that are based on evidence are not on a par with beliefs that one holds without evidence, and it is disrespectful of both science and religion to confuse these. To the faithful, having faith means sustaining belief despite the lack of evidence and sometimes even in the face of countervailing evidence. This accounts for the difference between a scientific test of a hypothesis and a theological test of faith. In the former case, we believe a proposed hypothesis only because it is supported by the evidence and has survived attempts to disconfirm it, and we reject it if the evidence opposes it. In the latter case, to survive a test of faith means to hold fast to one's belief even when everything goes against it. On the watered-down notion of faith implied in the criticism there would be no belief that is *not* based on faith—knowing one's birthday, that fluoride protects against cavities, and so on would all be matters of faith—so the concept would lose its important, distinctive meaning. These simple beliefs and other scientific conclusions are not based on faith but are inferred from the evidence of observation. To abuse the notion of faith so that it extends to such quotidian beliefs is to gut it of its theological significance.

Given other things creationists say, it is clear that they really do understand the stronger demands of faith. At ICR's Museum of Creation our guide directed our attention to the scriptural order of Creation, and she addressed the criticism that days are spoken of before the creation of the sun, by saying that this was in clear contradiction to the scientific account and was God's way of forcing the Christian to choose between faith in His Word and faith in mere scientific, human reason. These are not empirical tests of God, but tests of the believer. ICR's YEC view that God created the earth with the appearance of age is of a kind; we are told to trust the revealed truth of the youth of the earth in spite of the evidence of its "mature" appearance. Intelligent-design creationist Davis and Wayne Frair put the recommended view this way:

As Christians we believe that only God can know the universe as it *really is*. We are limited by our senses and our minds, and we know the universe only as it *appears* to us. . . . Truth as God sees it has been revealed in the pages of Scripture, and that revelation is therefore more certainly true than any mere human rationalism. For the Creationist, revealed truth controls his view of the universe to at least as great a degree as anything that has been advanced using the scientific method.[71]

As we have seen, the consistent creationist attacks not just biology, but also linguistics and almost every other science as well. The hard-won discoveries of astrophysics, geology, agronomy, medicine and on and on are all labeled false and are to give way to the "True Science" that is revealed by Scripture. But now we see that it goes even beyond this. The creationist holds not only that we must reject well-established conclusions of these sciences but also, as is evident in the quotation from Davis and Frair, that we must set aside the method of science itself. In the next chapter we will look in detail at this particular attack and how it is being pressed by the new creationists. This is a truly radical position, because science, far more than any of its specific conclusions, is fundamentally scientific method. Creationists would have us turn science on its head and replace scientific reasoning based on observable evidence with human interpretations of revealed truth. The confusion of human languages would be nothing compared to the great confusion that would result from such a program.

4

Of Naturalism and Negativity

It is easy to see why scientific naturalism is an attractive philosophy for scientists. It gives science a virtual monopoly on the production of knowledge, and it assures scientists that no important questions are in principle beyond scientific investigation. The important question, however, is whether this philosophical viewpoint is merely an understandable professional prejudice or whether it is the objectively valid way of understanding the world. That is the real issue behind the push to make naturalistic evolution a fundamental tenet of society, to which everyone must be converted.

—Phillip E. Johnson

Duels and Dualists

Dueling Models

There is no denying the visceral appeal of a classic courtroom showdown. One party stands charged with a crime. Are the allegations true or false? Is the defendant innocent or guilty? The battle between the lawyers who argue the case will determine the answers. This is an intellectual duel to the death. In the end one must win and the other must lose. As spectators we find the drama intense and enjoy the satisfaction of resolution when the verdict is read. It is no wonder that legal action plays so well on television. Occasionally, however, in real life, we are taken aback by an odd verdict and by the legal maneuvers and arcane objections and rulings that led up to it. "Was justice really served here?" we ponder. Was this contest between attorneys really the best method to uncover the truth?

Several years ago, I sat in on a course at the University of Texas School of Law on the legal rules of evidence. My own research is on the nature of scientific evidence and I thought there should be systematic commonalities that might be worthy of research. My colleague at the law school who taught the course shook his head when I told him of my intention. The legal rules of evidence, he said, vary from country to country and sometimes from state to state, and there is nothing systematic about them. They are the result of a combination of precedent, legislation, and political compromise, and they have little if any connection to scientific notions of evidence. He was right. Though there are some interesting connections to be drawn to science, by the end of the course I understood his point that the legal system is not concerned simply with optimizing the search for truth but rather incorporates a host of other interests that have been injected into the system historically and politically, and that the different sets of legal rules of evidence reflect these strange admixtures of goals. I came away rightly impressed by the delicately balanced way the legal rules of evidence attempt to satisfy these multiple interests but with the recognition that this was leagues apart from science. Given the significant difference between scientific and legal methods, it is ironic (and altogether telling) that the classic battles between evolutionary theory and creationism have taken place in the courtroom and that lawyers—from Bryan to MacBeth to Bird to Johnson—have been noticeably prominent in arguing creationism's case.

The legal model of a duel between two parties in which one wins by making the other lose has been the dominant way in which creationists have pressed their attack on evolution. The Institute for Creation Research has always tried to argue that creation-science and "evolution science" were the only two options, and Arkansas Act 590 was written to incorporate ICR's approach. The Oklahoma Republican Party included a creationist plank in their 1996 party platform that endorsed this explicitly: "We support the two-model approach to the teaching of origins in the public schools, giving balanced treatment to the view of evolutionists and Creationists." Creationists always try to set up debates following the same formula. A typical debate question might be: Does the scientific evidence better support the creation model or the evolution model of origins? In a scientific setting, this is not an unreasonable way to put the question, for science does require that hypotheses be compared with each other to see which better accounts for the evidence, but it is very misleading when posed in a debate. Fred Parrish, a biology professor at Georgia State University, wrote an amusing and revealing account of how a creationist group "suckered" him into a debate on this very question with Walter Brown, founder of the Center for Scientific Creation, and how the structure of the debate allowed the creationist side to obscure the real issues.[1] When the question is posed as though there are only two mutually exclusive options to choose from, creationists are able to simply argue negatively, pointing out holes, real or supposed, in evolutionary theory, and then claiming that since evolution has so many problems that creationism, as the only alternative, must obviously be correct.

Philosophers of science Michael Ruse and Philip Kitcher have previously pointed out the logical fallacy of the classic YEC version of this "dual model" argument—it is a false dilemma to argue that if evolution is wrong then a literal Genesis creationism is right. Christians who are theistic evolutionists have objected in an even more strenuous way from a theological point of view to drawing a line in the sand in this way. Mentioning ICR creationists like Morris and Gish, theistic evolutionist Howard van Till, who has written extensively on the relation of science and Christianity, chides those who "preach the 'gospel of either/or-manship.' They insist that the results of modern natural science stand in opposition to the Christian faith."[2]

Though the new creationists set up their dilemma in a slightly more subtle manner, they are preaching this same gospel. Now instead of creation-science they speak of theistic science or intelligent-design theory, and instead of "evolution-science" they rail against the "blind watchmaker thesis" and scientific naturalism. Ruse might have thought that he had seen the last of the dual model argument following his testimony at the Arkansas trial in 1982, but he encountered it again, this time dressed up in designer clothes, when he was asked to debate Phillip Johnson ten years later on the question of: "Darwinism: Scientific Inference or Philosophical Preference?" I want to suggest that this new dual model approach misunderstands the nature of the scientific issue in as fundamental manner as did the old version. In this chapter we will examine this new "philosophical" version of the dual model argument as it has been articulated by creationism's greatest new champion, Phillip E. Johnson.

Mr. Johnson for the Prosecution

Johnson is professor of law at the University of California, Berkeley, and prior to his entry into the creationism debate he was best known as the author of a popular textbook on criminal law. In his book *Darwin on Trial* (1991), Johnson renewed the creationist attack against evolution, and he is rightly credited with giving the movement a new lease on life. Christian creationist groups have been quick to recognize Johnson as an important asset, and they sponsor forums for him to present his arguments against evolution. The Ad Hoc Origins Committee, a group of professors and academic scientists from universities including Princeton and the University of Texas who describe themselves as Christian Theists, claims that Johnson has given a "penetrating and fundamental critique of modern Darwinism" and distributes free videotapes of one of his speeches.[3] Creationists have become increasingly well funded and well organized in the last two decades, but until now they have lacked an articulate spokesman with a high-profile institutional affiliation. Johnson fills this role and provides the movement with the measure of credibility it has longed for. Of course, Johnson is a lawyer who boasts that he is "entirely unprejudiced because [he has] no formal training in science past high school."[4] His credibility is thus not on the scientific side—indeed, William Provine and other biologists have called his descriptions of evolutionary theory a "crude caricature"[5]—but he knows how to draw upon his strengths and makes a classic courtroom move of shifting the locus of argument in a way that seeks to undermine the expert testimony of his scientist adversaries. His key argument is broadly philosophical, but Johnson also uses his considerable rhetorical skills to try to turn the tables on scientists by portraying them as naïvely doctrinaire and intolerant, while portraying creationists as rational and fair-minded skeptics. To meet Johnson's challenge, we must not only show how his argument fails on logical grounds, but also cut through his rhetoric.

One of Johnson's titles—"Evolution as Dogma: The Establishment of Naturalism"—neatly captures both his argumentative and rhetorical strategies. Unlike the creation-scientists, who try to put creationism on a par with the theory of evolution by claiming that creationism is scientific, Johnson tries to put them on a par by alleging that evolution is ideological. Darwinian evolution, he claims, "is based not upon any incontrovertible empirical evidence, but upon a highly controversial philosophical presup-position."[6] That presupposition is naturalism. Johnson argues that natu-ralistic evolution is not scientific but rather is a dogmatic belief system held in place by the authority of a scientific priesthood, and that without the naturalist assumption evolutionary theory would be rejected in favor of creationism. The charge that science is a "secular religion" is not new, but Johnson is the first to locate a basis for the charge in specific philosophical assumptions made by science, and to try to exploit this as a point of weak-ness in evolutionary theory to the advantage of creationism. Johnson's attack contains a kernel of truth—it is true that science makes use of a naturalistic philosophy—but Johnson has misunderstood naturalism's role in science in general and its implications in this instance. To show this, I will begin with a review of Johnson's main argument and discussion of its key concepts. In the course of discussion I will also highlight Johnson's prejudicial and misleading rhetoric, which serves to polarize the debate and undermine the possibility of peaceful coexistence between science and religion.

Johnson against the "Dogma of Naturalism"

Johnson's Argument
Johnson offers variations of the usual creationist arguments that try to poke holes in the broad fabric of scientific evidence for evolution, but we will focus upon his novel and strongest challenge, which is the whole-cloth charge that evolution is metaphysical dogma. We can summarize Johnson's main argument in the following three-step form: Evolution is a naturalistic theory that denies by fiat any supernatural intervention. The scientific evi-dence for evolution is weak, but the philosophical assumption of natural-ism dogmatically disallows consideration of the creationist's alternative explanation of the biological world. Therefore, if divine interventions were not ruled out of court, creationism would win over evolution.

This is not laid out formally as a deductive argument, but one recognizes at once a version of the familiar "dual model" tactic; the argument is presented as though evolution and creationism are the only alternatives, so if evolution gets knocked out, creationism wins by default. Creation-scientists, requesting "balanced treatment" of the issue in the public schools, used a very crude form of this type of argument structure, with Darwinian evolution on the one side and a thinly disguised biblical literalism on the other; let the children judge the evidence and decide for themselves which one is right, they asked in the name of fairness. (Of course, they did not plan to mention Mayan or Hindu or Asanti creation stories as alternatives.) Johnson is more sophisticated. He, too, wants to get his conclusion by means of a negative argument against evolution[7] but he tries harder to set up the dichotomy to logically exclude other alternatives by attempting to define the key terms of the debate—"Darwinism" and "creationism"—so that they are mutually exclusive and jointly exhaustive.

Johnson takes pains to distinguish his brand of creationism from the specific scripture-based commitments of creation-science[8] and to define creationism broadly. Here is the way he puts the definition in *Darwin on Trial*:

"Creationism" means belief in creation in a . . . general sense. Persons who believe that the earth is billions of years old, and that simple forms of life evolved gradually to become more complex forms including humans, are "creationists" if they believe that a supernatural Creator not only initiated this process but in some meaningful sense *controls* it in furtherance of a purpose.[9]

Elsewhere he reiterates this with a slightly different emphasis:

The essential point of creation has nothing to do with the timing or the mechanisms the Creator chose to employ, but with the element of design or purpose. In the broadest sense, a "creationist" is simply a person who believes that the world (and especially mankind) was *designed*, and exists for a *purpose*.[10]

A significant feature of Johnson's definitions is that they put no explicit restrictions on the manner of creation so long as God is involved in a significant way; guided evolution, special creation or any other mode of divine creation seems allowed. The definitions make no reference to the Bible, making it appear that Johnson countenances as creationist the cosmogonies of any other religious or cultural tradition. In *Evolution as Dogma*, Johnson is even more general:

[A] "creationist" is . . . any person who believes that God creates.[11]

Such apparent open-mindedness makes the defender of evolution look narrow-minded in contrast with the tolerant creationist. It also serves to enlarge Johnson's constituency, for most people will identify themselves as creationist in the minimal sense of commitment to the idea that God creates.[12] Additionally, the broad definition helps bolster Johnson's claim that evolution is necessarily at odds with religion, for he contrasts this mild-mannered creationism with a view of evolutionary theory that makes the latter essentially atheistic.

Johnson defines "evolution" very narrowly. He does not deny that evolution by natural selection occurs if all one means by that is that "limited changes occur in populations due to differences in survival rates."[13] Even creation-science allows microevolution, he claims—God created "kinds" but thereafter individuals can diversify within the limits of the kind.[14] On the other hand, the important thesis of evolutionary theory, he says, is the further one about macroevolution: that evolutionary processes also explain "how moths, trees, and scientific observers came to exist in the first place."[15] Most of *Darwin on Trial* attacks this claim, but Johnson narrows his sights still further to set up his general argument. Since it is possible that God did not create creatures suddenly, but used instead a gradual evolutionary process, even macroevolution does not contradict creationism unless it is "explicitly or tacitly defined as *fully naturalistic evolution*— meaning evolution that is not directed by any purposeful intelligence."[16] This is the form of evolution that Johnson sets up as his target. Here is his positive definition:

By "Darwinism" I mean fully naturalistic evolution, involving chance mechanisms guided by natural selection.[17]

Johnson's main argument hangs on his conception of the role of naturalism in this scheme, which we will examine shortly, but his central point is that in naturalistic evolution God's intervention is excluded.

Taking these two definitions together, we see how the argument is supposed to work. Creationism holds that God plays a role in Creation (however it occurs) and Darwinism denies the same. Though closer inspection makes it clear that Johnson's definitions do not establish the logical dichotomy he needs (as we will discuss in the last section), on the surface it looks as though he has set up the major premise of a valid dilemma that will then allow creationists to rely solely upon negative argumentation. This is Johnson's first innovation. His second is that his characterization of the terms of the debate allows evolution to be attacked not only on scientific grounds, but also on philosophical grounds. He spends seven chapters in *Darwin on Trial* on the first task, trying to cast doubt upon the wide range of empirical evidence for evolutionary theory so that he can claim that creationism is a better theory which would be accepted if not for the "powerful" and "doctrinaire" naturalistic assumption that rules it out by definition.[18] As we shall see, however, the philosophical charge does the real work.

Johnson is well aware that scientists have not and will not now find creationist criticisms of the evidence for evolution to be persuasive,[19] but this matters little for he is playing to the jury. When the scientific expert witness rebuts his negative appraisals of the evidence for evolution, Johnson will argue that biologists are "[unable] to make any sense out of creationist criticisms of their presuppositions" because of their "philosophical naiveté"[20] and their "blind commitment to naturalism."[21] The evidence cited for evolution, Johnson claims,

looks quite different to people who accept the possibility of a creator outside the natural order. To such people, the peppered-moth observations and similar evidence seem absurdly inadequate to prove that natural selection can make a wing, an eye, or a brain. From their more skeptical perspective, the consistent pattern in the fossil record of sudden appearance followed by stasis tends to prove that there is something wrong with Darwinism, not that there is something wrong with the fossil record. The absence of proof "when measured on an absolute scale" is unimportant to a thoroughgoing naturalist, who feels that science is doing well enough if it has a plausible explanation that maintains the naturalistic worldview. The same absence of proof is highly significant to any person who thinks it possible that there are more things in heaven and earth than are dreamt of in naturalistic philosophy.[22]

Again we see how Johnson's rhetoric tries to make the creationist appear to be the rational "skeptic" who merely accepts the "possibility" of a Creator, and the biologist the "blind" and "naïve" ideologue who dogmatically rejects that possibility and thereby misjudges the evidence. Thus, he concludes, it is not the evidence but the ideology that supports evolution. Here is the conclusion the reader is supposed to draw:

Victory in the creation-evolution dispute therefore belongs to the party with the cultural authority to establish the ground rules that govern the discourse. If creation is admitted as a serious possibility, Darwinism cannot win, and if it is excluded *a priori* Darwinism cannot lose.[23]

The claim that evolution is held up solely by "metaphysical assumptions"[24] and "speculative philosophy"[25] allows Johnson to ignore the weakness of his negative scientific arguments. In his public lectures, Johnson follows the same pattern, usually taking a few token swipes at the empirical evidence for evolution and then moving quickly to his philosophical indictment of its naturalistic metaphysics.

Although this philosophical criticism of naturalism has to carry the weight of his conclusion, Johnson fails to provide any philosophical analysis of the concept that he charges scientists have uncritically accepted. Neither does he support his thesis that the concept is inherently dogmatic, or provide evidence that scientists do subscribe to it in the way he claims. Let us now briefly review the history of naturalism, and then evaluate Johnson's characterization and his application of the concept to the biological case.

Varieties of Naturalism

The generic meaning of "naturalism" is a philosophical view based upon study of the natural world, with an implicit contrast to the supernatural world, but this leaves room for a wide range of specific variations. Since the time of the ancient Greeks, naturalism has often been associated with various forms of secularism, especially epicureanism and materialism, but it has also been used as a label for religious views such as pantheism, as well as the theological doctrine that we learn religious truth not by revelation but by the study of natural processes. In the centuries leading up to the twentieth century, concomitant with the rise of the natural sciences, the term became associated more directly with the methods and fruits of the scientific study of nature. One spin-off at the turn of the century was the naturalist movement in literature, epitomized by Zola but continuing in a form through Steinbeck, which featured "scientific" portrayals of human characters playing out predetermined roles as amoral creatures governed by natural law. Another extreme expression was Auguste Compte's philosophy of *positivism*, the scientific stage of philosophical development which society purportedly reached after progressing beyond theological and metaphysical conceptions of the world. Positivism concerned itself only with regularities of observable phenomena, so naturalism at that time became associated with phenomenalism. This version of naturalism was carried forward into the philosophy of science in the early twentieth century by the influential logical positivists, who restricted knowledge to propositions with a determinable truth-value—if a proposition was not verifiable then it was taken to be meaningless. The so-called verifiability criterion of meaning turned out to be unworkable, and its collapse was one of several reasons for the demise of the logical positivist view in the middle of the century. Since then, in philosophy at least, the naturalist view of the world has become coincident with the scientific view of the world, whatever that may turn out to be. Many people continue to think of the scientific world view as being exclusively materialist and deterministic, but if science discovers forces and fields and indeterministic causal processes, then these too are to be accepted as part of the naturalistic worldview.[26] The key point is that naturalism is not necessarily tied to specific ontological claims (about what sorts of being do or don't exist); its base commitment is to a method of inquiry.

Of course one could choose to take some set of basic ontological categories from science at a particular time and then claim that only these things exist. The seventeenth century mechanistic materialists, who held that the world consists of nothing but material particles in motion, did just this, and there are any number of other ways that one could decide to fix base ontological commitments. This type of view is known as *metaphysical* or *ontological naturalism*. The ontological naturalist makes substantive claims about what exists in nature and then adds a closure clause stating "and that is all there is." A thorough historical review of positive formulations of ontological naturalism could fill an article in itself, but amidst this variety many do agree on a common negative claim: because God standardly is assumed to be supernatural, the typical ontological naturalist denies God's existence. It is possible, however, for an ontological naturalist to allow God in the picture, provided God's attributes are appropriately constrained to conform to the regimen of the given natural ontology. Hobbes and Spinoza were ontological naturalists who thought they found room for God (indeed, for a Judeo-Christian God) in this way. Some traditional theists, however, were not willing to countenance their naturalized conceptions of the deity; Hobbes was branded an atheist and Spinoza a pantheist. The problem of trying to naturalize theology is that traditionalists want God to be able to control nature from outside nature; they take God to be supernatural by definition. Probably the main reason for the strong secularist strand among the varieties of naturalism is that many naturalists also have tended to take for granted this traditional conception of God and have found it difficult to square with their other ontological commitments.

Ontological naturalism should be distinguished from the more common contemporary view, which is known as *methodological naturalism.* The methodological naturalist does not make a commitment directly to a picture of what exists in the world, but rather to a set of methods as a reliable way to find out about the world—typically the methods of the natural sciences, and perhaps extensions that are continuous with them—and indirectly to what those methods discover. An important feature of science is that its conclusions are defeasible on the basis of new evidence, so whatever tentative substantive claims the methodological naturalist makes are always open to revision or abandonment on the basis of new, countervailing evidence. Because the base commitment of a methodological naturalist is to a mode of investigation that is good for finding out about the empirical world, even the specific methods themselves are open to change and improvement; science might adopt promising new methods and refine existing ones if doing so would provide better evidential warrant. Understanding the nature of scientific evidence is critical for answering Johnson's charge, but let us postpone examination of that concept and how it relates to the question of God's existence and creativity until we have seen the details of Johnson's philosophical claims that naturalism is assumed dogmatically and that its ideology alone supports evolutionary theory.

Although it is the linchpin of his argument, Johnson provides only a cursory discussion of the concept of naturalism. Taken individually, his few statements do pick out versions of naturalism, but taken together they suggest a biased and misleading picture. In *Darwin on Trial*, Johnson defines naturalism as follows:

Naturalism assumes the entire realm of nature to be a closed system of material causes and effects, which cannot be influenced by anything from "outside." Naturalism does not explicitly deny the mere existence of God, but it does deny that a supernatural being could in any way influence natural events, such as evolution, or communicate with natural creatures like ourselves.[27]

This is a good definition of a common form of ontological naturalism; the "causal closure of the physical" is another way this idea is expressed. The acknowledgment that naturalism does not "explicitly" deny the "mere existence" of God, however, is significant, for it is another indication that Johnson is not as tolerant and ecumenical as his definition of creationism might initially lead one to believe. The clear implication here is that, because naturalism rejects continuing divine intervention, it does *implicitly* deny God's existence, but this conclusion follows only if one has a particular conception of divine power. We see this view expressed again as Johnson immediately follows the above definition by introducing a specific form of naturalism that he calls "scientific naturalism."

Scientific naturalism makes the same point by starting with the assumption that science, which studies only the natural, is our only reliable path to knowledge. A God who can never do anything that makes a difference, and of whom we can have no reliable knowledge, is of no importance to us.[28]

Note that in such statements Johnson is dismissing views such as Deism that do allow God to influence natural events, to make a difference and conceivably even to communicate with us, by setting up the world in the appropriate way at Creation but thereafter not intervening in the natural order. He is also rejecting views that hold that God is concerned with our spiritual rather than our material being and thus intervenes only at a spiritual level. He is also ignoring religious views that do not posit a personal God, but conceive of God as a universal life force or a mystical unity. Also unimportant, apparently, are views that say we can have "no reliable knowledge" of God; this restriction leaves out even many Judeo-Christian thinkers who hold that the nature of God is unknowable to the human mind. These spiritual views Johnson excludes are prevalent worldwide, so we should not be misled by his attempt to portray his form of creationism as generically tolerant. Such views, however, *are* compatible with varieties of both ontological and methodological naturalism and belie Johnson's attempts to conflate naturalism and atheism.

Returning to the definitions, one may think at first that "scientific naturalism" is Johnson's term for methodological naturalism, but in light of his other comments we see that he mixes in elements of ontological naturalism. He says, for example, that in the present context he considers scientific naturalism to be equivalent to evolutionary naturalism, scientific materialism, and scientism:

All these terms imply that scientific investigation is either the exclusive path to knowledge or at least by far the most reliable path, and that only natural or material phenomena are real. In other words, what science can't study is effectively unreal.[29]

By ignoring distinctions among such positions[30] Johnson again is able to associate evolution with (godless) materialism and to portray naturalism as monolithically dogmatic. "Scientism," for example, is a term of derision coined by hermeneutic critics of science to label those who wanted to apply the methods of the natural sciences "inappropriately" to the human sciences, for which they thought the literary model of hermeneutic *interpretation* should reign as the proper method. Their target was specifically the followers of the logical positivists, but, as was noted, the exclusionary positivist view that only the scientifically verifiable was meaningful has not held currency for several decades. Contemporary methodological naturalists would not recognize themselves in this description, yet it is just this sort of view that Johnson insistently portrays as the essence of scientific naturalism.[31]

When he applies naturalism to evolution Johnson says that one gets:

... a theory of naturalistic evolution, which ... absolutely rules out any miraculous or supernatural intervention at any point. Everything is conclusively presumed to have happened through purely material mechanisms that are in principle accessible to scientific investigation, whether they have yet been discovered or not.[32]

Here it is clear that Johnson is describing a form of ontological naturalism—besides the reference to mechanistic materialism, the terms "absolutely" and "conclusively" emphasize supposed dogmatic commitment to the substantive ontological claims. Johnson claims that evolutionary biologists assume this sort of positivistic philosophy, but certainly evolutionary biology as a science does not have to do so, and it is hard to believe even that any scientist who has kept abreast of developments in philosophy of science would affirm this form of ontological naturalism.

Indeed, it seems clear that the two biologists that Johnson most often decries—George Gaylord Simpson and Stephen Jay Gould—do not endorse such a view, but are instead methodological naturalists. Simpson discussed naturalism as part of his review of the principle of uniformitarianism in geology and biology and is explicit that the scientific postulate of naturalism is "a necessity of method" and that the rejection of appeal to preternatural factors must be made on "heuristic grounds."[33] When one

looks for Gould's view on the matter, one finds in his discussion of uniformitarianism that he used precisely the distinction reviewed above to disam-
biguate "substantive uniformitarianism" (a descriptive hypothesis holding
that the history of life was uniform) from "methodological uniformitarianism."[34] Gould uses the latter term to label the assumption in geology that
natural laws are invariable—a position that implies absence of supernatural intervention. The name Gould gives this presupposition tells us just
how he views it; he recommends that the special term be dropped because
it follows from the fact that geology is a science. Clearly, both Simpson and
Gould understand that science does not affirm naturalism as a substantive
ontological claim but rather as a methodological assumption.

Methodological Naturalism and Evidence

We have seen how Johnson misleadingly inserts terminology with connota-
tions of dogmatism into the definition of naturalism. He regularly refers
to naturalism using such terms as "extravagant extrapolation, arbitrary
assumptions, and metaphysical speculation,"[35] but such name-calling is
no argument. Johnson provides no analysis to show that science assumes
the naturalistic principle dogmatically; he simply asserts this. We have now
seen that naturalism is not properly put forward as an ontological claim
about what conclusively does or does not exist, but rather as a methodological rule that states a valid way for investigation to proceed, so clearly it is
not dogmatic in the sense Johnson claimed. But is the methodological rule
itself dogmatic? To say that a belief or principle is dogmatic is to say that
it is opinion put forward as true or valid on the grounds of authority rather
than reason. Does science put forward the methodological principle not
to appeal to supernatural powers or divine agency simply on authority? Is
it just an extravagant, arbitrary, speculative assumption? Certainly not.
There is a simple and sound rationale for the principle based upon the
requirements of scientific evidence.

Empirical testing relies fundamentally upon the lawful regularities of
nature which science has been able to discover and sometimes codify in
natural laws. For example, telescopic observations implicitly depend upon
the laws governing optical phenomena. If we could not rely upon these

laws—if, for example, even when under the same conditions, telescopes occasionally magnified properly and at other occasions produced various distortions dependent, say, upon the whims of some supernatural entity—we could not trust telescopic observations as evidence. The same problem would apply to any type of observational data. Lawful regularity is at the very heart of the naturalistic worldview and to say that some power is supernatural is, by definition, to say that it can violate natural laws.[36] So, when Johnson argues that science should allow in supernatural powers and intelligences he is in effect saying that it should allow beings that are above the law (a rather strange position for a lawyer to take). But without the constraint of lawful regularity, inductive evidential inference cannot get off the ground. Controlled, repeatable experimentation, for example, which Johnson explicitly endorses in his video "Darwinism on Trial" (1992), would not be possible without the methodological assumption that supernatural entities do not intervene to negate lawful natural regularities.

Of course, science is based upon a philosophical system, but not one that is extravagant speculation. Science operates by empirical principles of observational testing; hypotheses must be confirmed or disconfirmed by reference to empirical data. One supports a hypothesis by showing that consequences obtain which would follow if what is hypothesized were to be so in fact. As we have seen, Darwin spent most of *The Origin of Species* applying this procedure, demonstrating how a wide variety of biological phenomena could have been produced (and thus are explained) by the simple causal processes of the theory. Supernatural theories, on the other hand, can give no guidance about what follows or does not follow from their supernatural components. For instance, nothing definite can be said about the processes that would connect a given effect with the will of the supernatural agent—God might simply say the word and zap anything into or out of existence. Furthermore, in any situation, any pattern (or lack of pattern) of data is compatible with the general hypothesis of the existence of a supernatural agent unconstrained by natural law. Because of this feature, supernatural hypotheses remain immune from disconfirmation.[37] Johnson's form of creationism is particularly guilty on this count. Creation-science does include supernatural views at its core that are not testable, and it was rightly dismissed as not being scientific because of these in the Arkansas court case, but it at least was candid about a few specific nonsupernatural claims that are open to disconfirmation (and indeed that have been disconfirmed), such as that the earth is less than 10,000 years old and that many geological and paleontological features were caused by a universal flood (the Noachian Deluge). Johnson, however, does not provide any creationist claim beyond his generic one that "God creates for some purpose," and, as a purely supernatural hypothesis, this is not open to empirical test. Science assumes methodological naturalism because to do otherwise would be to abandon its empirical evidential touchstone.

Finally, allowing appeal to supernatural powers in science would make the scientist's task just too easy, because one would always be able to call upon the gods for quick theoretical assistance. Johnson wants us to accept the claim that "God creates for some purpose" as an explanation of the biological world, but there would be no reason to stop there. Once such supernatural explanations are permitted, they could be used in chemistry and physics as easily as creationists have used them in biology and geology. Indeed, all empirical investigation beyond the purely descriptive could cease, for scientists would have a ready-made answer for everything. Obviously, science must reject this kind of one-size-fits-all explanation. By disqualifying such short-cuts, the naturalist principle also has the virtue of spurring deeper investigation. If one were to find some phenomenon that appeared inexplicable according to some current theory one might be tempted to attribute it to the direct intervention of God, but methodological naturalism prods one to look further for a natural explanation. Clearly, it is not just because such persistence has proven successful in the past that science encourages this attitude.

Johnson claims that "If the possibility of an 'outside' intervention is allowed in nature at any point . . .the whole naturalistic worldview quickly unravels."[38] He intends by this only that atheistic Darwinism will lose in a head-on comparison with theistic creationism once the "ideological" restrictions are removed but, as we have seen, the consequences would be far more serious. Johnson wants to make an exception to the law in this one area, but it would infect the entire enterprise. Methodological naturalism is not a dogmatic ideology that simply is tacked on to the principles of scientific method; it is essential for the basic standards of empirical evidence.

Creationism's Evidence

With his attack upon naturalism, Johnson is arguing that science abandon a sound methodological principle and reintroduce miraculous "explanations." We have seen that science has good reasons for retaining this principle—without it, standard inductive evidential inferences would be undermined—but we have also admitted that rules of scientific inquiry are themselves open to change or modification if a better method of evidential warrant is found. Does Johnson have something better in mind? It seems that he does, for he regularly claims that the supernatural theory of creationism is a better theory than Darwinism, and he constantly complains that "Creationists are disqualified from making a positive case, because science by definition is based upon naturalism."[39] Such statements lead one to expect that Johnson will supply what he says the scientific priesthood has suppressed, but one will look in vain for this positive evidence. Amidst all the negative arguments one finds only two small hints of what type of positive evidence the creationists have to offer—revelation and the design argument.

The first occurs only as a passing remark following an (inadvertently self-undermining) acknowledgment that empiricism is a "sound methodological premise."[40] Johnson writes:

Science is committed by definition to . . . find[ing] truth by observation, experiment, and calculation rather than by studying sacred books or achieving mystical states of mind. It may well be, however, that there are certain questions . . . that cannot be answered by the methods available to our science. These may include not only broad philosophical issues such as whether the universe has a purpose, but also questions we have become accustomed to think of as empirical, such as how life first began or how complex biological systems were put together.[41]

The sly implication here is that the "sacred books" and "mystical states of mind" could indeed be appropriate ways to answer empirical as well as teleological questions. Is this Johnson's new source of positive evidence for creationism? I asked Johnson just this question following one of his public lectures and he replied that he was not defending this position. However, neither did he deny that such appeal to scriptural authority or mystical experience would count as positive empirical evidence. Johnson seems to be pleading the Fifth on this important issue. He cannot reject these methods without alienating his constituency, for the biblical account, perhaps supplemented by religious experiences, is the prime motivation for Christian creationists. On the other hand, he cannot endorse the "method" of supernatural revelation without abdicating his claim of expertise as a lawyer, for anyone would be laughed out of court who argued that one could help establish an empirical fact (say, that the defendant set off the explosion) by reference to the authority of psychic or spiritual testimony.

The second hint of a source of positive evidence for creationism is in the following statement:

[F]rom a creationist point of view, the very fact that the universe is on the whole orderly, in a manner comprehensible to our intellect, is evidence that we and it were fashioned by a common intelligence.[42]

This is the only instance where Johnson makes an explicit commitment to any type of positive evidence the creationist can provide, but what we have here is nothing but a version of the old *argument from design*—the world appears to exhibit a designed arrangement so we should infer the existence of a designer—which relies on, at best, only a weak analogy from the human case to the divine. This is not the place to review the vast literature on the design argument, so I will confine myself to a few remarks on Johnson's specific formulation of it, and we will return to look at it in more detail in the next chapter.

What sort of order do we find in the universe, and what can it tell us? Examples of design which we attribute to a human designer include such things as a house, a formal well-manicured garden, or Paley's famous pocketwatch, but I would argue that we draw the inference in these cases precisely because the kind of order we see in them is so *unlike* what we typically find in nature; their simple geometrical forms and periodicities are strikingly different from the complexities and irregularities that the surface structure of the world presents. And when we do discover an underlying order in nature that is "comprehensible to our intellect," it is order to which we have been able to give natural, scientific explanations. Such order thus does not provide good reason to infer a supernatural creator. It is rather those features of the world that, for the time being at least, are *incomprehensible* to our intellect that are more likely to lead us to think of a higher power. At one time this might have been the awful power of a thunderstorm, leading us to suppose an angry god. Or we might have been struck by the wondrous beauty of a rainbow after the storm and interpreted it as a sign that God was appeased. We typically appeal to supernatural agency to explain that which we cannot explain otherwise. But when these phenomena are eventually accounted for in terms of natural electrical and optical properties, they lose their persuasiveness as indications of the literal presence of God and at most retain only an emotional or symbolic force.

Similarly, the version of the design argument that appeals to the adaptedness of organisms was persuasive only when this adaptation was mysterious and the idea of purposeful creation seemed the only possible explanation. But Darwin showed how simple natural processes could explain such adaptations. Johnson's "God creates for a purpose" view can say nothing about the supernatural processes by which the Creation was accomplished or what divine ends it serves; such an "explanation" starts and stops with the will of God. To pick just one of any number of common examples, Darwin's theory also accounts for those organisms that are not properly adapted to their environment—random variation produces both fit and unfit individuals, and natural selection is more likely to eliminate those that cannot compete as well in the given environment—but why would God create the world in such a way that the vast majority of individual organisms die because they are maladapted? The design argument has always been criticized on this sort of point even before Darwin; if God designed the world, why did He do such a poor job of it? Isn't the evident waste and sloppiness actually an argument *against* the existence of God? Such questions show the flip side of the design argument and highlight its weakness, for if it is applicable at all it is applicable for both the theist and the atheist. The creationist answer to such impious questions is that God must have had His reasons. Period.

There are two ways a creationist might become more specific about the will and methods of God. The first would be to appeal to revelation through mystical experience or scripture, but we have already seen that this does not in itself count at all as empirical evidence. The other is to revert to an earlier form of naturalism we mentioned—natural theology—and to try to judge the nature of God by looking at the book of nature. Johnson indulges explicitly in this approach just once, concluding that the elaborate-tailed peacock and peahen are "just the kind of creatures a whimsical Creator might favor, but that an 'uncaring mechanical process' like natural selection would never permit to develop."[43] If we are to take Johnson seriously, such "creationist explanations" in terms of divine "whimsy" postulated on the basis of peacock tails, are better accounts than those given by evolutionary theory and are thereby supposed to favor creationism.

If this is the "positive case" that purportedly has been suppressed, it is no wonder that creationists rely exclusively upon negative argumentation and why Johnson labors so mightily to legitimize it. The creationists' insistence upon viewing the issue as a simple either/or choice mistakenly leads them to think that their negative arguments against Darwinian evolutionary mechanisms directly prove creationism. However, negative argument will not suffice to establish the creationists' desired conclusion even using Johnson's nonstandard definitions of "Darwinism" and "creationism" that were quoted earlier. Johnson's definition of Darwinism mentioned only the classical evolutionary mechanism of natural selection operating upon tiny chance variations, and omitted other possible processes. So even if negative argument were to rule out this sort of mechanism (something that Johnson certainly has not done), one could not thereby accept creationism since there are other alternatives that do not rely upon divine intervention. His definitions of creationism are similarly problematic. For example, as mentioned above, one definition seems to rule out the sort of Deist who believes that God created the world and set it going in the way He wanted, but then no longer intervenes. Another definition rules out supernatural but nontheistic views that would stand in contrast to naturalistic evolution. Johnson would have us believe that the logical situation is the following:

Creationism (C_1) or Darwinism (D_1)

as though these were the only candidates. If this were the case, and a negative argument disproved the latter, then the former would follow as a deductive logical conclusion, but the logical situation is rather:

C_1 or C_2 or ... or C_n or D_1 or D_2 or ... or D_n or X_n

and in this case, even if D_1 were rejected, a variety of options remains besides Johnson's preferred C_1, so purely negative argument is not sufficient to establish it. Furthermore, these positions are not necessarily mutually exclusive. Direct conflict occurs when evolutionary theory is confronted with *specific* creationist stories (like the creation-scientists' literal 6-day instantaneous creation), but Johnson claims not to defend any such view.

"Darwinism," as Johnson defines it at least, is not a single thesis but rather the conjunction of at least two different theses: (D_m) the specific evolutionary mechanism of the modern synthesis (which we discussed in chapter 2), and (D_a) the atheistic denial of divine intervention. But all of Johnson's and other creationists' negative arguments are directed at undermining (D_m), so even if they were successful they would leave the possibility of (D_a) untouched. The existence or nonexistence of some particular evolutionary process is independent of the question of whether or not there is a creative deity. Johnson himself must grant this for he claims to allow the possibility that "He might have created things instantaneously in a single week or through gradual evolution over billions of years."[44] Thus, negative arguments against evolutionary processes are irrelevant to the key question of divine creative power when stated in this general way. Indeed, if Johnson cared only about his broad, ecumenical sort of creationism as he claims, then he should have no reason at all to argue against (D_m) and could confine himself to arguing against (D_a). Only someone who has a specific conflicting creationist scenario in mind, such as the one-week instantaneous Creation story, need worry about the evolutionary mechanism. In any case, negative argumentation is not going to establish either the general or the specific creationist thesis. As prosecuting attorney for the new creationists, Johnson needs to provide positive evidence for his and his clients' preferred conclusion, but, as we have seen, he has none to offer.

Methodological Naturalism versus Theistic Science

Unsurprisingly, Johnson and other IDCs are as unhappy with methodological naturalism as they are with metaphysical naturalism. When Johnson published a response to my argument of the previous two sections he argued that the former is just a slick sales ploy to sucker the unwary into buying the latter. He wrote:

That distinction implies that Darwinists do not claim to make ontologically true statements about the history of life, but only statements about what inferences can be drawn from a naturalistic starting point. If the Darwinists were really as modest as that, there would be little to argue about. For example, I agree with Richard Dawkins that the "blind watchmaker" mechanism is the most plausible naturalistic hypothesis for how complex organisms might have come into existence. We disagree only over whether the theory is true.[45]

He goes on to say that:

We who know how this game of bait-and-switch is played just look for the "switch" that turns innocent "methodological" naturalism into the real thing. In Pennock's version, the switch is the argument that naturalism and rationality are virtually identical—because he thinks that attributing the design of organisms to an intelligence would imply that all events occur at the whim of capricious gods, so that there would be no regularities for scientists to observe. No doubt this caricature explains why no science was done in the century of Newton.[46]

But Johnson's statements about truth and rationality are misleading. Johnson wants truth about reality—indeed, he wants absolute truth—but he neglects the more basic issue that truth claims must be justified by some method. He says it is irrational to act in conflict with true reality and asserts that God is truly real, but he fails to provide a method by which we can justify claims about God. Instead, he challenges scientists to say that they know, rather than just dogmatically assume, that God did not create us. But we have seen that biologists do not assert by fiat that God played no role in the development of life forms; they simply proceed, as all scientists must, to search for purely natural mechanisms. When they find evidence for a natural explanation (and Johnson admits that the Darwinian mechanism is not only possible but somewhat plausible), they can legitimately say that they have discovered something true about the natural world. To be sure, this is an approximate and tentative scientific truth, not an ontological (metaphysical) truth, in the sense that it cannot rule out the possibility that a supernatural Creator is involved in the process. (On the other hand, it does rule out one version of the teleological argument: that God is necessary to explain this development.)

Consider for comparison the geneticist who, applying methodological naturalism, searches for a natural explanation for hypertrichosis. People with hypertrichosis grow hair all over their faces and upper bodies, and were once thought of as werewolves. Finding evidence for the X-linked gene and an evolutionary explanation of the trait, the geneticist might reassure a patient that his disorder is "the result of a purposeless and natural process that did not have him in mind," the phrase of G. G. Simpson that creationists find so offensive. Surely we may accept that statement as true, even though as a merely naturalistic scientific truth, it does not rule out the possibility of an intelligent supernatural cause—a "curse of the werewolf," say—so it cannot be said to be absolutely true in the ontological (metaphysical) sense. Similarly, the creationists' supernatural story might be a metaphysical truth—God might have created the world 6,000 years ago but made it look older as mature-earth "appearance of age" creationists hold—but it is not a scientific truth.

In the most important section of *Reason in the Balance* Johnson proposes "theistic realism" and tries to support the creationist hypothesis from the viewpoint of its "theory of knowledge."[47] He begins by citing John 1:1–13 as "the essential, bedrock position of Christian theism about creation,"[48] and he goes on to make his central argument that it is "obvious" to all who have not had their reason clouded by the "drug" of naturalism that living beings are the products of intelligent creation: "Because in our universal experience unintelligent material processes do not create life. . . ."[49] Weigh that reason in the balance! Genetic engineering might one day allow humans to create life, but so far we do not have a single case of intelligent creation of life; rather, our universal experience to date is that *only* unintelligent material processes do so. We will return to look at this argument more closely in the next chapter, for it is the intelligent-design creationists' central argument. Here the important point is that the idea of a theistic science is set out in opposition to methodological naturalism.

Intelligent-design creationists unite in this attack and on their insistence in the viability of theistic science. Johnson continues to write as though methodological naturalism is essentially synonymous with metaphysical naturalism, but others acknowledge that methodological naturalism is a distinct view and attack it directly. Notre Dame philosopher of religion Alvin Plantinga, for example, joins the IDCs in opposing evolution and in rejecting methodological naturalism. As he puts it, "a Christian academic and scientific community ought to pursue science in its own way, *starting from* and taking for granted what we know as Christians."[50] In arguing against methodological naturalism and in championing such "Augustinian science," Plantinga admits that his suggestion "suffers from the considerable disadvantage of being at present both unpopular and heretical" but argues that it has the "considerable advantage of being correct."[51] Plantinga suggests that a theistic scientist could reason as follows:

God has created the world, and of course has created everything in it directly or indirectly. After a great deal of study, we cannot see how he created some phenomenon P (life, for example) indirectly; thus probably he has created it directly.[52]

But we have already considered this form of argument. When God is brought in to explain what we find unexplainable we have no more than a God of the gaps, and this form of reasoning is little different from the negative argumentation of the dual model. Plantinga takes pains to distance his epistemology from the God-of-the-gaps view, but he can do so only by explicitly giving up the idea that God is meant to be an explanatory hypothesis and by appealing to revelation.

First, the thought that there is such a person as God is not, according to Christian theism, a hypothesis postulated to *explain* something or other, nor is the main reason for believing that there is a such a person as God the fact that there are phenomena that elude the best efforts of current science. Rather, our knowledge of God comes by way of *general* revelation, which involves something like Aquinas's general knowledge of God or Calvin's *sensus divinitatis* and also, and more importantly, by way of God's *special* revelation, in the Scriptures and through the church. . . .[53]

Johnson's "theological science"[54] might emphasize different Scripture and Plantinga's "Augustinian science" might have a more sophisticated theology, but with regard to the creationism debate neither is too different from the standard "creation-science" of Henry Morris' Bible-based young-earth creationism in the sense that all these begin with some assumption of what Christians supposedly "know" and can take for granted in their science. But the battles we witnessed within the Tower over what may be presumed as "True Christianity" with regard to the "theological facts" should give us sufficient reason to doubt whether revelation could possibly supply the purported unified basis for such a science. Broaden one's view to observe battlegrounds of theological disputes among competing religious traditions and even angelic scientists would fear to tread there.

We also hear echoes of Morris in Johnson's comment about Newtonian science, in which he follows a trend among creationists to pick up on studies in the history of science that show how some natural philosophers of the scientific revolution supported their notion of natural law by reference to a conception of God as an orderly law-giver. Creationists are now eager to credit Christianity with the origin of modern science (or at least of the "True Science" that supports their views) and to anachronistically call theistic scientists like Newton fellow "creationists." However, they want Newton as one of their own because of his scientific successes, not because of his less enduring theological work on "ancient wisdom," interpretation of the Book of Revelation, and numerology regarding "the number of the beast." God might have underwritten the active principles that govern the world described in the *Principia* and the *Opticks,* but He did not interrupt any of the equations or regularities therein. Johnson and other creationists who want to dismiss methodological naturalism would do well to consult Newton's own rules of reasoning, especially his first that says not to admit unnecessary causes when explaining phenomena, and his fourth that says to regard the conclusions of inductive methods as "accurately or very nearly true"[55] and to eschew contrary hypotheses until new evidence requires them. Are such rules metaphysical dogma? Commentators note that Newton's *theology* sometimes led him to regard these dogmatically, but that in his scientific passages he took them as ". . . a matter of method merely, to be used tentatively as a principle of further inquiry."[56]

In Newton's day, many Christians thought atomism was tantamount to atheism in much the same way that Johnson and other creationists now say that evolution is, and Newton engaged in spirited exegesis to combat this view. He traced the atomic idea back through pagans such as Lucretius, Epicurus, Democritus, Thales, Pythagoras, and finally to Mochus, whom he identified with Moses to make the link to Christianity. Today Christian theists rightly find such contrivances unnecessary, for theism is not threatened by atomic theory. Nor do Christians still feel troubled by the heliocentric and geokinetic theories of the solar system, though an earlier generation went through a great turmoil over that conflict with the Bible. Such theories have been confirmed under the methodological assumptions of naturalistic science and we may properly call them true and factual in as strong a sense as empirical justification allows; they are not metaphysical but simple mundane truths. Evolutionary theory is of a kind and, again after some turmoil, most religious groups have come to accept it as a scientific truth about reality that is fully compatible with their faith and with a mature understanding of Scripture. Johnson's, Plantinga's and other IDCs' attempts to turn back the clock do a disservice to both religion and science.

I certainly agree that believers can be good scientists; what I cannot agree with is that they must be believers of the creationist variety who want not just absolute truth, but their unjustified, antiscientific version of it. Scientific naturalism is no dogma. It is a sound principle of method. We will return in chapter 6 to look in greater detail at why it is unwise to introduce the supernatural into scientific reasoning, but we have already seen enough to give a preliminary assessment of what is going on.

Revolutions and Revolutionaries

Johnson, Plantinga, and the other intelligent-design creationists see themselves as revolutionaries. They contend that evolutionary theory is so shot full of anomalies that it should be abandoned completely. Michael Denton expressed this attitude exactly when he titled his book *Evolution: A Theory in Crisis* (1985). Jonathan Wells suggested in the discussion following one of his talks that biologists should just stop investigation into evolutionary matters and leave it to history in the same way that physicists left behind the notion of the ether. They see themselves as leading the final charge that will bring the edifice of evolution crashing down and with it the reigning paradigm of naturalism. They aim at nothing less than to resurrect the creationist paradigm and to return the crown to theology as the queen of the sciences.

With many of these notions, the IDCs are drawing upon the work of philosopher and historian of science Thomas Kuhn, whose 1962 book *The Structure of Scientific Revolutions* is probably the most broadly influential work of this century in the field. Johnson calls upon Kuhn's name explicitly in the central, key chapter in *Darwin on Trial,* in which he lays out his criticism of the naturalistic "rules of science" and calls for the overthrow of this philosophical paradigm. To understand what is going on here we must know a little about Kuhn's philosophy of science and how it has been interpreted.

Kuhn focused on the process of scientific progress and pointed out what he saw as a pattern in the way scientific revolutions occur. In most periods, he claimed, scientists carry on their research within a paradigm, that is, using the core set of concepts of a theoretical structure. The "normal" work of the scientist is research that works within the paradigm to solve puzzles—phenomena that one would expect to be accounted for by the theory, but that are still unexplained. There are always many such puzzling holes in a theory and scientific research advances by slowly filling them in. A science progresses as long as scientists continue to be able to solve puzzles successfully within the paradigm. At any given time, scientists have a sense of what the interesting problems are and which would be practical to pursue. However, some puzzles may resist solution. If these are not ones whose solutions would reasonably have to await, say, improvements in technology or knowledge from another area, but rather are puzzles that should be solvable under the current state of the art, then they are put into the class of anomalies. Holes in a theory are not of concern; anomalies are. A paradigm can tolerate a number of anomalies (every rule has its exceptions, after all, and what appears to be an anomaly may turn out to have been simply a hard puzzle), but if too many anomalies start to accumulate the paradigm will reach a crisis. In a crisis stage, scientists might start to look seriously at radically different approaches, and some will call for a theoretical revolution and a shift to a new paradigm.

So what happens at the point of revolution, as scientists move to a new paradigm? How do they pick from among available alternatives? Kuhn shocked many of the philosophers of science of his day with his answer. Breaking with the received view that held that theory choice was a matter of logic alone, Kuhn made the bold claim that picking which alternative to move to is not a matter of *rules* that *determine* a choice but rather of *values* and *norms* that *influence* choice. Theory choice, said Kuhn, "cannot be resolved by proof."[57] Even more shocking, he suggested that individual scientists might be drawn to one alternative rather than another in part for "subjective" or "nonrational" reasons, and he compared the transfer of allegiance to a new paradigm as being like "conversion." The concepts of different paradigms could be "incommensurable"—so radically distinct in meaning—to the degree that scientists working in different paradigms simply do not understand one another. Finally, Kuhn claimed, since our view of the world depends upon our paradigm, scientists working in different paradigms might be said to actually live "in different worlds."

Such statements might easily lead one to believe that Kuhn's point was that truth is subjective and relative to the individual. Indeed, this is the lesson that many drew from Kuhn, concluding that what we call scientific "knowledge" is not objective and that the scientist's story of how the world is has no special claim over anyone else's story. In later works, Kuhn tried to explain explicitly that this was not the proper conclusion to draw from his work. For instance, the notion of "conversion" was not meant to suggest that a paradigm is like a religion, but rather to highlight Kuhn's idea that paradigms are "gestalts" and that, like viewing the classic picture of the duck/rabbit or the goblet/facing profiles, one cannot conceptualize both views simultaneously or move smoothly between them, but must psychologically "jump" from one view to the other. Nor is this switch irrational. When Kuhn had said that theory choice was "subjective" he had not meant that it was a matter of taste, but rather that it involved *judgment* as scientists weighed how well alternative theories fared in achieving the values of accuracy, consistency, broad scope, simplicity, and so on. Furthermore, he held that such values were not relative to but were invariant between paradigms.

Within much of academia, however, it was the superficial relativist interpretation of Kuhn's work that spread. Most significantly, it was picked up by a new movement that had its start in the philosophy of literature and began as a theory of literary interpretation. Known as "deconstructionism," this approach had its roots in the true observation that literary works are typically open to various readings and the reasonable suggestion that it might not be possible to discover which reading is the correct interpretation of the author's intentions. Is the character's obsession with smoking stogies supposed to be an indication of his strong, aggressive masculinity, or does it represent anxieties and self-doubts that he must struggle to repress? Is it a phallic symbol or is this one of those cases in which a cigar is just a cigar? As one reads on, other elements in the story could lead the reader to settle on one or another of these interpretations, but that could lead in turn to a revised reading of other symbols, suggesting a new overall interpretation. This is the "hermeneutic" process, a concept with a long history that as we saw in chapter 1 goes back to theories of interpretation of holy texts. Deconstructionists take this notion of interpretation several steps further. One thing they claim is that authors might not even realize what is going on in their own works, and that we can deconstruct the pictures they have constructed in ways they might not have intended. Perhaps the cigar was consciously intended to be a way to emulate the rich, but when deconstructed it really suggests an ironic devaluing of wealth, an unintentional assertion of the disdain of such values by the lower class. But if even the author's own intentions do not determine the true interpretation, then what is really true is just subjective, being left to each reader. The reader becomes the writer and no one reader's interpretation is truer than any others. Truth on the deconstructivist or, more generally, the "postmodernist" view is completely relative in that we can never escape our own viewpoint; we have no way to reach and see truth from what they call a "God's-eye view." If we happen to think that there is in fact some real truth this is only because one or another particular group—because of their position, prestige, or power—has been able to establish and enforce their own view. Postmodernists believe that Kuhn showed that even science is a narrative and interpretive activity of just this sort, and so that its "truths" are not objective but are constructed by power relations and prejudices. That Kuhn himself believed that his work did *not* undermine objective scientific truth is irrelevant from the deconstructive point of view.

Intelligent-design creationists often talk about truth, and even more often about "Truth, with a capital T."[58] Johnson acknowledges that Darwinism might be the best explanation of the data and then asks rhetorically "But is it *true?*" He applauds a Christian campus group called *Veritas*, which lends its financial and organizational support to IDCs. Indeed, he illustrates the Kuhnian notion of incommensurability with an example from his own experience debating evolutionists. He writes:

[I]t is pointless to try to engage a scientific naturalist in a discussion about whether the neo-Darwinist theory of evolution is *true*. The reply is likely to be that neo-Darwinism is the best scientific explanation we have, and that this *means* it is our closest approximation to the truth. . . . To question whether naturalistic evolution itself is "true" . . . is to talk nonsense.[59]

By now, the reader should be sensitized to notice Johnson's illegitimate slide from neo-Darwinism to "naturalistic evolution," so I need not repeat the explanation of this error. Johnson is also slightly but significantly in error about the relation of explanation to truth, but I will hold off tackling that complex issue until later. Here let me keep the focus on truth. The IDCs' use of the postmodernist interpretation of Kuhn is a red flag that should alert us to look carefully at the notion of scientific truth that they have in mind.

Although they are forever calling out "Truth! Truth!" their actions betray a very different view. This is another interesting difference between the new creationists and their predecessors. The old creation-science is founded on the idea that science does provide truth and that the truth it uncovers fits with that revealed in Scripture, so the strategy of YECs has been to attack evolution head-on by confronting it with their Genesis-based alternative. The IDCs, on the other hand, are relativists about natural human knowledge, and they therefore think science is rotten to its core because it claims that its naturalistic method can discover objective empirical truths. Their strategy, therefore, is to be quiet about the specifics of their own alternative and to seek out scientific discontents, inciting them to a political revolution—an overthrow of scientific naturalism itself—claiming that conditions will be improved once "theistic realism" is the ruling paradigm and "theistic science" is in control of knowledge. This is the classic postmodernist approach, for which truth is just politics.

Johnson revealed his allegiance to this philosophy at a talk at a political science department during his 1995 world speaking tour, and he repeated it in one of the series of detailed letters describing the tour that he sent to his circle of supporters:

Sept. 20, Wednesday. After a morning of writing I met Political Science professor Patricia Boling who hosted a noon colloquium for the department faculty and grad students. I told them I was a postmodernist and deconstructionist just like them, but aiming at a slightly different target.[60]

In his writings and speeches, Johnson usually makes his points using conditional statements or rhetorical questions or by reference to others so it is usually difficult to pin him down to any specific positive view. Given how rare it is for him to give his own position in a straightforward declarative statement, this explicit revelation of his view and aim is particularly significant. However, it is not an isolated instance. In a newspaper interview he said that his plan is "to deconstruct" the philosophical roadblocks set up by materialist biology, and to "relativiz[e] the philosophical system."[61] Indeed, Johnson's original title for *Darwin on Trial* had been *Darwinism Deconstructed.*[62] Once one is sensitized to this, one finds postmodernist language throughout Johnson's work. When he claims, for example, that scientists are attracted to naturalism because "It gives science a virtual monopoly on the production of knowledge,"[63] he is echoing the deconstructionist charge that knowledge is not discovered but rather is fabricated by the intellectual capitalists who own the factories of the knowledge business. When he equates scientific naturalism with "scientism" he is repeating the name-calling led by antiscientific cultural relativists. When he says Darwinism is science's "creation story" he is echoing the social constructionist charge that science simply delivers narratives that are epistemically on a par with other myths and stories. When he says that "Darwinist religion" is forced upon the public through "a program of indoctrination in the name of public education,"[64] he is following the lead of cultural relativists who hold that science is somehow a "Western" construct and that science education is merely propaganda. When he describes the scientific community as a "priesthood" that "guards the door" of knowledge, he is making the central postmodern point that knowledge is simply that story whose authors have the power to suppress other stories. One could cite many similar examples from Johnson and from others among the IDCs.

But, one might ask, what about other passages in which relativism is denounced and creationism is put forward as truth? Such passages have to be understood in light of the postmodernist view and keeping in mind creationists' other beliefs. Postmodernists talk about truth as much as anyone else does, and they believe that one may advocate the truth of one's view in as strong a manner as one can. However, they also hold that the notion of truth involved here is narrative truth—it is the truth that characters in any fictional story hold about their own story relative to their own subjective view, and thus it should not be confused with objective truth. Scientists might claim that scientific conclusions are objective because they are based on empirical evidence, but according to postmodernism what scientists call evidence is just a special form of rhetoric. Given this view of science, it is not surprising that we find one of the new IDCs, University of Memphis rhetorician John Angus Campbell, telling us that Darwin's *Origin of Species* was little more than clever rhetoric.[65] Johnson's whole argument against evolution by way of his attack on scientific naturalism is of a kind with this view. You might think that basing scientific explanations on observation and experiment and uniform natural law is good evidence for objective truth, but Johnson says you have simply been duped by the smooth-talking perpetrators of a philosophical conspiracy who have garnered the political power to "establish" this "naturalistic religion." *Defeating Darwinism* is essentially a revolutionary creationist manifesto that incites believers to "step off the reservation" and escape the "oppression" and "domination" of the "materialist rules" to which the Darwinian "intellectual elite" have forced them to "submit."[66] The postmodern revolutionary relies on negative arguments and rhetoric, and Johnson's expertise as a lawyer has made him a master of the art. We even hear the postmodernist strains in Plantinga's call for a multiframework "Augustinian science," in which Christians pursue their separate approach to science based on "what they know" and leave others to their own sciences. Such a balkanized science is at one with the radical multiculturalists' calls for feminist science or Hispanic mathematics.

The IDCs are in lockstep with postmodernism's skeptical contention that human truths, including scientific truths, are merely subjective narratives. Both hold that what passes for objective knowledge depends simply on which narrator is in political power, and both think that science has been in power long enough and seek to overthrow its epistemic privilege. Both hold that human knowledge is necessarily relativistic. There is, of course, one significant additional belief that takes the IDCs a step beyond postmodernists—or perhaps I should say a step beyond and two steps backwards. Postmodernists accept relativism and seem happy to dispense with the notions of objective truth, embracing instead the rich plurality of subjective human viewpoints. Creationists, however, as we have seen, believe that although human reason by itself is impotent, there remains one way to get a "God's-eye view" of the world, namely, from God himself.

God's divine revelation saves us from relativism by providing us with absolute truth in Scripture.

Though Johnson has so far refused to endorse divine revelation publicly as a source of empirical evidence, it is obvious that he must accept this given the rest of his view and especially his statements that God's word should be the basis for theistic science. Indeed, he closes *Defeating Darwinism* by returning to the question of "Truth with a capital T" and rhetorically rejecting the truth of science whose "foundation is materialism," while reminding us that Jesus referred to himself as the Truth and "warned against the scoffers who build their house on a foundation of sand."[67] Plantinga, at least, is forthright in stating his view that warranted knowledge requires a supernatural basis. On this point—that all knowledge must in the end rest on divine knowledge—these new creationists reunite with their predecessors, who hold that "True Science" must be based on what is revealed in the Bible. This should come as no surprise to anyone for despite creationists' protests to the contrary, it is patently obvious that their alternative hypothesis of intelligent design is not a scientific conclusion but a religious one. Since their positive view is based on revealed truth, and since they cannot publicly reveal this religious basis and still hope to claim a place in the public school science classroom, it is perfectly clear why they rely solely on negative argumentation and why they complain of the rules of science that require them to state clearly the details of their "theory" and support it with positive evidence.

Would-be revolutionaries are notorious for the great detail in which they will express their dissatisfaction with the "establishment" but also for the vagueness with which they will describe the new system that they plan to establish in its place. We should think carefully before joining the IDCs' revolutionary movement. Would it be rational to institute a system in which all hypotheses are to be given the same "balanced treatment" in the science classroom irrespective of whether they have earned a place there by virtue of the empirical evidence? Is it wise to abandon scientific standards for assessing the truth of hypotheses in favor of a system in which some hypotheses may be held up as absolute truth by right of special revelation? Or should we rather stick with science's successful method that gives all hypotheses equal opportunity to prove themselves but allows none to be "above the law"? Creationism held sway for centuries, but by the mid-nineteenth century, even before Darwin proposed his superior alternative account, it was already collapsing in the face of the data. Having lost the battle within the system, creationists are now trying to regain their position by overturning the system, changing the rules of evidence, and returning to special revelation as the only source of truth. Reason should tell us to resist such a revolution.

5

Chariots of the Gods

Innumerable suns exist; innumerable earths revolve about these suns in a manner similar to the way the seven planets revolve around our sun. Living beings inhabit these worlds.
—Giordano Bruno

Swing low, sweet chariot
Comin' for to carry me home
—Traditional

God's Charioteers in the Stadium of Science

The centerpiece of the movie *Ben Hur* is the thrilling chariot race that pits Ben Hur against the charioteers of the Roman emperor. When Charlton Heston, playing the heroic title character, pulls his white chariot to the starting line the audience knows that he is about to do battle in the name of God against those who care nothing for God. Creationists see themselves as being in such a race against the proponents of evolutionary theory. They conceive of their work simultaneously as a form of ministry and as a holy struggle. They are carrying the message of God (or at least their understanding of it) to the world and stepping into the stadium of science itself to defend it against what they see as its greatest enemy. When talking among themselves, they ride God's own chariot, the Bible, and arguments among creationists typically involve citing Scripture and going back to the Hebrew terms to support one view over another. As we noted in the first chapter, which view of the physical world they are willing to accept is driven by their preferred interpretation. When they go on to try to promote these views as scientific in opposition to evolution, their argument strategy necessarily changes. As we documented in several previous chapters, their primary form of argument is negative. Rather then emulating Ben Hur, they follow the example of his black-suited opponent; by trying to poke holes in the evidence for evolution and overturn scientific method, they hope to break the spokes of the wheels of the scientific chariot and win the competition by default. In a few cases, however, some creationists do offer some positive evidence for their views, hoping to claim scientific warrant. These positive arguments, or purported positive arguments, are the subject of this chapter.

Young-Earth Creation-Science

The classic form of creation-science claims two main positive novel predictions, namely that the world is only about six-thousand years old and that its major features are the result of a catastrophic global flood. They try to make the view scientific by putting forward these claims without mentioning their basis in Genesis and without making any reference to Scripture in their arguments. This is the version of creationism that was at issue in the federal court cases that struck down attempts to legislate balanced treatment for creationism at the state level, and it is the form that continues to be the predominant view that activist creationists promote at the local school level today. Just as scientists have time and again answered creationists' negative arguments against evolution, they have also repeatedly addressed and refuted the supposed positive arguments for a young earth and a global flood. Nevertheless, young-earth creationists (YECs) continue to flog the same dead horses. I will not take the space here to rehash the details of this debate, but will simply mention a few of the "evidences" for a young earth that I hear most frequently, to give a flavor of the "positive" arguments of creation-science for those who are unfamiliar with them.

When Man and Dinosaur Walked in Paluxy
Following out the implications of the Genesis account of Creation, young-earth creationists claim that dinosaurs and humans lived together on the earth at the same time. Of course evolutionary theorists think that is true as well in the sense that contemporary birds are likely the descendants of and are classified in the same taxonomic group as the dinosaurs. But this is

not what creationists have in mind. YECs hold that the picture of cavemen living in the Lost World with T-Rex and the other terrible lizards is no mere schoolchild's or Hollywood fantasy, and that the Bible tells us so.

Dinosaurs, they claim, are mentioned in the Bible as the Behemoth and the Leviathan. Institute for Creation Research (ICR) scientists say that the former was probably a dinosaur because of its Scriptural description. They give away posters of a seated man observing what appears to be an Apatosaurus with the scriptural passage from Job: "Behold now behemoth, which I made with thee; he eateth grass as an ox. Lo now, his strength is in his loins, and his force is in the navel of his belly. He moveth his tail like a cedar: the sinews of his stones are wrapped together" (40:15–17). At the Museum of Creation and Earth History, our guide drew the children's attention to the phrase "he moveth his tail like a cedar," noting that no animal we know of besides dinosaurs had a tail so large. Scholars of biblical Hebrew would have to stifle a chuckle if they heard this exegesis, for the King James translation utilizes the term "tail" as a common euphemism for the male genital member.[1] Stephen Mitchell's authoritative translation of the book of Job removes the linguistic fig-leaf and renders the passage somewhat differently: "Look now: the Beast that I made: he eats grass like a bull. Look: the power in his thighs, the pulsing sinews of his belly. His penis stiffens like a pine; his testicles bulge with vigor."[2]

Obviously, this is not the "proof text" that it might have appeared to be on its face. Similarly, the supposed physical evidence that creationists have pointed to that humans and dinosaurs lived contemporaneously has proven to be not quite what they purported it to be.

In Glen Rose, Texas, about halfway between Ft. Worth and Waco, is the site of the infamous "Paluxy mantracks." Creationists claimed to have found in the bed of the Paluxy river several lines of human footprints alongside dinosaur tracks. Here at last was something that looked like it could count as positive evidence for one part of the YEC view. Indeed, this was the best evidence that creationists ever had, and in the mid-1980s the fossils garnered national and even international attention. If true human and dinosaur prints were in Cretaceous strata, then the evolutionary chronology would be false in a significant manner. The original claim had been made by Clifford L. Burdick in 1950 in the Seventh Day Adventist periodical *Signs of the Times*, in an article titled "When GIANTS Roamed the Earth: Their Fossil Footprints Still Visible!" Henry Morris published some of Burdick's photographs in the seminal *The Genesis Flood* in 1961. In 1968, creationist Stanley Taylor was convinced by the Paluxy prints, and in 1970 he returned with a film crew to produce *Footprints in Stone* for Films for Christ, bringing greater attention to Burdick's claims. The Institute for Creation Research began its own investigations of the Glen Rose area in 1975, and shortly thereafter John Morris published his data and conclusions in a book *Tracking Those Incredible Dinosaurs and the People Who Knew Them* (1980), which touted the import of the prints. However, the most recent and purportedly best findings were announced by the Reverend Carl Baugh in 1983, in a Bible-Science Association Audio Tape of the Month.[3] Baugh reported uncovering 44 "human" footprints in four trails that crossed and in a couple of cases even overlapped prints of "*Tyrannosaurus rex.*" The image Baugh conjured of human beings digging for clams and crossing paths with the king of the thunder lizards was compelling. And, indeed, footprint trails are especially significant for paleontologists, for they can provide significant information about stride length as well as other features.

Of Baugh's "spectacular" tracks Henry Morris wrote in the second edition of his book *Scientific Creationism* (which ICR still advertises as being the most widely used and comprehensive creationist reference text) that "the only possible escape from the conclusion that man and dinosaurs were contemporary is to say that the human tracks were not really human but were made by some unknown two-legged animal with feet like human feet!"[4]

A team of scientists came to a very different conclusion as they observed Baugh's excavations at Paluxy as well other alleged mantracks at a couple of other Texas sites, during repeated visits in 1982 and 1983. Led by Dr. Laurie R. Godfrey, an anthropologist specializing in paleontology and anatomy at the University of Massachusetts, Amherst, the team reviewed the creationists' accounts and compared them against each other and against their own observations. They reported their findings in a series of papers.[5] Of the creationists' own published data, Godfrey concluded:

We consistently found these data to be shoddy and, even when taken at face value, to lead to absurd conclusions about the stride length, foot length, and foot shape of the "humans" that presumably made them.[6]

Not only were the tracks not shaped in the way they would have been had they been made by human beings, they were also much bigger than human prints. This brings me to one additional point that I have not yet explicitly mentioned about the creationists' claims. Although the ICR poster I mentioned depicts a man of ordinary build observing the dinosaur, and the "mantracks" were usually described by creationists as "human" footprints, in fact most of the prints were so large that they might have been made by Goliath himself. The set of tracks with the largest number (28) of prints had an average length of 16 inches, and Dr. Baugh calculated that the individual would have been eight and a half feet tall. Another set averaged 22 inches in length, and Baugh noted that this person had been thirteen feet tall and weighed 600 pounds. If these were manprints then these were no ordinary men! They must have been giants. And, indeed, this is just what creationists concluded, citing Genesis 6:4: "There were giants in the earth in those days. . . ." and similar passages from Joshua, Deuteronomy, and I Samuel. Baugh dubbed the "man" who made the prints *Humanus Bauanthropus* and engraved the name of it on a bronze plaque at the site to commemorate the discovery. He also included on the plaque the Job 40:15 reference, apparently interpreting "Behemoth" as referring to this race of giants.

But even if we were willing to accept these creationist tales of giant men this would not save their case. Summarizing the results of their study, Godfrey wrote:

[W]e found that the alleged Cretaceous mantracks consistently failed the test of human origin but often passed the test of dinosaur origin. Indeed, some were quite clearly portions of dinosaur footprints. Others—those most responsible for the Paluxy mantrack legend—turned out to be inept carvings. Although these were definitely in the minority, had they never existed it is doubtful that creationists would have focused on the Glen Rose area in the first place.[7]

The scientists also noted that many of what creationists had identified as "toe prints" were nothing more than erosional pits or invertebrate burrow casts of *Thalassinoides*. These are just a couple of examples of a host of problems that were found. One member of the team, Ronnie Hastings, who lived near Glen Rose and observed the creationists' work over many years, provided a revealing first-hand account of Baugh's and the other creation-scientists' appallingly unprofessional research methods.[8]

One other line of evidence that was important in revealing that all the tracks had been made by dinosaurs involved the changing patterns of coloration of the prints, which were the result of secondary infilling of the original track depressions by an iron-containing sediment. The pattern was first noticed in 1984 by Glen Kuban, a self-described Christian believer but neither a creationist nor an evolutionist, who had been investigating the area on his own and with others, including Hastings, since 1980, and who subsequently presented his findings at the first International Conference on Dinosaur Tracks in 1986. Kuban reported that the material had begun to oxidize after it was exposed to the air, progressively changing from a blue-gray to a rust color, and increasing its contrast with the surrounding substrate. When the coloration patterns were observed in addition to the indentations they clearly revealed the shape of dinosaurian digits on the "mantracks."[9] Kuban went back to look at early photographs and Taylor's film and was able to point out previously unnoticed coloration patterns recorded then, refuting John Morris's implication that they might have been fraudulently painted or stained on. Kuban's findings hammered the final nail in the coffin of the Paluxy prints, even for many former promoters. Taylor withdrew *Footprints in Stone* from the Films for Christ list.

The "mantracks" were thus made by neither normal- nor giant-sized human beings, and a 1988 article "The Rise and Fall of the Paluxy Mantracks" in *Perspectives on Science and Christian Faith*, the journal of the evangelical ASA, concluded that it was "improper for creationists to continue to use the Paluxy data as evidence against evolution."[10] Nevertheless, such claims continue to be made and were even given network television airtime in a notorious NBC broadcast *Mysterious Origins of Man* in 1995, hosted by none other than Charlton Heston himself. The program's main message was that the evolutionary account of human origins and much of the scientific chronology was false. In an ironic twist, however, it turned out that some of the material for the program was taken from the book *Forbidden Archaeology*, which forced some YECs to furiously back-pedal. In a review in the February 1996 issue of the *Answers in Genesis* newsletter, young-earth creationist Ken Ham warned that *Forbidden Archaeology* "is dedicated to 'His Divine Grace A. C. Bhaktivedanta Swami Prabhupade.' It appears that the authors are Hare Krishna adherents! . . . Everything cycling continuously over millions of years fits well with Krishna philosophy! That seems to be what this program is all about!" Thus, we see now, it is not only fundamentalist Christians whose religious cosmological story leads them to oppose scientific conclusions. Though he promotes the teaching of creationism in the public schools, Ham would not likely support giving equal time to the Hindu version of creation-science. Regarding the erstwhile vaunted mantracks, Ham wrote, "According to leading creationist researchers, this evidence is open to much debate and needs much more intensive research. One wonders how much of the information in the [NBC] program can really be trusted!"[11] For some reason it seems easier to recognize the danger of reading religious beliefs into science when that religion is not one's own.

It may yet be a while, however, before the creationists' give up their most prized "evidence." Henry Morris's conclusions about the "spectacular" tracks remain uncorrected in the March 1996 printing of *Scientific Creationism*, though we should note that he and John Morris quietly avoid any *direct* mention of Paluxy in their new work, *The Modern Creation Trilogy*.[12] What evidence do they substitute in its place for their conclusion? They cite mythological accounts of dragons, sightings of the Loch Ness Monster, and "reports brought out by pygmies and other natives" of "a large, living, lake-dwelling dinosaur-like animal in an almost inaccessible swamp area in the central African rain forest."[13] With such evidence as the only alternative, it is no wonder that some creationists refuse to yield the manprints. If you go to Glen Rose today you can still visit the Creation Evidences Museum of Carl Baugh, see his casts of the prints, and hear how they prove evolution wrong by showing that humans and dinosaurs once roamed the plains of Texas together.

Moon Dust and Landing Pads

One of the most memorable positive empirical arguments that YECs have given in favor of their 10,000-minus year chronology and against the scientific view involved interplanetary dust infall, or accumulation, and predictions from these two models of the depth of dust on the surface of the moon. If the world were really billions of years old as scientists say, then, given the rate of infall, there should have been a layer of interplanetary dust on the moon over 180 feet deep. It was supposedly because of fear that the astronauts would sink into this dust that they equipped the lunar landing module with round landing pads on its feet. But astronauts were surprised to find no dust layer on the moon. This observation thus supports the view that the universe is only a few thousand years old, as the creationists' model holds.

I first heard this argument a few years ago from a student in one of my philosophy classes. On other topics we had covered in class I had given a fair presentation of the different arguments, the student wrote me in a letter, so why was I not giving a balanced account of creation-science's proofs, such as this one that the earth was young? I had to admit that I did not know of this argument and that on the surface it looked quite reasonable. However, as is typically the case when one is confronted with creationist negative arguments against evolution or, as in a few cases like this, with purportedly positive evidence for their view, it sometimes takes a bit of digging to discover the basis of the argument, and invariably it turns out to be not quite what it was purported to be.

One key factor in the argument is the rate of infall of interplanetary dust. It turns out that creationists' calculations of the expected dust depth used data from the first attempts at measuring the influx, done atop the Hawaiian volcano Mauna Loa and published by Petterson in *Scientific American* in 1960. But there were several weaknesses with this preliminary estimate, problems that Petterson himself mentioned in his report, and subsequent improved measurements showed his initial estimate to be too high by at least one or two orders of magnitude. Also, calculations of the infall based on observations of crater-size distribution on the moon gave an independent measurement that agreed well with direct measurements taken from satellites. Indeed, even before measurements were available from lunar probes, scientists at a 1963 conference on the Lunar Surface Layer agreed that the rate of infall should be quite low, and their estimate that the rate for masses smaller than 10 kilograms was about 1 gram per square centimeter during the past 4.6 billion years agreed with subsequent infall measurements.[14] The Apollo astronauts were not at all surprised to find only a shallow layer of dust. As further confirmation of the scientific view of the age of the moon, radiometric dating of rock samples brought back by the astronauts showed no age less than 3.1 billion years. The greatest age measured was 4.5 billion years of small particles enclosed in fragments of igneous rocks. These measurements agree with independent radiometric measurements of the ages of meteorites found on earth.[15]

On the other hand, if one accepts the YEC timescale, then the rate of infall must have been very much larger than the current rate to even begin to account for the observed craters on the moon. It also implies that there should be a very large number of craters on the earth's surface, since there would not have been nearly sufficient time, in the six to ten thousand years they postulate, for them to have eroded. But, of course, we do not observe such extensive cratering on the earth, and even those large ones that we do find—such as the recently discovered 180-kilometer diameter crater off the Yucatan coast of an asteroid impact 65 millions years ago which might have caused the extinction of the dinosaurs at the end of the Cretaceous period—are eroded and buried, detectable only by measurements of tiny variations in the strength of gravity. Neither does the creationist view explain the patterns of lunar erosion, for example, that the exposed surfaces of half-buried rocks are rounded and covered by small pits made from space dust impacts whereas the buried surfaces are angular and unmarked, or the observed pattern of crater erosion which suggests a prolonged slow bombardment. Reviewing the creationist argument from interplanetary dust, biologist Frank Awbrey briefly considered Henry Morris's hypothesis that the battered appearance of the moon is the result of "continuing cosmic warfare" between Michael's angels and Satan's minions. Not surprisingly, Awbrey concluded that the creationist view is "nothing more than a set of miracles offered up in place of a simple, natural explanation that accords very well with the moon's features and with actual measurements of space dust."[16]

In short, the moon dust argument was based on outdated and incomplete data and simply ignored a variety of countervailing evidence. Howard van Till, the Calvin College physicist and Christian theist whom I mentioned in the first chapter, eventually took YECs to task for persisting in their use of this argument, writing:

The claim that a thick layer of dust should be expected on the surface of the moon, and the claim that not more than a few inches of dust were found on the surface of the moon, are contradicted by an abundance of published evidence. The continuing publication of those claims by young-earth advocates constitutes an intolerable violation of the standards of professional integrity that should characterize the work of natural scientists.[17]

Despite such reprimands, the moon dust argument remains uncorrected in the latest (March 1996) printing of Henry Morris's *Scientific Creationism*. And in their newest work, *The Modern Creation Trilogy*, which they advertise as "the standard-bearer in its field" and "the key reference resource and evangelistic tool for years to come,"[18] the Morrises still refer the reader back to the relevant pages of that earlier book for its "scientific discussions of evidences for recent creation."[19]

Be Fruitful and Multiply

As a final example of positive evidence for the young-earth creationist view, I choose the first one described in detail in *The Modern Creation Trilogy*. This is the argument from human population growth, and it is especially characteristic of creation-science. Henry and John Morris write:

It is easy to show mathematically that, starting with just one man and one woman, it would take only about 1,100 years of exponential growth to produce the present world population of around six billion people, if the population were increasing by 2 percent each year. This cannot have been going on for very long in the past or the world would long ago have been overrun with people.[20]

They continue along this line of reasoning to try to undermine the time estimates given by evolutionary theory and the rest of science and also to try to support their own young-earth model.

Is it really plausible to believe that for *almost a million years*, populations increased annually by only 15 people per million, and now they have suddenly exploded to 20,000 *each year* per million? Or that the doubling time has suddenly decreased from 30,000 years to 35 years? This is evolutionary philosophy at its most preposterous, and is contrary to all known history and human nature.[21]

Since this argument is included in the "scientific" volume which expressly avoids referring to Scripture, the Morrises leave this man and woman unnamed and say they simply assume this small initial population in order "to be . . . conservative."[22] They then calculate that this couple would have been able to produce the present world population in about 4,000 years with just an annual population growth rate of half a percent, conservatively within their preferred 6,000-year time frame.

The first thing to notice about this argument is that to get their desired conclusion the Morrises are willing to assume for human population growth what they deny for radioisotopes and other physical processes, namely, that the relevant rates are constant. This gets the appropriate relation backwards. While application of the principle of uniformity makes good sense when speaking of radioactive decay or random mutation, it does not work so simply given what we know of the history of human population size. Despite the implications for population growth from what we know of "human nature," there are other mitigating factors that we need to take into account. Data on other animal species in nature reveal that population size is typically highly variable, with cycles of increase and decrease that average to a growth rate of zero, which is what scientists believe held for most of the early history of the human species as well. It was only the advent of agricultural production, the development of permanent settlements and cities, and the introduction of mechanization that allowed the rate of human population growth to depart significantly from this norm to achieve exponential increase. Morris's calculation fails to take into account such background information as well as data we have of periods when, because of famines or plagues, human populations actually declined. Analyzing Morris's calculations, evolutionary biologist David Milne concluded: "Only by ignoring these contrary indications and by assuming that the growth rate of the pre-Industrial Revolution years was somehow typical of all of human history can creationists arrive at the conclusion that two human individuals living in 4300 B.C. could in actual reality have produced the entire world population of today."[23]

The second point to note is that the final figures mentioned above from *The Modern Creation Trilogy* are exactly those that Henry Morris presented in the original 1974 edition of *Scientific Creationism* and that are retained in the second edition.[24] In the *Trilogy*, the Morrises do not address or even mention the devastating criticisms that scientists made of that earlier argument. Besides pointing out various factual and mathematical errors, scientists showed that if one accepts Morris's assumptions about population one is led to a host of absurd conclusions. For example, using "data" on the length of time from Creation to the Flood, the number of children per family, and the duration of each generation taken from Morris's publications, Central Michigan University geology professor James Monroe applied Morris's reasoning to calculate the size of the population at the time of the Flood and showed that the population density of the earth would have been 13,000 persons per square foot for the entire earth's surface and that their total mass would have exceeded that of the earth. The purported evolutionists' problem of a world overrun with people is nothing compared to this! On the other hand, for the post-Flood population, using Morris's model Monroe calculated that there would have been only "eighty-six persons in the entire world in 1300 B.C., the time of the exodus, or 354 persons to witness the judgment at Babel."[25]

We could continue to review such examples of purported positive evidence for creation-science, but this representative sample is more than enough. In *The Modern Creation Trilogy*, Henry and John Morris attempt to support their view of a young earth and young universe by repeating other arguments that they have made over the years—the purported decay of the earth's magnetic field, the decay of comets, and the influx of salt and other solids into the oceans—and referring to earlier books for other arguments. In each instance, when scientists have examined the purported evidence, they have found nothing of substance.[26] ICR-type young-earth creationism continues to be the dominant view among creationists rooting in the stands, but with the weakness of its arguments and its series of defeats in the courts, other creationist leaders are trying to take up the reigns. Let us now return to our consideration of these new champions.

Intelligent-Design Theory

Reviewing the literature of scientific creationism, one writer reflected the opinion of many, concluding that its arguments "often involve tortured logic, a stubborn denial of the evidence, a shallow understanding, or a reckless disregard for truth."[27] Given such a track record, it is no wonder that many of the new creationists are trying to distance themselves from their YEC progenitors. But can they do any better? We have already seen a bit of the struggle going on within the creationist Tower and we have also looked at many of the negative arguments, particularly the central argument against scientific naturalism, that are the mainstay of the new attacks on evolution. What more do they have to offer?

What Intelligent-Design Theorists Won't Say

Intelligent-design theorists have learned a few lessons from the failures of their predecessors and have devised a more sophisticated strategy to compete head on with evolution. One of the main things they have learned is what not to say. A major element of their strategy is to advance a form of creationism that not only omits any explicit mention of Genesis but is also usually vague, if not mute, about any of the specific claims about the nature of Creation, the separate ancestry of humans and apes, the explanation of earth's geology by a catastrophic global flood, or the age of the earth—items that readily identified young-earth creationism as a thinly disguised biblical literalism. Occasionally an old-earth intelligent-design creationist will even explicitly reject one or another element of the young-earth view and endorse the standard scientific models. Behe even says that he endorses the common descent thesis. In either case their emphasis is on the minimal thesis that "life, like a manufactured object, is the result of intelligent shaping of matter."[28] That is, whether or not they believe the manufacturing process was gradual or abrupt, and making no claim about how long it took, they concur on the basic point that the biological world was *designed* by an *intelligent agent*.

To the extent that it sticks to the minimal, generic thesis and is coy about other points, intelligent-design creationism is able to rally a wide variety of creationists to its banner. One of the reasons that Phillip Johnson is the most widely accepted creationist champion is that, as we saw in the previous chapter, he defines creationism as simply the minimal positive thesis that God creates for a purpose, and confines his arguments to opposing evolution and scientific naturalism. He claims that issues of the timing of Creation are relatively unimportant and in his major writings declines to state his own view on the issue. Others, like Hugh Ross and Fred Heeren, while continuing to reject evolution, explicitly accept much of the standard scientific picture and openly criticize the young-earth view. Heeren says that not only is the Bible not incompatible with the standard geological views, but that many devout Christians recognize that it fits better with the idea of lengthy geological ages, and he tells his readers not to be "bullied" by the "my-way-or-no-way" views of the Institute for Creation Research.[29] For their part, young-earth creationists support the generic thesis of intelligent-design, but of course they disagree strongly with the claim that the question of age is unimportant, so there is an ideological gulf that separates the groups.

Intelligent-design creationists' most carefully crafted game plan appears in the textbook *Of Pandas and People,* which was supposedly written for secondary school biology students, but which really looks like it was written to try to circumvent Supreme Court rulings against young-earth creation-science. However, almost all of the arguments that appear are exactly the same negative ones as those of the young-earth creationists: they recite the same litany of supposedly insuperable problems with evolutionary theory, with the aim of showing that it does not have the resources to account for the origin of life, the Cambrian Explosion, and biological complexity and adaptedness. Again, the most important difference involves what they leave out. They conclude that if the world looks designed and evolutionary theory cannot explain that fact with its natural mechanisms, then we must conclude that the world is designed. This is just the same failed strategy we saw before: they offer two models and try to support the one by negative arguments against the other. If this were all there was to intelligent-design theory then we could dismiss it without further ado. However, IDC does offer one important new positive argument.

The SETI Analogy
The authors of *Pandas* are keen to reject the notion that IDC is necessarily religious or supernatural. There is nothing unscientific, they say, about investigating intelligent design. The reason some might think otherwise, they claim, is that many scientists have confused the notion of intelligence with the idea of the supernatural.

[S]cientists from Western culture failed to distinguish between intelligence, which can be recognized by uniform sensory experience, and the supernatural, which cannot. Today, we recognize that appeals to intelligent design may be considered in science, as illustrated by the current NASA search for extraterrestrial intelligence (SETI).[30]

It is hard to believe that scientists really did not understand the difference between intelligence and the supernatural, but, be that as it may, it is worth examining the IDCs' claims in more detail. It is certainly an interesting idea that the creationist hypothesis could be investigated on the model of the search for extraterrestrial intelligence. To be able to evaluate their analogy, we must begin with a bit of background about the SETI project.

The modern search for extraterrestrial intelligence began with the work of Frank Drake, a radio astronomer who realized that microwave radio waves might be a way to send interstellar messages. He reasoned that if there were intelligent beings on other planets with that level of technological advancement, we might be able to pick up their transmissions. In the spring of 1960 he made the first search for such microwave signals, training an 85-foot antenna on two nearby sun-like stars. Drake's receiver operated at only one frequency and picked up no signs of life, but his project inspired others. In the early 1970s, NASA sponsored Project Cyclops, a feasibility study of SETI science and technology issues, and its positive assessment led the agency to commit further resources to the project, establishing SETI programs at Ames Research Center and at the Jet Propulsion Laboratory. Intelligent-design creationists fail to mention that the main reason scientists think there might be intelligent life elsewhere in the universe is the plausibility of cosmic evolution—that is, it is reasonable to think that evolution, as a natural process, occurs regularly throughout the cosmos where conditions are favorable. Of course, it is hard to estimate how likely it is that evolutionary mechanisms would produce intelligent beings similar enough to us for us to be able to hear and understand them. Drake devised an equation to help assess the probabilities. It begins with the rate of star formation and then progressively whittles this number by factoring in the proportion of stars that could be expected to have a planetary system, the proportion of these that have suitable conditions for life, and so on. At first, putting numbers into the Drake Equation was mostly blind guesswork, and most of the factors in the equation still have wide margins of error, but in the last few years the first observations of planets orbiting other stars and of possible microfossil evidence of bacterial life on Mars have actually increased estimates of what was admittedly a long shot. After more than a decade of preliminary design work, NASA began observations in 1992, but within a year Congress terminated funding. Part of the NASA project continued as Project Phoenix with private funding, pursuing a Targeted Search of 1,000 nearby sun-like stars. Several other SETI projects that began in the 1970s at Ohio State University, the University of California (SERENDIP), and the Planetary Society (META) continue observations. The SETI Institute has a home page on the Internet where one can keep up with the latest developments.

The one development that has yet to be reported is any sign of intelligent life, though over the decades a few candidate signals have briefly tantalized researchers. Even during Drake's original search, one brief and unusual signal that appeared to emanate from Epsilon Eridani caused momentary excitement but it could never be reestablished. Drake reports that some UFO enthusiasts still believe that he made contact with extraterrestrials that day but that, because of a government conspiracy, the information had to be kept secret.[31] Drake also describes the excitement caused in 1967 when Jocelyn Bell, a graduate student in astronomy at Cambridge University, picked up a beacon-like radio source that pulsated every 1.3 seconds. A blinking astronomical source was completely unprecedented at that time, and the scientists wondered whether this might be a signal from intelligent civilization. Some astronomers spoke of the "Morse code" nature of the signal and others called the sources "LGMs," as a joking reference to little green men. Lacking computer programs that would have been needed to analyze the pulses for patterns that might suggest an intelligent origin, Drake writes:

I made long recordings of the pulse intensities, on tape and on chart paper, and then sat scrutinizing the charts, trying to discern signal patterns in them. I stared at them for hours at a time, but even in my eagerness to find an alien message, I never saw any evidence to make me think these tracings were of intelligent origin.[32]

No little green men were sending the signal after all. Actually, this was the discovery of the first pulsar, a term that Drake coined for these radio pulses that astronomers eventually showed to be emissions from rotating neutron stars. The discovery of pulsars was worth a Nobel Prize, but of course if an intelligent signal were ever to be discovered that would be worth even more. In 1997, the movie *Contact* gave SETI researchers a fictionalized success, but unless one believes the conspiracy theorists, real-life SETI scientists have yet to make contact with any extraterrestrials.

The Talking Pulsar

Intelligent-design theorists argue that just as the scientists of the SETI Project seek evidence of intelligence beyond the world, so too do they. In a clever rhetorical move, they frequently quote the late astronomer and SETI pioneer Carl Sagan to show that even a confirmed skeptic such as he admitted that such investigation is scientifically legitimate. But are IDCs really sincere in their protestations that theirs is not a religious hypothesis? Can they use the SETI model to ground the hypothesis of intelligent design as a truly scientific alternative to evolution? To begin to examine their claims, let us take an extended representative passage that sets out the main lines of their argument. There is one important item of information that we will want to watch for in anyone who promotes ID creationism as a scientific theory. We need to check out their ID card and determine the identity of the intelligent designer!

Intelligent-design creationist Fred Heeren is up-front about the identity of the designer in his recent book, *Show Me God: What the Message from Space is Telling Us About God.* In it, he summarizes the key points of the IDC argument in a fictional dialogue between Margaret, a SETI project researcher, and the Sultan of Brunei, who is considering helping to fund the project. The Sultan begins with a question:

"How will you be sure when you've received a signal from an intelligent source? How do you tell the difference between a natural, pulsating signal, as from a pulsar, and a signal from an intelligence?"

"If you get a message in Morse code," explained Margaret, "you know there's intelligence behind it. Nature can't duplicate that. In the same way, if we get a signal containing encoded information, even if we can't break the code at first, we'll know it's coming from an intelligent source."

"You mean like the encoded information we find in DNA."

Margaret didn't hear this interruption and continued her speech: "Nature can't duplicate specified complexity. The chance of nature creating a pattern that has meaning is almost infinitely small."

"Like the specified complexity of hemoglobin. . . . Are you familiar with the calculations of Hoyle and Wickramasinghe? . . . Did you know that the chance that amino acids would line up randomly to create the first hemoglobin protein is one in ten to the 850th power? Which is in the realm of infinity, considering the fact that there are only 10 to the 80th power atoms in the entire universe."

Mark and Margaret could only stare at each other. This man obviously had more than a casual interest in science.

The Sultan continued: "There's an even smaller chance that the DNA code could have randomly reached the required specificity: one chance in 10 to the 78,000th power even for the DNA of a simple microorganism. This must be a signal of intelligence, wouldn't you say?"[33]

As the story continues, the star Epsilon Eridani begins flashing out a signal in Morse code that turns out to be some familiar words in Hebrew: "In the beginning God created the heavens and the earth." The signal continues and, over the course of a year and a half, transmits the entire Hebrew Bible, and then it stops. Following the SETI protocol for post-contact procedure, the nations of the world meet and decide how to respond. The UN sends back a reply asking "What does this mean? What do you expect us to do with this message?" Epsilon Eridani is eleven light years away, so they expect to wait 22 years for the signal to reach there and for a response to get back, but instead the star begins immediately to transmit the New Testament in ancient Greek: "The beginning of the good news about Jesus Christ, the Son of God." Then scientists find that the message is beaming from all stars and from all parts of the sky. Margaret is finally convinced that this is a message from God, but Mark is not and asks why doesn't God just appear before him if he wants to be known. Margaret replies:

"If God were to personally appear before you, in any form that was spectacular enough to convince you that it was really Him, then you wouldn't have any real *choice* about following Him—you'd know you *had* to. And when there's no choice, there's no opportunity for love. That's what He wants, He doesn't want to *make* you follow Him."

Mark looked for holes in the argument but found none.[34]

Of course, no Bible-quoting pulsars have yet been observed, so the last part of the story is not meant to provide any positive evidence, but it is important in that it lets us know how intelligent-design creationists want us to interpret the first part. Because it serves so well to connect the ideas of information, divine intelligence, and the SETI project, the talking pulsar has become a regular character in IDC arguments. The first talking-pulsar story I heard was in a presentation by William Dembski at an IDC conference in Dallas. In his version, the pulsar says, again in Morse code, that it is the "mouthpiece of Yahweh" and proceeds to prove it by answering any question that is deposited in an ark on Mount Zion, predicting future events, giving cures for diseases, and providing answers to provably hard math problems.[35] After a discussion in which he argued that this science-fiction case would be an indication of a super-intelligence because of the extreme improbability that the correct answers could have been found, Dembski then moved on to the improbability of the specific sequences of nucleotides in the genome of biological organisms. In discussion and in other papers Dembski also makes the connection to the SETI project. We are supposed to think that the information in DNA molecules is a sign from God in the same way as would be a transmission of the Bible in Morse code beamed from a pulsar. The SETI project looks for intelligent life in the universe by searching for a signal with information, and IDCs claim that we find such an information signal in the DNA of every cell, which therefore indicates the existence of an intelligent designer who put it there. The *Pandas* text substitutes a love note scrawled in the sand for the talking pulsar and speaks only of a generic intelligent designer instead of specifying that it is God, but the argument is otherwise the same. The lessons we are asked to take from the SETI analogy are two-fold. First, we are supposed to agree that intelligent-design creationism is not religion but good science. Second, we are supposed to conclude that intelligent design is the best scientific conclusion to draw from the DNA data.

As we will see, the intelligent-design creationists' SETI analogy is a space-age version of the classic philosophical argument from design for the existence of God. To evaluate it we will therefore not only have to look a bit at scientific work on information theory and probability and their applications in evolutionary biology, but also return to deeper philosophical questions about the nature of evidence, to help us differentiate science from religion and from pseudoscience. We will also have to consider a bit of theology to understand the importance of Heeren's concluding remarks about why God doesn't just make a spectacular appearance. As in our discussion of linguistic evolution and the Tower of Babel, it will be useful to first investigate these issues in a context that allows us to consider them afresh and without prejudice. Moreover, for philosophical analysis of an issue, it is often very helpful to have a contrasting case that can highlight conceptual features that we might otherwise be blind to. For these reasons I now want to introduce one further sort of anti-evolutionary viewpoint.

Extraterrestrial Intelligent Design

Suppose we adopt, for the sake of argument, the intelligent-design creationists' claim that the evidence supports their hypothesis that life is the result of intelligent creation. Let us also take them at their word that they mean to provide scientific support on the SETI model. If so, then it looks as though the conclusion we should draw is that we were designed and created by intelligent extraterrestrials. Interestingly, Heeren writes that when the first edition of his book was published he was "barraged with calls from UFO watchers and skimming readers who wanted to know more facts about this reported contact with extraterrestrials," and he says that he "felt like Orson Welles must have felt after his radio broadcast of War of the Worlds in 1939."[36] My original intention at this point was to devise an alternative extraterrestrial intelligent design (ET-ID) view of this kind and to show how ID creationism's arguments compared to those of the UFO enthusiasts whom we would expect to believe such a story. As it turns out, I did not need to concoct such a tale, for once again truth is stranger than fiction. A large international group—the Raëlian Movement—advocates just this ET-ID view. Like creationism, this is a religiously based movement that rejects evolution. Unlike creationism, Raëlianism denies supernatural divine creation. Raëlians promote a third view—that intelligent aliens landed here millennia ago in spaceships and formed all of life on earth, including human beings, using highly advanced genetic engineering. I think that if we investigate the question of intelligent design in this context it will be easier to see why the IDC conclusion is not scientific.

The Raëlian Movement originated in the early 1970s in France and now claims over 35,000 members in over eighty-five countries around the world. Outsiders think of it as a New Age religion, and in many ways that is a fair characterization, but Raëlians think of their religion as being directly linked to Christianity and the other great world religions, and moreover, as the final religious form now that the world has entered the time of the Apocalypse. The beliefs of Raëlians are based upon messages our extraterrestrial creators conveyed to the world through the "Guide of Guides," Claude Vorilhon, a French journalist and race-car enthusiast. Vorilhon, who later adopted the name "Raël," claims to have been twice contacted by an alien in a flying saucer who revealed to him the true story of the creation of life on earth. Here, in brief, is the story, as described in *The Book That Tells The Truth* (1986 [1974]).

Eons ago, on a distant planet, these aliens—the Elohim—had reached an advanced level of scientific understanding and technical ability that enabled them to create primitive living cells in the laboratory. Some of them were fearful, however, of creating some new life form that might prove dangerous to life on their own planet, so they sought a lifeless planet where they might pursue their experiments safely. That planet was the earth. They began with simple cells but as research progressed they soon were able to engineer seeds, grasses, and a wide variety of vegetation. As their technical abilities developed, the scientists collaborated with artists to produce beautifully decorative and scented plants. Plankton, small fish, and then bigger fish came next, and as the scientists created new species they worked to see that each fit well into the ecology of the whole system. Eventually, they began to create animals—dinosaurs, sea and land creatures, herbivores and carnivores. The artists had a big hand in the creation of birds, going to such wild aesthetic extremes in some cases in their design of plumage that the creatures could hardly fly. Finally, the scientists were ready for their greatest technical challenge, creating beings like themselves. Several teams set to work. After producing a series of prototypes (some of the skulls and bones of which we have found), the scientists successfully created *Homo sapiens*. Their final forms were all slightly different—the various human races—but all were alike in being made in the image of their creators.

Evolution is a myth, according to Raëlianism. Biologists are correct that life forms of increasing complexity appeared over time, but that is because the alien scientists began with simple cells and then progressively modified these to produce more complex life forms as their techniques improved. Biologists are also partially correct in saying that humans descended from earlier primates; according to the Elohim scientist who contacted Raël, "Human beings . . . are only an improved model of the monkey, to which we added that which makes us people."[37] However, do not think that this happened by chance with purely natural mechanisms! No natural need could produce the beautiful curled horns of certain wild goats or the wild exotic colors of tropical fish. It is incredibly improbable that accidental evolution could produce the wide array of life forms. No, the complex biological world is all the result of the intentional design work of the Elohim scientists and artists.

It is important to understand that Raëlianism fulfills all the elements that intelligent-design creationists set out to characterize their view, at least in the minimal version they propose for public consumption. The only difference is that Raëlianism specifies that the intelligent designers were not supernatural spirits or gods, but alien beings from another planet. We thus have a real example of an ET-ID view. In all other respects, Raëlianism is very similar to standard Creationism. They have been far less organized and productive than creationists in arguing against evolution, because they are more interested in proselytizing their positive message, but when they do talk about it they tend to follow the same negative argumentation strategy.

Raëlians make many of the same bad arguments against evolutionary theory as creationists. The Raëlian who explained their view at an introductory meeting I attended began by noting that evolution is "simply a theory," and she suggested that radiometric dating of fossils is inaccurate because radiation decay is not constant. Raëlians follow creationists in proclaiming that evolution "lies in direct contradiction" to the second law of thermodynamics.[38] We also hear an echo of Philip Johnson's allegation that evolution is akin to religious dogma; Marcel Terrusse, Raëlian Bishop Guide for France, writes that scientists "indoctrinate" us with the evolutionary view, and that people would need "the courage to be seen as heretics"[39] to question openly these professors. Raëlians also question how random mutations could ever be advantageous and improve survival or increase organized complexity. Terrusse makes the point as follows:

An accident of this sort can never increase organisation but will only result in damage in the same way as throwing a watch on the ground can never increase its precision, nor hitting a computer with a spanner endow it with extra calculating properties. And the time factor will not change anything since what was impossible yesterday will also be today.[40]

Yet Terrusse had already inadvertently given counterexamples to this claim, noting that mutations in *drosophila* have been observed to change their size, color, and wing shape. He claimed that none of these mutations confers a survival advantage, but we already know of ways that individuals with increased size can better compete with smaller rivals for scarce resources. We already know that having one color rather than another could improve an individual's camouflage in specific environments. And it doesn't take an aerospace engineer to know that some wing shapes are aerodynamically better than others. Variations of just these sorts can give one individual a competitive advantage over another and thereby improve its relative chances of survival and reproduction, and over time, inheritance of such incremental improvements generation after generation can lead to better and better adapted organisms. So whence the claim of impossibility?

Raëlians speak of whacking a computer with a wrench in the same way that creationists speak of a tornado whipping through a junkyard. Could such random processes improve the programming of a computer or create a working automobile? they ask rhetorically. What evolutionists want you to believe is absurd, they conclude. But, as we saw before, such analogies are highly misleading. They misrepresent evolutionary processes, which are not random in this way at all. In fact it is not only theoretically possible to improve a computer's programming using Darwinian mechanisms, it has already been done. Nathan Myrhvold, director of the Advance Technology Group at Microsoft, developed a "software breeder" in the late 1980s using genetic algorithms such as we discussed in chapter two. Discussing this work, Stuart Brand explains, "A human programmer comes at a problem with a mind-set that causes him or her to solve it one way; in fact, there's a large space of other solutions—and evolution can find them. . . ."[41] For instance, Myrhvold's software breeder found an efficient way to translate files from binary into ASCII form.

Another interesting point about the Raëlian creation story is that it seems it might be only partially antievolutionary. The passage that describes the process that the Elohim scientists used to form increasingly complex life forms is ambiguous, but one reasonable interpretation is that they produced new life forms by modifying earlier ones. Raël writes, "In reality the first living organism created on earth was unicellular, and afterwards it produced more complex life. But not by chance!"[42] Together with the previously quoted passage of how humans were made by modifying monkeys, this suggests that Raëlians could accept a form of Darwin's view of descent with modification, namely, that organisms form an interconnected ancestor-descendant tree of life. If so, it looks like they might agree with the basic fact of evolution, though they would no doubt disagree about the pathways and pace of the diversification. Their major disagreement, however, is the same one that Johnson and the intelligent-design creationists have, namely, over the "blind watchmaker" mechanisms: They reject evolution by natural selection and insist upon intelligent design. Of course the bio-engineering process used by the Elohim is no doubt not standard biological reproduction, but if it involved manipulating genetic material taken from each species of organisms to produce the next in sequence, then there is an important sense in which Raëlians accept the fact that evolution occurred. Even if the Elohim did not directly manipulate genetic material from each species to create the next, but instead used a process that was more indirect, this still fits with a form of the common descent thesis.

Suppose they started with the "genetic recipe" for individual cells, say, in the form of a digital computer program that specified the organism's genome, and had a machine to churn out DNA molecules like pasta noodles once they added the chemical ingredients and entered the recipe for the desired sequences. (This may not be so far-fetched a possibility; already microbiologists have machines that allow them to build short DNA strands with the exact sequence of nucleotides they choose just by pressing the buttons A, T, G, and C in the desired order.) And suppose they designed subsequent cells by using this recipe as a model and simply adding to it and changing bits here and there, the way a cook modifies a recipe to produce a variation of an old favorite. They could continue in this way to produce, in turn, more complex organisms. For each new species they would have their machine crank out the prototypes to be released into the wild. This would certainly be a nonstandard method of reproduction, but it would maintain the basic idea that organisms are connected in a relationship of descent with modification at just a slightly higher level of abstraction.

This is probably just one of those odd hypothetical science-fiction scenarios that only a philosopher could love, but it does lead to a very interesting conceptual possibility when we connect it back to the view of Genesis creationists. Perhaps God created the biological world in such a manner. As we noted, some of the creationists with whom Darwin argued held that God created biological organisms following "ideal plans," and who is to say these were not master genetic recipes? A creationist who thought that special creation worked in this way would thereby accept the basic fact of evolution (though in an nonscientific form). Indeed, it would even be open for such a creationist to accept the more direct method—after all, perhaps God was extracting genetic information when he created Eve from Adam's rib![43] I mention this odd scenario because it leads to a further ironic twist in our comparison. Let us return a moment to the Raëlian creation story. Didn't that line about the Elohim creating humans in their image sound rather familiar? Isn't this just a corrupt retelling of Genesis? Almost, but not quite, explains the Raëlian, and here is where the story really begins to get interesting. Actually, it is the other way around; the Bible contains a partial and somewhat corrupted version of the true story. Now that the truth has been revealed to Raël, we can correctly interpret the Bible at last. According to Raëlians, evolution is a myth but so is creationism.

The Bible was written by the created beings on earth, but they did not know the complete truth about their creation, and what they did know they were unable to fully comprehend. Arthur C. Clarke, dean of science-fiction writers, who predicted the use of telecommunication satellites, wrote that any sufficiently advanced technology would be indistinguishable from magic. Think of what primitive peoples would have thought had they encountered modern people with our technology—they would have viewed us as sorcerers and magicians. According to Raëlians, the Elohim, with their even greater scientific technologies, were thought of as supernatural beings—as gods. Indeed, we translate the Hebrew term "Elohim" in the Bible as "God," but it is not a singular form but a plural term, which Raëlians claim means "those who came from the sky." There are no divine beings and there was nothing supernatural about the creation of life. We should respect and be thankful to the Elohim who are our creators but they are natural beings as we are, and everything that is supposedly miraculous in the Bible can be explained scientifically. With the new knowledge given to Raël, they say, mysterious passages in the Bible suddenly make sense.

Who were the "sons of God" who married many of the "daughters of men" after human beings once again spread over the world after the Flood, causing Yahweh to feel disgrace (Genesis 6:1–4)? Raëlians claim they were aliens who had been exiled to earth because they had loved their creations too much and violated the home planet's ruling that they be kept in ignorance of scientific knowledge. (Compare that to ICR's Henry Morris's explanation that they were Satan's evil angels coming down to earth to violate human women.) What was the fire and brimstone that rained from the sky, destroying Sodom and Gomorrah and the surrounding plains (Genesis 19:23–28)? Atomic bombs dropped on those cities, say the Raëlians. The pillar of cloud and fire that led Moses and the Israelites out of Egypt? A rocket ship with repellent beams that parted the Red Sea. Indeed, Moses had many encounters with UFOs including the one that descended upon Mount Sinai to deliver the tablets of the Decalogue.

The mountain of Sinai was wrapped in smoke, because Yahweh had descended on it in fire. Like smoke from a furnace the smoke went up, and the whole mountain shook violently. And the voice of the trumpet sounded long and waxed louder and louder. (Exodus 19:18–19)

Purported references to UFOs and alien encounters are not confined to Genesis but occur throughout the Bible. Some are not at all obvious (Raëlians believe that the Tower of Babel was not a ziggurat, but a spaceship that would literally allow them to reach heaven—the home planet of the Elohim), but others are startlingly realistic. Think of those most striking and bizarre passages in which Ezekiel gives extended descriptions of when he saw the chariot of Yahweh, that begin "I looked; a stormy wind blew from the north, a great cloud with light around it, a fire from which flashes of lightning darted, and in the center a sheen like molten bronze," and speak of strangely garbed and helmeted beings that moved in vehicles of wheels within wheels that glittered like crystal (Ezekiel 1:4–28 ff.). Raëlians outdo even literalist creationists in their interpretation of such passages, insisting that they are factual descriptions rather than symbolic mystical "visions."

In talking with Raëlians and reading their literature, one is struck not only by their similarity to creationists but also to UFO enthusiasts. Raëlian books typically appear in bookstores on the same shelf as books on UFOs. A classic of the UFO genre is Erich von Däniken's aptly titled *Chariots of the Gods.* Von Däniken was the first to point out several passages in the Bible that could be interpreted as referring to extraterrestrial visitation. Indeed, he cites the Ezekiel passage in a chapter titled "Was God an Astronaut?" and rhetorically inquires why an omnipotent God who can be anywhere he pleases would "come hurtling up from a particular direction" with such "noise and fuss."[44] Who were the four living beings with the likenesses of men who emerged from the fiery landing of Yahweh's chariot to speak to Ezekiel? Von Däniken concludes that:

They were certainly not "gods" in the traditional sense of the word, or they would not have needed a vehicle to move from one place to another. This kind of locomotion seems to me to be quite incompatible with the idea of an almighty God.[45]

Von Däniken notes similar references in the mythological literature of other cultures as well. He discusses ancient carvings, illustrations, monoliths, and other physical artifacts that he believes support the idea that the earth was visited by ancient astronauts. He also looked for physical evidence and, lending new credence to the aphorism about birds of a feather, sent a cameraman in 1975 to Glen Rose, Texas, to film the Paluxy river prints, which he supposed not to be "manprints" but footprints of giant extraterrestrials. Raëlians point to similar facts as well as eye-witness sightings of UFOs as evidence of the truth of their view. The Raëlian who gave the introductory talk I heard explained that she had written to von Däniken about the Raëlian Movement, and she showed me the letter she had received from him expressing interest but polite skepticism about some of Raël's claims. Raëlians are used to having their views dismissed out of hand and mocked, so this lukewarm response from someone who might be thought to be naturally sympathetic was accepted with equanimity. *The Book That Tells the Truth* ends with a plea for open-mindedness. Raël writes:

Now that you have read this book . . . in which I have tried to reproduce as clearly as possible all that was said to me [by the extraterrestrial], if you will think perhaps that I have a great imagination and that these writings were simply to amuse you, I shall be profoundly disappointed. Perhaps these revelations will give you confidence in the future and allow you to understand the mystery of the creation and the destiny of humanity, thus answering the many questions that ever since childhood we pose during the night, asking why we exist and what is our purpose on this earth.[46]

According to Raëlians, the evidence that this story of creation is true exists before our eyes. Others can bear witness to the reality of their encounters with aliens and flying saucers, and you might have the chance to observe one yourself, Raël tells his readers. He concludes:

If you still have doubts, read the newspapers and look to the sky where you will see that the appearances of the mysterious crafts will be more and more numerous. . . .[47]

Raëlianism offers the possibility of a meeting with the extraterrestrials when they arrive in their spaceships as the ultimate potential observational evidence of its truth. The Raëlian I spoke with, however, admitted that she herself had never seen a UFO, let alone had a close encounter with any extraterrestrial. Like most Raëlians, she bases her belief in the return of the Elohim upon faith in Raël's account of his transformational encounter, in much the same way that early Christians had to take the word of the Apostles that they had encountered the risen Jesus. (Actually, the parallel is particularly strong on this point, for Raël claims to have met Jesus as well when he was transported for a brief visit to the Elohim homeworld.) Also like the early Christians, she says she fully expects that the hoped-for return will occur within her lifetime. Of course, the history of religion is filled with Christian sects who believed that the Second Coming was imminent and some even hazarded a guess about when it would happen, but these invariably saw the predicted dates come and go without the expected rapture. Recent history has seen groups of UFO believers who have been similarly disappointed when the Mother Ship failed to land at the time of an anticipated arrival. Supposing that Elohim in either the Christian or Raëlian form do eventually return in a convincing fashion, we would then have reason to consider seriously the respective creation account. But given that such an appearance remains but a hope, the pertinent question is whether we currently have scientific evidence of the existence of an intelligent designer of the world or whether that belief is religious.

Redesigning the Argument from Design

New Bumpers for the Teleological Argument
Philosophers over the centuries have considered three main sorts of argument for the existence of God. The ontological argument is the most philosophically intriguing, attempting to provide an *a priori* proof of God's existence from the very idea of God as a perfect being, but as a purely conceptual argument it does not pretend to be empirical. The cosmological argument, on the other hand, does appear at first to be empirical. In its basic form, as put forward by St. Thomas Aquinas, it starts from the observed fact of motion in the world, asserts that no motion occurs without something to put it in motion, reasons to a necessary "First Mover" to initiate this change, and claims that this is what all men understand to be God. Some commentators have interpreted this as involving a temporal sequence with God as the origin. At the Museum of Creation and Earth History, our guide gave an informal version of this argument, concluding that God also created time itself in the Genesis Creation event. Some old-earth creationists particularly like this interpretation, because it makes the

cosmological argument appear to be an empirical argument and they can then point to the scientific account in which the universe had a beginning in time at the Big Bang and argue that this fits with Christian cosmogony. Raëlians, on the other hand, reject the cosmological argument. As they see it, the universe is infinite in all directions. They agree with Pascal's idea that there are infinite numbers of universes all nested within one another. For Raëlians, the universe is also temporally infinite and, if there is no beginning, there is no need for a First Mover. Severe philosophical criticisms have been made of the empirical interpretation of this argument, and the argument rapidly becomes more complicated once one moves to a more sophisticated interpretation (which is arguably what Aquinas intended in the first place) and takes God not to be a First Mover situated at the start of a temporal causal series "within" the universe, but instead as being situated at the apex of a hierarchy that in some sense holds up the temporal succession of events at all points. This interpretation avoids the logical problems but, in so doing, takes the argument out of the realm of science.[48] However, we need not delve further into the details of this or other versions of the cosmological argument, because it is the third sort of argument—the teleological or design argument—that is the mainstay of Creationism. The classical version of this argument begins with the claim that things in the world act "to achieve the best result" and, assuming that things without knowledge cannot move toward such an end without the help of a thing with knowledge (that is, without an intelligent being to set the goal and guide them towards it), concludes that some intelligent being must exist by whom all natural things are directed to their ends. That being we call God. Aquinas gave this version of the design argument in the thirteenth century, but it was greatly elaborated upon in the eighteenth century by Cotton Mather in his *The Christian Philosopher* (1721) and, especially, by William Paley. As we noted in chapter 2, Paley's sustained development of the design argument was particularly influential on Darwin and his contemporaries, and it set the standard for subsequent discussion of the subject.

Paley is most remembered for his brilliantly stated watchmaker analogy, which opens *Natural Theology*: as we may infer a watchmaker from the observation of the complex workings of a watch, so may we infer a maker of the world by observation of *its* complex design. After impressing upon his readers the intricate construction of real watches, Paley goes on to imagine hypothetical watches that contained within themselves additional mechanisms that enabled them to reproduce. How infinitely greater would have had to have been the skill of a watchmaker who could craft such an instrument! And yet it is just such reproductive ability, and still greater complexities that we see in biological organisms, so we know that God is great indeed. The watch analogy is Paley's driving argument, but it accounts for only a fraction of the 352 pages of his book. In the bulk of the work Paley describes in great detail a vast array of specific illustrations from the natural world that he claims proves the existence and wisdom of a designer who created in each case means perfectly tuned to ends. Paley finds benign utility and thoughtful design everywhere he turns in the world. Of water, for instance, he writes that "besides maintaining its own inhabitants, [it] is the universal nourisher of plants, and through them of terrestrial animals; is the basis of their juices and fluids; dilutes their food, quenches their thirst; floats their burdens." But this is only a partial list, for happily its properties allow us to boil, freeze, evaporate, or "run or spout [it] out in what quantity and direction we please."[49] On Paley's view, God designed everything in the end for the sake of human beings, but one might well ask how anyone could know that God had these or any other specific purpose in mind.

Water is also to be admired for "the constant *round* which it travels; and by which, without suffering either adulteration or waste, it is continually offering itself to the wants of the habitable globe."[50] Paley describes how the sea water evaporates to form clouds that drift over the land to fall in showers upon the earth, penetrating the crevices of the hills to feed springs and then streams that finally unite in rivers that rejoin the sea. A drop of water, he writes, "takes its departure from the surface of the sea, in order to fulfill certain important offices to the earth; and having executed the service which was assigned to it, returns to the bosom which it left."[51] Of course, this complex circulation and the multiple functions it performs cannot be accomplished by the water droplets alone, but requires the aid of air, soil, sun, and more. Paley points out that water and all the other elements "bear not only a strict relation to the constitution of organized bodies, but a relation to each other. Water could not perform its office to the earth without air; nor exist as water, without fire."[52] In such passages Paley is trying to impress upon his reader the interdependent arrangements of parts that are necessary for the functions he identifies. By highlighting complexity of this sort he hopes to strengthen the analogy to the watch.

Most of Paley's examples are drawn from the biological world, and of these, his strongest example by far is the amazing complexity of the human eye. Indeed, a series of anatomical drawings of the eye is the first thing one sees upon opening *Natural Theology*. Darwin himself pointed out that the eye, as an "organ of extreme perfection" appeared to be a serious difficulty for his theory, writing in the *Origin*:

To suppose that the eye, with all its inimitable contrivances for adjusting the focus to different distances, for admitting different amounts of light, and for the correction of spherical and chromatic aberration, could have been formed by natural selection, seems, I freely confess, absurd in the highest possible degree.[53]

A lesser intellect might have given up in the face of this apparent difficulty, but Darwin pressed on to show that natural selection could indeed have formed even so complex an organ by a series of incremental modifications from simpler organs. He listed varieties of intermediate eye-types found in nature—starting from merely a pigmented optic nerve followed by other forms with their "numerous gradations of structure"[54]—to show their viability and thus the possibility of gradual transitions. In light of such evidence, the various complexities that Paley and others cited no longer seemed to be so dauntingly inexplicable. Richard Dawkins has elaborated Darwin's argument for the evolution of the eye, and for other biological complexities, in considerable detail in several books.[55] Dawkins is an important figure in the contemporary debate about the creationists' use of the argument from design, for he agrees with Paley that the biological world has features that appear as if designed, but he always goes on to show how the Darwinian mechanisms can produce these same results without intelligent design.

However, not everyone agrees that the world even appears to be designed. Challenging Paley, the skeptic asks: what about the needless waste, the freaks and monstrosities of nature, and the many examples of apparently slipshod construction? Richard Owen, recall, a creationist of a different sort, matched Paley's list of biological perfections with his own list of biological imperfections. It is important to remember that such criticisms of the design argument in terms of apparent imperfections were made well before Darwin arrived on the scene, but they became especially significant in the wake of evolutionary theory. Evolutionary biologists often cite examples of odd constructions as one line of positive evidence for evolution. Stephen Jay Gould's most famous example is the pseudothumb of the giant panda, which is built from its wrist bone. Natural selection operates on the materials at hand, and so we find organisms with adaptations that seem makeshift since they evolved from parts that might originally have had other purposes. In his popular writings, Gould also has used this example, which provides the title for his book *The Panda's Thumb*, not only to explain some of the positive evidence for evolution but also as an argument against the creationist view:

If God had designed a beautiful machine to reflect his wisdom and power, surely he would not have used a collection of parts generally fashioned for other purposes. . . . Odd arrangements and funny solutions are the proof of evolution— paths that a sensible God would never tread but that a natural process, constrained by history, follows perforce.[56]

Creationists, of course, have tried to defend the Design Argument against such criticisms. Intelligent-design creationist Paul Nelson, writing under the pseudonym Peter Gordon, has attacked Gould specifically on this point, arguing that the pseudothumb is not imperfect at all, but appears to be used rather efficiently by the Panda. More recently Nelson has also tried to question Gould's and other biologists' use of the imperfection argument as a positive argument for evolution. He claims that using the argument as positive evidence presupposes the intelligibility of theological assumptions about what God would or would not have done and is thus "inconsistent with the doctrine of methodological naturalism."[57] Nelson concludes that biologists either should give up citing the argument as favoring evolution or should abandon the rule imposed by methodological naturalism. Obviously, he most favors the latter option, since the main goal of intelligent-design theorists is to legitimize theistic hypotheses and supernatural interventions in science.

However, Nelson's argument improperly conflates the positive argument for evolution with Gould's negative argument against creationism.

Nelson's confusion is not surprising given that on the dual model view creationists presuppose that rejecting the one is equivalent to proving the other. If Gould's comment about what a sensible God would never do were meant to be taken as part of a scientific argument, then Gould has indeed improperly violated methodological naturalism in his negative argument, for science has no way to test what would or would not be sensible from God's point of view. But ruling out this argument does not affect the positive argument. The reason that odd arrangements and makeshift adaptations are evidence of evolution is that these constitute a kind of pattern that is an expected effect of evolutionary processes, which must work at any given moment within the constraints imposed by previous evolutionary development. This positive argument need not refer to God or any theological hypothesis. That it traditionally has been associated with the argument from imperfection against the design argument is simply the result of the contingent historical fact that it originally was articulated in the context of an earlier debate regarding Paley's highly specific claims about perfection. However, the slipperiness of the theology makes it impossible to tell what is or is not an "imperfection" and thereby makes it impossible to test the design argument. This is highlighted by the fact that some have argued that imperfections actually *support* the design argument and evolution as well.

Historically, a common reply to the argument from imperfections has been to suggest that apparent imperfections in the world are indeed only apparent and would disappear were our knowledge complete. In 1884, the Bishop of Exeter, Frederick Temple, in his fourth Bampton Lecture on the perceived conflict between religion and evolution, suggested that this answer to the skeptic is given even greater force by accepting evolution.

But what force and clearness is given to this answer by the doctrine of Evolution which tells us that we are looking at a work which is not yet finished, and that the imperfections are a necessary part of a large design the general outlines of which we may already trace, but the ultimate issue of which, with all its details, is still beyond our perception! The imperfections are like the imperfections of a half-completed picture not yet ready to be seen; they are like the bud which will presently be a beautiful flower. . . .[58]

Temple's position is a kind of theistic evolutionism so, of course, it is anathema to intelligent-design theorists. As they see it any accommodation to evolution is unacceptable. However, since the only positive argument they have for their view is the design argument, they need some other answer to the problem of imperfections. Some, like Nelson, doggedly retain something close to Paley's vision and argue that we simply don't appreciate or understand the perfection of the world. Davis and Frair offer the following explanation to a child who asks why her pet rat has scales on its tail: ". . . God had put the scales there for reasons He knew to be perfectly good ones but which may take us a lot of research to discover, since He has not told us what they are."[59] However, others among the intelligent-design creationists are taking stronger measures to protect the design argument against the argument from imperfections.

Michael Behe offers three arguments against the imperfection argument that take the foregoing defenses to their logical conclusion. I'll have a couple of criticisms of them, but in the end I will want to say that Behe is essentially correct. Here are his arguments:

The most basic problem is that the argument demands perfection at all. Clearly, designers who have the ability to make better designs do not necessarily do so.

I do not give my children the best, fanciest toys because I don't want to spoil them, and because I want them to learn the value of a dollar. The argument from imperfection overlooks the possibility that the designer might have multiple motives, with engineering excellence oftentimes relegated to a secondary role.

Another problem with the argument from imperfection is that it critically depends on a psychoanalysis of the unidentified designer. Yet the reasons that a designer would or would not do anything are virtually impossible to know unless the designer tells you specifically what those reasons are.[60]

The first complaint is a bit unfair, of course, since the notion of perfection was introduced by Paley and others who were forthright and specific about their conception of how the Designer was supposed to have operated. Paley assumed that God was perfect and would necessarily have created a perfect Creation with every creature suited in all its parts to its situation. It was this kind of panglossian notion that Voltaire so effectively satirized in *Candide*: "Everything is made for the best purpose. Our noses were made to carry spectacles, so we have spectacles. Legs were clearly intended for breeches, and we wear them." Does Behe really mean to deny that the perfect designer did a perfect job, or is he just making a virtue of vagueness? Whatever the case may be, he is certainly correct that we have no way of knowing that a supposed designer of the world intended, or was even capable of, engineering perfection. In the second objection one suspects that Behe hopes his comments about his own motives as a father will be seen as an analogy that might suggest possible motives of God the Father, but if we leave that aside he is quite correct that we cannot tell whether a designer had one or many motives or what they might have been.

As for his third objection, it provides the design argument with a virtually impenetrable shield. Although we are occasionally able to infer some of the design objectives of fellow human beings, provided that we may assume substantive background knowledge about shared biological needs and cultural values, even this can be extremely problematic. Move on to trying to psyche out the design intentions of extraterrestrials and the task becomes even trickier. I know of no Raëlians who have written about the imperfection objection explicitly as such, but we do know that they hold that the Elohim genetic engineers who designed terrestrial life-forms had to experiment before they improved their technique. "Imperfections" would thus be quite expected on their view. In this case, Raëlians could say that they are able to avoid the problem Behe mentions because the extraterrestrials did tell Raël specifically what their design intentions were, at least in part, as well as something of the causal means by which they tried to accomplish them. Lacking substantive background information of this sort, the ET-ID theorist would have nothing to go on. We will take this point up again shortly. Finally, if we think of the Designer with a capital "D," as advocates of the design argument of course want us to, the task becomes quite impossible, for we are in no position whatsoever to psychoanalyze God or any other supernatural intelligence.

I conclude that Behe has successfully insulated the design argument against the imperfection argument. Equipped with such bumpers it can now withstand any impact. What this means, however, is that IDCs, unsurprisingly, are in exactly the same boat as YECs who make the Omphalos "mature earth" defense. As we will see, it also further supports my earlier contention that intelligent-design theorists have no positive evidence to offer for their position. Their so-called design inference still relies upon negative argumentation and is just another form of the God of the gaps argument.

Let Fly the Arrow

We now return to the intelligent-design creationists' SETI analogy and the argument from information. Heeren outlined the argument which in one form or another is the mainstay of intelligent-design theory. Johnson asks us to recall the opening line of the Gospel of John, that "in the beginning was the Word," and he links this to an argument against materialism: "Matter," he says "is important, but secondary. The Word (information) is not reducible to matter, and even precedes matter."[61] That organisms contain highly complex information is purportedly the proof of God's Creation "because highly complex information that is independent of matter implies an intelligent source that produced the information."[62] We have seen this idea at work as the organizing idea of *Pandas*, where authors Percival Davis and Dean Kenyon suggest that organisms contain complex information that can come only from intelligence. Intelligent-design creationist William Dembski develops this idea by drawing out the notion of complex specified information (CSI). He gives an archery example to elucidate the notion intelligent-design creationists have in mind and the inference we are supposed to be able to draw from it: Suppose an archer shoots an arrow from 50 paces away from a huge wall and it hits a particular point. Hitting any particular point on the wall might be as highly improbable as hitting any other, but we would sit up and take notice if the point hit was the center of a single bulls-eye on the wall. We would not likely attribute that hit to chance because that highly unlikely point had been specified in advance. It must be that the archer hit the bulls-eye intentionally, that is, by design. This illustration makes it clear that Dembski intends to link up with the classical design argument for the existence of God, for Aquinas had put forward his original teleological argument in just this way, giving as one premise that: "Things lacking knowledge move towards an end only when directed by someone who knows and understands, as an arrow by an archer."[63] Dembski argues that this special feature distinguishes intelligent design; CSI is what SETI scientists are purportedly looking for. Finding it would suffice to prove intelligent design because, the argument goes, only intelligent agents generate CSI. Since DNA contains CSI, this can only mean that life did not evolve by natural means but was put together by an intelligent designer. Furthermore, say intelligent-design theorists, the only reason scientists don't recognize this conclusion is that they are wearing the blinders of scientific naturalism.

Dembski and the authors of *Pandas* are not the first to make this sort of argument. Norman L. Geisler is the first I know of who outlined all its elements. He put forward the argument in a written debate that appeared in *Creation/Evolution* in the mid-1980s. After pointing out that we would infer that the faces on Mount Rushmore were created by an "intelligent primary cause" because they convey "complex information," Geisler notes that "DNA has specified complexity in its message . . . [and] it is our regular, uniform experience that specified complexity results from an intelligent cause."[64] Furthermore, he argued that "since there admittedly was no human intelligence before life arose, then it would be necessary for this intelligence to be superhuman," and since the whole universe had a beginning, "there is no reason why this cause could not be a supernatural one." To reject the supernatural cause, he writes, "only reveals one's philosophical bias in favor of naturalism." Finally, to bolster this he cites the SETI idea that "we could posit the existence of superintelligent beings in outer space upon the receipt of the very first short message from them."[65] This is almost the whole of intelligent-design creationism in a nutshell. In his review of *Of Pandas and People,* Geisler compliments the authors for not siding with either the young- or old-earthers, but criticizes them for "appeasing [our] enemies" by "avoiding the word 'creation' like the plague" and for not clearly distinguishing their view from that of "naturalistic (pantheistic) 'creationists'" who see the creator within the universe."[66] Despite such reservations, it is not surprising that Geisler praises *Pandas* as "the first credible attempt of its kind to offer the scientific case for a creationist view in a text intended for public schools" and "the best book in its category,"[67] given that he is credited as one of the critical reviewers of the manuscript and the text essentially hews to the line of argument he previously laid out.[68]

It is fitting that Geisler was the first to articulate this argument, since we know that he takes the SETI analogy very seriously indeed. Geisler, then professor of systematic theology at Dallas Theological Seminary, was one of the star expert witnesses on theological matters in defense of creation-science at the Arkansas trial, and he made headlines with his testimony on the reality of UFOs. However, he took them not to be evidence of extraterrestrials, but as evidence of the existence of a personal devil. UFOs, he declared under oath, are "Satanic manifestations for the purposes of deception" and "represent the Devil's major, in fact, final, attack on the earth."[69] At that time, however, he offered more direct proof of their existence than the complexity of DNA. When asked by the prosecuting attorney how he knew that UFOs existed, he said their existence was confirmed by an article he had read in *The Reader's Digest*.

Raëlians would of course concur with Geisler about the existence of UFOs, but they would strenuously object to calling them Satanic manifestations. Recalling Morris's writings and our trip to the Museum of Creation and Earth History, we know ICR's official stand is that all religions, besides Judaism and Christianity, were created by Nimrod and Satan at the Tower of Babel, so we can expect them to take Geisler's side against the New Age Raëlian Movement on this point. Should we dismiss the scientific evidence for evolution, conclude that all these views are "just theories," and be sure to give each "balanced treatment" in the school biology curriculum? Should we abandon scientific methodology and "keep an open mind" about such supernatural and extraterrestrial "alternatives" and present them as though they were on an evidential par with the well-tested scientific account? Should we, to be "fair," provide *Of Pandas and People, Chariots of the Gods,* and a Raëlian supplemental text, and "allow students to decide for themselves"? Surely this would be a terrible mistake. If we wish to avoid being taken down the garden path in this way, we need to watch carefully to see where creationists' arguments go wrong and how they can be misleading.

Science humorist Sydney Harris drew a cartoon that might well have been the inspiration for the IDCs' SETI analogy. In the first of a series of four panels he depicts two astronomers looking up into the dark night sky. In the second panel, we see a pattern of bright dots in the sky that form the shape "EAT." In the third panel, the sky is again dark, but EAT flashes once more in the fourth panel. One astronomer growls to the other, "I don't care what it looks like. I say they're pulsars." The analogies that IDCs put forward are supposed to convince us that in denying their conclusion of intelligent design, scientists are being obstinate in just this way. It does seem as though one would have to be perverse to deny that this was not a flashing sign from a cosmic diner. But what if the flashing dots formed the shape PAIN? We English speakers might take this as a foreboding sign of doom. French speakers, on the other hand, might now be thinking of a cosmic bakery, since this is the word for "bread" in French. The point is that we cannot necessarily tell what a signal means just by looking, but must judge it in light of other assumptions we make. But what if someone argued that even if we didn't understand the message we could nevertheless know in this case that there was an intelligent signaler simply because the signal contained information? Here is where things begin to get tricky.

We need to be particularly wary whenever we see IDCs appeal to information signals since the term "information" is used in a variety of ways. It can be rather confusing to hear talk about information theory, since someone unfamiliar with this specialized, technical subject might think that it dealt with information in the same way that we understand the term in everyday contexts. "Information" as used by information theorists is a technical term that involves the resolution of uncertainty. Claude Shannon, the founder of information theory, chose the binary digit, or bit, as his unit of information. One bit of information resolves the uncertainty between two equally probable alternatives, say, whether a coin landed heads or tails. If the alternatives are not equally probable then it takes more information, more bits, to resolve the uncertainty of the improbable alternatives than it does the probable ones. A "message" in information-theoretic terms is just a series of bits. Thus to say that a signal "contains information" is not necessarily to say that it contains a message that is indicative of an intelligent agent. Keep this in mind as we continue. We would have to find a certain sort of message in a signal to say that we have found evidence of intelligence.

SETI project scientists are not looking for a signal that simply contains information in this technical sense. They already have plenty of information-bearing signals—light and radio waves from stars provide all sorts of information about their physical properties. The point is to look for that sort of information that indicates intelligence. We have already seen something of how difficult this might be. The pulsed radio signal discovered by Jocelyn Bell initially set SETI scientists' hearts a-pounding because it was unprecedented as an astronomical phenomenon and because it was suggestive of a beacon. But to describe something as a "beacon" is already to suggest that there was some intention involved to draw attention to it, and how are we to distinguish a beacon from a mere pulse? The answer, of course, is that it won't be easy. Remember the days that Drake spent looking for a message in the varying intensities of the pulsar signal. The world is full of complexities of all sorts. To identify something as an intelligent signal from among these will require not only that it stand out in sharp contrast to ordinary complexities but that it be recognizable by reference to a pattern that we already know to be characteristic of our own intelligent signaling. Unless we can discern a message pattern that matches in this way, we will completely overlook signals that are right before our face. The SETI Institute's internet site points out just this problem:

How will we know what the signal means?: If the signal is intentional, it is likely to be easy to decode. In order to send or receive a signal over interstellar distances, a civilization must understand basic science and mathematics. Hence, a message from another civilization would probably use a language based on universal mathematical and physical principles. Signals that a civilization uses for its own purposes may be difficult to decipher. Such emissions may have no detectable message content.[70]

This is generally true for inferences about purposeful design. Unless the purposes are of a kind with our own ordinary (that is, not only natural but recognizably human) purposes, they will be invisible to us. This is one reason the love notes in the sand and the talking pulsar examples are misleading, for in each case we infer an intelligent signaler not because these are cases of complex specified information in a generic sense, but because the pattern of information matches a previously known pattern that we associate with intelligence. To draw an analogy between such patterns of information and the patterns of information found in DNA is to make an unwarranted leap. Information-rich as DNA might appear, this is not sufficient to conclude that it is a sign of intelligence. The information in DNA can be attributed to natural processes.

Intelligent-design creationists deny this outright about DNA and sometimes even seem to make the stronger claim that it is impossible for natural processes to generate information at all. But chance can generate information. It can even generate complex information. And natural selection can preserve and channel it. As we saw in the excerpt from Heeren, creationist claims about DNA are often accompanied by probability calculations that purport to show that it is unbelievably improbable that one or another protein molecule could have arisen by chance. The only explanation, they want to conclude, is that it was intentionally designed. Let us briefly look into this argument.

An article titled "Meeting Darwin's Wager,"[71] which describes Michael Behe's challenge to evolution, carries an illustration of Behe seated at a card table and playing a game of cards with Charles Darwin, while Stephen Jay Gould, Richard Dawkins, Phillip Johnson, and others observe their play. Darwin certainly appears to be scowling as he considers the card he has drawn, while Behe is sketched with a thoughtful poker face, and we are able to see that he is holding what appears to be the ace of hearts in his hand. Games of chance like cards are a standard setting in which to discuss probability, and this is an appropriate place to dismiss a common misconception about what we should conclude upon encountering a highly unlikely event. Improbable as it is to get a royal flush, players are occasionally dealt that hand. Moreover, and this is the key thing to remember, every other card sequence of the same length is improbable to exactly the same degree. In a fair lottery it might be highly unlikely that a given person will win, but it is very likely that someone will. Thus, even if it is true that something is highly improbable, that fact by itself does not tell us that we can rule it out.

This point has been made many times before to bring out one of the fallacies of the creationists' ubiquitous arguments that purport to show the extreme improbability that some biological molecule arose by chance. But there is another lesson to learn from the card-game analogy that is particularly relevant to the way that the new creationists make the argument by focusing upon the special functional nature of biomolecules. The lesson is that the specific sequence that is a royal flush is a winner only in certain games. *Any* specific sequence could "function" as a winner in some game. We have no way to evaluate a given hand unless we know what game is being played. For instance, the ace in Behe's hand might give him the advantage over Darwin if they are playing a game in which they have specified beforehand that aces are high, but it might give him a losing hand if aces are low. Intelligent beings can play all sorts of games with each other, and without a lot of additional information and assumptions we are in no position to judge whether and how one hand "functions" until the players tell us the object of the game and we know something about the causal properties of the "game pieces." Even Dembski's archery example works only because we know in advance the relevant features of the game and something about the physics of arrows and targets. It is by no means clear that biological information is analogous to the sort of information pattern that allows us to infer design. It is certainly not a function that is specified in advance in the same sense that the card players or the archers specify their goal in advance.

The general point is the same as before—the functionality of something is not seen on its face. Things are causally interacting constantly, but it is not possible to say that they are "functioning" in some way unless there is reasonable sense to be made of what it would be for them to malfunction, and both notions require some prior specification of purpose. Is a tree functioning or malfunctioning when it topples to the forest floor and becomes a source of nutrition for mushrooms and a shelter for foxes? Who can say? Is the earth as a whole itself a functioning life-form with its myriad homeostatic feedback mechanisms, as James Lovelock's Gaia hypothesis suggests? Most scientists rightly dismiss this as romantic nonsense. What about Paley's claim of the complex, interconnected functionality of water? Is a rock-slide functioning or malfunctioning when it blocks a stream?

Biologists do indeed speak of the functions of organs, and functional explanations occur in biology in a way that they do not occur in any other of the natural sciences. Philosophers of biology have even argued that such ascriptions of functions make sense, though just as a shorthand for a longer causal explanation that would depend critically on the evolutionary history of the feature in question. It is only because some property has been "selected for" by natural selection that it is legitimate to speak metaphorically of its having a function. But this sort of specification of function depends necessarily upon the evolutionary processes, so the biological notion of function is not a specification that would serve to license a design inference. Quite the opposite. Genetic information is natural, not indicative of intelligence.

Intelligent-design theorists try to contest this in several ways. Dembski tells us that pure chance is:

. . . incapable of generating CSI. Chance can generate complex unspecified information, and chance can generate noncomplex specified information. What chance cannot generate is information that is jointly complex and specified.[72]

Notice that Dembski is doing just what we saw other creationists doing earlier, namely, simply asserting that something is impossible in principle. He gives no argument to support this critical bold premise. Since complexity and specification are independent properties, as Dembski explains them, if chance can generate complex unspecified information or noncomplex specified information, then what mechanism makes chance in principle "incapable" of generating information that is jointly complex and specified?[73] There seems to be no such mechanism that applies to the biological case. Dembski is wrong to say that chance is incapable of generating CSI. It might be highly unlikely, to be sure, but not impossible. He has overstated the case to try to make a hard problem sound like a profound one.

In any case, as we have seen, the Darwinian mechanism does not appeal to chance alone, but chance in combination with natural selection. Dembski tells us, however, that this process also is incapable of generating information:

The Darwinian mechanism of mutation and natural selection is a trial-and-error combination in which mutation supplies the error and selection the trial. An error is committed after which a trial is made. But at no point is CSI generated.[74]

Again, this is overstated. For example, Dembski tells us that one's telephone number constitutes CSI: "the complexity ensures that this number won't be dialed randomly (at least not too often), and the specification ensures that this number is yours and yours only."[75] However, I know that I am not the only one who is acquainted with some poor soul whose home phone rings off the hook with wrong numbers at certain hours because it is just one digit different from a local pizza parlor. The mechanisms of DNA replication and repair make it likely that the random mutations that occur will be "nearby" to the original, similar to the way that a random mistake in dialing a phone number is more likely to reach someone still in the same calling area. So, if one's phone number qualifies as CSI, then chance can indeed generate CSI under certain conditions. Dembski and other intelligent-design theorists seem to apply the notion of a specification in a vague and *ad hoc* manner so that it is difficult to say for sure what would or would not count as CSI, but since I have here used Dembski's own example of CSI this should be a legitimate counterexample to his claim.

We could even modify the scenario slightly to make it model a Darwinian process. My office phone is part of the university system that requires one to press "9" to get an outside line, while internal phone numbers require pressing only the five digit extension, all beginning with either a "1" or a "5." If you start to dial an outside number without first pressing the "9" you are quickly cut off by a recording saying that the call cannot be completed as dialed. If you do press "9" first, but then get a few digits into a long-distance number, another recording interrupts to say that an authorization code must be entered first. Suppose that the entire phone system was set up so that it would cut one off in this manner if dialing hit illegitimate prefixes. We might imagine a phone system that did this at first as I described, but then also had cut-offs for domestic versus international prefixes, say, then respectively for either working area codes or country codes, then for local calling areas, and so on, until the only legitimate calls were those of five digits that were individuals' numbers. Suppose now that we start with simple mechanical phone dialers that begin by dialing just a single digit. Those that are cut off "die" and those that "survive" then replicate, with some chance of random variation. That is, in the next generation of dialers, most will dial the number that survived but a few will dial a randomly different digit, or perhaps add one or a few additional random digits. With this sort of replicating dialer and our hypothesized phone environment acting as a natural selector, we have a simple Darwinian system that could explore the space of phone numbers, and it would soon turn out that most of dialers of the population were reaching one or another working individual number, improbable though reaching any specific one of them might have been. This is a rough illustration of the searching power of naturally selected, randomly varying replicators. This process, as a kind of genetic algorithm, can generate complex information and, assuming that phone numbers are indeed "specified" in the relevant sense, also generate CSI.

Dembski's and the other intelligent-design theorists' argument is essentially a variation of one made by astronomer Fred Hoyle in the early 1980s.

(Recall that Heeren had the Sultan cite Hoyle in his SETI story.) Hoyle calculated the probability of getting the exact sequence of amino acids for a specific small protein at random and pronounced it so unlikely as to be not worth considering. There was thus simply not enough time for chance to have produced such proteins on earth.[76] Geisler had cited Hoyle's calculation and a similar one made by Hubert Yockey[77] in his original article, mentioned previously.

Again, one problem with this argument as a criticism of Darwinism is that it makes the same mistake that Behe made when he tried to impress upon us how unlikely it was for a groundhog to cross the Schuylkill expressway. It calculates the probability of successfully crossing a given gap under the assumption that it must be done all at once. Indeed, the odds against a successful single step can be mind-bogglingly enormous. However, as we saw at the expressway, evolution does not take giant steps in ten-league boots but rather takes many small steps, with natural selection operating cumulatively each step of the way. Redo the calculations under the assumption of cumulative selection and the enormous improbabilities shrink away. Because his improbability argument completely misses the most basic point about Darwinian cumulative selection, it is sometimes referred to as "Hoyle's Howler." In *The Blind Watchmaker*, Richard Dawkins beautifully illustrates the power of cumulative selection with an example that considers the probability that a monkey banging at a keyboard would type out a line from Shakespeare at random.[78] The chance of our monkey hitting upon the line "METHINKS IT IS LIKE A WEASEL" from Hamlet is tiny if we require him to get all 28 characters right in a single step. But switch now to a Darwinian monkey who begins with a random string of 28 characters, produces multiple replications of this sequence with some chance of a copying error each time, and then repeats the process starting for the next set of copies with whichever of the copies is closest to the target sequence as the original. If he continues in this way, in a surprisingly small number of generations he hits the target. Philosopher of biology Elliott Sober explains the example using a combination lock;[79] my phone dialer is another variation.

Intelligent-design creationists are smart enough not to question the math here, but they do try to avoid the conclusion by changing the subject. Here is Behe's reply:

What is wrong with the Dawkins-Sober analogy? Only everything. It purports to be an analogy for natural selection, which requires a function to select. But what function is there in a lock combination that is wrong? . . .

Evolution, we are told by proponents of the theory, is not goal directed. But then, if we start from a random string of letters, why do we end up with METHINKSITISAWEASEL instead of MYDARLINGCLEMENTINE or MEBETARZANYOUBEJANE? Who is deciding which letters to freeze and why? Instead of an analogy for natural selection acting on random mutation, the Dawkins-Sober scenario is actually an example of the very opposite: an intelligent agent directing the construction of an irreducibly complex system.[80]

With a final swipe at Sober he concludes: "The fact that a distinguished philosopher overlooks simple logical problems that are easily seen by a chemist suggests that a sabbatical visit to a biochemistry laboratory might be in order."[81]

Johnson made the same move in criticizing a somewhat different point in *Evolution and the Myth of Creationism*,[82] where the author, Ohio State University zoologist Tim Berra, illustrates a point about the nature of an evolutionary sequence using a series of photographs that show the development of the Corvette over several decades. Because automobiles are designed by human beings, Johnson wants to disqualify the example, charging that "[t]he Corvette sequence . . . does not illustrate naturalistic evolution at all. It illustrates how intelligent designers will typically achieve their purposes by adding variations to a basic design plan."[83] Like Behe, Johnson adds a swiping flourish:

I described the credentials of Professor Berra and named the publisher so nobody could accuse me of attacking a "straw man." A distinguished university press would not publish such a book without obtaining professional reviews certifying that its scientific explanations were reliable. Evidently the reviewers saw nothing wrong with equating automotive engineering and biological evolution. I am not surprised, because evolutionary biologists typically do not understand that sequences resulting from variations on common design principles (as in the Corvette series) point to the existence of common design, not its absence. I have encountered this mistake so often in public debates that I have given it a nickname: "Berra's Blunder."[84]

Despite his protestations to the contrary, Johnson has indeed attacked a "straw man," a fallacy that, by the way, has nothing to do with whether or not the argument one is attacking is made by someone with "credentials." Rather, the straw-man fallacy involves setting up and knocking down a false caricature of someone's position. This is just what Johnson has done in his treatment of Berra's example, which occurs in the context of a discussion of transitional fossils leading up to modern human beings. Johnson regularly accuses biologists of playing fast and loose with the term "evolution," but it is Johnson who here is performing the terminological sleight-of-hand. Or perhaps he simply fails to understand Berra's point. The Corvette sequence is not meant to be an illustration of natural selection, so it is a fallacy to set it up and knock it down as such. Berra explicitly says that this is an illustration of a kind of descent with modification. He wants here to emphasize how small changes, where the relatedness of intermediate forms is easily recognizable, can add up to differences such that the endpoint is distinct from the starting point to the degree of being nearly unrecognizable. For this purpose, Berra's using artificial rather than natural selection works perfectly well. (Darwin himself used examples of artificial selection as the first step in his argument for descent with modification in the *Origin*, though his real biological cases demonstrated the point a bit better.) Furthermore, given that creationists continue to believe in the immutability of species, and to insist that the cumulative selection of small variations doesn't add up to large ones, it is an important, basic point to make with a familiar example. It is thus Johnson, not Berra, who has blundered.

Behe attacks a straw man as well, for Dawkins and Sober had *not* meant their examples as analogies for natural selection on random variation, but rather as illustrations of the power of *cumulative* selection on random variations, which is just the aspect of the Darwinian process that is relevant to the improbability argument. It is completely misguided of IDCs to accuse Dawkins and others of the supposed logical blunder of overlooking the teleological aspect of the "WEASEL" argument. Indeed, Dawkins himself was careful to warn his readers that the analogy is misleading in just this way in the very paragraph following his original presentation of it. He explains that in each generation of selective breeding, "the mutant 'progeny' phrases were judged according to the criterion of resemblance to a *distant ideal* target . . . Life isn't like that. Evolution has no long-term goal. . . . In real life, the criterion for selection is always short-term, either simply survival or, more generally, reproductive success."[85] The fact that Dawkins used an example with a distant functional target rather than one where an immediate selection criterion operated is irrelevant to making the pertinent point of how cumulative selection answers the improbability argument. To give an illustration that is one step closer to the biological process, Dawkins goes on to construct a more complex model on a computer using randomly variable line patterns which he terms "biomorphs" because selection can turn them into shapes that could bear an uncanny resemblance to biological forms. Having used this software in one of my classes, I can attest to the surprise one feels at the forms that can develop from the combination of chance mutation and selection even without any long-term goal in mind. Here too, of course, the model is not perfect, but Dawkins is again careful to point out the limitations of the biomorph program, noting that "it still uses artificial selection, not natural selection. The human eye does the selecting."[86] He then says exactly what would be required to get a realistic model of natural selection:

To simulate natural selection in an interesting way in the computer . . . we should concentrate . . . upon simulating nonrandom death. Biomorphs should interact, in the computer, with a simulation of a hostile environment. Something about their shape should determine whether or not they survive in that environment. Ideally, the hostile environment should include other evolving biomorphs: "predators," "prey," "parasites," "competitors." The particular shape of a prey biomorph should determine its vulnerability to being caught, for example, by particular shapes of predator biomorphs. Such criteria of vulnerability should not be built in by the programmer. They should *emerge*, in the same kind of way as the shapes themselves emerge. Evolution in the computer would then really take off, for the conditions would be met for a self-reinforcing "arms race."[87]

Dawkins wrote this in 1986. Less than a decade later, we saw the beginnings of just such a model in Tom Ray's Tierra program, which I described in chapter 2. And, as we saw, an arms race of the sort that Dawkins predicted is exactly what occurred, with novel and unexpected digital "organisms" emerging just as he had expected.

The intelligent-design theorists' responses to evolutionary models from computer science are always along the same lines: Sure, computers can create or exhibit intelligent action, but that is because intelligent human designers programmed them. Johnson, for example, says that citing the decision-making capability of computers "is another instance of Berra's Blunder. Those computers are intelligently designed. Unassisted matter never made a computer, nor did naturalistic evolution."[88] Once again, Johnson's criticism is wide of the mark, for the origin of the computer itself is irrelevant to examples of artificial life except in the trivial sense that you need the hardware to run the software. Any computer hardware could "hold" a digital "environment," and certainly hardware designers never planned computers with such a use in mind. Nor could he level his criticism against the software, for Tierra was intentionally designed only in the sense that allowed it to instantiate elements of the Darwinian process—a replicator, a source of random variation, an unsupervised selection function. The properties that evolved—parasitism, mutualism, and so on—had not been "designed in" but truly emerged naturally.

Johnson's misapplied complaint reveals one of the basic problems with his view. He claims that Berra's example actually is an illustration of the design inference. He says that "sequences resulting from variations on common design principles (as in the Corvette series) point to the existence of common design" and that it "illustrates how intelligent designers will typically achieve their purposes by adding variations to a basic design plan."[89] But if this Corvette analogy is correct as an instance of the design inference, then it looks as though we should infer that human beings are simply "this year's model" in a long line of primates that the "designer" has put out over the years. Actually, this is essentially what Raëlians want us to believe; they claim that other primates and fossil hominids served as "prototypes" for human beings as the Elohim genetic engineers worked through their design process. Well, maybe so, but can we legitimately infer that there was a Designer with such intentions for the biological world from an analogy with automobile design of the good folks at General Motors? Raëlians might claim to see the work of a designer in their observations of the world, but isn't a better explanation, rather, that they are simply reading their own design wishes into the world in the same way that Paley did in seeing design in the properties of water? More could be said about these problems, and we will consider them further in the next chapter, but let us now turn to the final version of the intelligent-design theorists' attempts to revitalize the argument from design.

Mike Behe and His Irreducibly Complex Mousetrap

Darwin was quite forthright in identifying possible observations that, if shown to be true, would falsify his theory. In addition to admitting those mentioned earlier, he also noted that:

If it could be demonstrated that any complex organ existed, which could not possibly have been formed by numerous, successive, slight modifications, my theory would absolutely break down.[90]

Of course, identifying such an organ would not necessarily undermine the common descent thesis, and other evolutionary hypotheses would remain unaffected as well, but it would be a devastating counterexample to the gradualist mechanism of natural selection. Intelligent-design theorist Michael Behe has said that he has no reason to doubt the truth of common descent, but he does doubt the power of natural selection to shape the full range of biological complexities. In *Darwin's Black Box* he claims to have found a number of such "complex organs" to prove his case. This is clearly an important claim. Fellow IDC Tom Woodward, a professor of religious studies at Trinity College of Florida, eschewing any pretense of subtlety, described Behe's work as being of earth-shaking magnitude and Behe himself in terms that would befit the returning Messiah himself. Reporters, he said, were making their "pilgrimage to Bethlehem, Pennsylvania" to interview the "master-teacher," and when Behe gave a talk in Florida, an audience of biologists, students and schoolteachers "braved rains from an approaching hurricane to hear him."[91]

So what does Behe have to say? We already have a fairly clear idea given our earlier discussions of critical passages from *Darwin's Black Box*— Behe hopes to show the impotence of Darwinism by pointing out purportedly profound explanatory gaps. Trying to do this is nothing new. ICR's Duane Gish has tried to do this by pointing out gaps in the fossil record. (As George Gaylord Simpson wryly observed about this sort of challenge, the argument from absence of transitional types boils down to the striking fact that such types are always lacking unless they have been found.) Indeed, as we saw, almost every creationist attack proceeds in the same way, by citing something that Darwinism supposedly cannot explain. Behe takes his list of unexplained facts from biochemistry. He says, correctly, that for the Darwinian theory of evolution by natural selection to be true, it has to be able to account for the functional molecular structure of life, and he finds no such explanations in the literature for the systems he cites. Behe is also correct that the systems he mentions are incredibly complex, though whether they are more or less complex than the physical structure of the eye is hard to say. But we have already seen that complexity by itself is not necessarily a problem for Darwinism. What is new here? Behe claims that his molecular systems exhibit one property that the Darwinian mechanisms cannot explain, namely, that they are what he defines as *irreducibly complex*:

By irreducibly complex I mean a single system composed of several well-matched, interacting parts that contribute to the basic function, wherein the removal of any one of the parts causes the system to effectively stop functioning.[92]

Behe describes the complex molecular machinery involved in the visual cascade, blood clotting, vesicular transport, and the movement of flagella as examples of irreducible complexity. He says that there is a suspicious silence in the literature with regard to the evolutionary development of these systems and that the reason is that biologists do not have the foggiest idea how evolutionary mechanisms could have produced them. He tells us that they are overlooking the obvious explanation:

There is an elephant in the roomful of scientists who are trying to explain the development of life. The elephant is labeled "intelligent design." To a person who does not feel obliged to restrict his search to unintelligent causes, the straightforward conclusion is that many biochemical systems were designed. They were designed not by the laws of nature, not by chance and necessity; rather they were *planned*. The designer knew what the systems would look like when they were complete, then took steps to bring the systems about. Life on earth at its most fundamental level, in its most critical components, is the product of intelligent activity.[93]

At this point, Raëlians are no doubt breaking in with hearty cheers of support. But is the cheering warranted? We will return shortly to consider this elephant that Behe says scientists are ignoring, but let us first look a bit more closely at his notion of irreducible complexity.

Behe calls upon Paley as the inspiration for his idea of irreducible complexity but he recognizes the danger of doing so. After all, Paley's example of the interactions of water with the other elements seems to count as an irreducibly complex system, and yet we can explain those causal interactions by means of physical laws without postulating a designer who set it up specifically so that water could fulfill its supposed "offices." Knowing full well the absurdities that his predecessor was led to by his enthusiastic eye for design, Behe protects his flank by taking a few potshots of his own by pointing out some of Paley's many silly examples—"The short unbending neck of the *elephant* is compensated by the length and flexibility of his proboscis. . . ."—and admitting that these can provide "a rich source of comedy material."[94] Nevertheless, Behe argues, if we don't let Darwinists sidetrack us with Paley's bloopers, we will see that he got the argument from irreducible complexity exactly right in his central analogy of the watch. The reason an irreducibly complex system is supposed to cut the wheels from under Darwinism is that such a system "cannot be produced directly (that is, by continuously improving the initial function, which continues to work by the same mechanism) by slight, successive modifications of a precursor system, because any precursor to an irreducibly complex system that is missing a part is by definition nonfunctional."[95]

Philosophers always prick up their ears when someone says something is true "by definition," so be sure to keep this explanation in mind as well as the earlier quoted definition of irreducible complexity as we consider Behe's argument. For the argument to work, Behe needs to be able to show that there are indeed irreducibly complex systems and that they are unreachable by Darwinian means. Let us take these claims in turn.

Behe tells us that "[t]he first step in determining irreducible complexity is to specify both the function of the system and all system components," and that the second step is to "ask if all the components are required for the function."[96] However, Behe has not done this for any of his examples. In part, this is simply because not enough is known about them. For example, he writes: "The bacterial flagellum, in addition to the proteins already discussed, requires about forty other proteins for function. Again, the exact roles of most of the proteins are not known. . . ." Continuing on, he writes that "[i]t is very likely that many of the parts we have not considered here are required for any cilium to function in a cell."[97] However, the very complexity of these systems makes it extremely difficult to tell whether they are indeed irreducibly complex, and the data Behe gives are insufficient to support his claim for in no case have all the components of these systems been identified and studied. Conclusions about the functional properties of biological systems are only preliminary, for in a complex interacting causal system even if we know that knocking out any single factor results in a loss of a given function we cannot thereby say that knocking out two, for instance, might not restore it by some other means. Behe tells us that he cannot imagine any way around the apparent irreducibility in his examples, but even just upon reviewing his book other biologists have been able to suggest some specific biochemical pathways that might work. Only future research will tell. Thus, it is simply premature to say that these systems are irreducibly complex.

But perhaps this response is unfair to Behe. Even if the examples he cites turn out upon further investigation not to be irreducibly complex as he defines the notion, at this point scientists don't know the answer, and it at least seems possible from a conceptual standpoint to suppose that such systems could exist, if not these then perhaps others. But there is a more serious problem with the argument, which we can see if we look at Behe's central analogy. Behe illustrates the notion of irreducible complexity with the idea of a household mousetrap. He can specify its function—crushing mice—and its components: the ordinary mousetrap has a platform, spring, hammer, holding bar, and catch. He can also argue persuasively that each of those components is needed for the trap to perform its function; remove the hammer and a mouse could dance all night on the platform. It would be similarly ineffective were any other part missing. Thus it looks as though the trap is indeed irreducibly complex and that no precursor system missing a piece could fulfill its specified function.[98] With this simple example Behe aims to crush any Darwinian mouse who attempts to take the bait. Indeed, the subtitle of Woodward's laudatory article is "How Biochemist Michael Behe Uses a Mousetrap to Challenge Evolutionary Theory."

But I'm afraid that Behe is just a bit too swift; his argument is just the same bait-and-switch tactic that IDCs accuse others of using. Look back now to the critical passage where he explains that irreducible complexity can't be reached by a Darwinian process because any precursor which is missing a part is "by definition nonfunctional." He baits us in that passage with a conceptual argument: were the mousetrap to lose any of its parts it would not function as a mousetrap, which is indeed true by definition. But he then switches to the *empirical* conclusion, that any irreducibly complex system that lost a part would be nonfunctional. Behe needs this stronger premise to rule out the Darwinian mechanism and to conclude that there was no way to have gotten to it except "in one fell swoop." However, even if a system is irreducibly complex with respect to one defined basic function, this in no way implies that nearby variations might not serve other nearby functions. Behe claims that there could never be any functional intermediates that natural selection could have selected for on the way to *any* irreducibly complex system, but he can't get the empirical conclusion from his "by definition" conceptual argument. The strong empirical premise he needs is false.

Consider Paley's pocket-watch. It is irreducibly complex with respect to a highly specific function, say, to hang at the end of a chain in a gentleman's pocket without being damaged by his daily routine so that he can quickly check the passage of the hours and minutes without having to open a lid. Our gentleman might find the watch unacceptable, should it be damaged in a manner such that even one aspect of his specific set of needs is compromised, and toss it aside on the heath, whereas someone with different needs who picked it up might find it perfectly functional.

One notices that Behe tries to head off this sort of objection by specifying the function he wants by definition. He argues, for example, that the clear face of the watch was "not necessary for the function of the watch" but just an added "convenience" that is not part of the irreducibly complex system itself. Nor, he says, is it relevant that the wheels are made of brass to prevent rust, because "the exact material, brass, is not *required* for the watch to function," even though it might help. Paley, he says, gets himself into trouble "when [he] digresses from systems of necessarily interacting components to talk about arrangements that simply fit his idea of the way things ought to be."[99] But functionality, even of interacting components, must be judged relative to some standard, and when Behe dismisses possible standards as "peripheral" he is simply making the system fit *his* idea of the way things ought to be. A rust-proof material is indeed required for the gears, if the function is to have a watch that won't rust. Even a clear lid could be absolutely necessary for someone's purpose. Nor does narrowing the requirements to *interacting* components help, even assuming Behe could somehow define "interaction" in such a way that keeps out the causally connected parts that he wants to exclude above. Let me give a slightly different example to show why.

The case I have in mind involves the extraordinarily difficult scientific problem of how to determine one's longitude when at sea and out of sight of land. Checking one's latitude was fairly easy, but not being able to measure longitude caused such serious losses of human life and commercial goods as sailors became lost at sea that in the early eighteenth century the

British Parliament offered the incentive of a large monetary reward to anyone who could devise a workable method. Isaac Newton suggested that the longitude problem could be solved with a timepiece that could maintain accuracy on board ship. However, he noted that severe design constraints made this an almost impossible goal:

[B]y reason of the motion of the Ship, the Variation of Heat and Cold, Wet and Dry, and the Difference of Gravity in different Latitudes, such a watch hath not yet been made.[100]

A watch built to fulfill this specified function would be an irreducibly complex system in Behe's sense. If it could not maintain accuracy to within 3 seconds every 24 hours in this demanding environment it would fail in its basic task (as defined by Parliament). Moreover, failure to meet this minimal function could result in total destruction, for being off course by more than half a degree of longitude caused the wreck of many a ship. A watch that lost but one of its necessary interacting parts would indeed by definition be nonfunctional *as a solution to the longitude problem.* But it is not true that it would not be functional at all. For example, the chronometer that John Harrison built had two dumbbell-shaped bar balances and four helical balance springs, each of which helped to compensate for the motions of the ship. If one or another of these were to have broken, the clock would no longer have fulfilled its specific function aboard ship, but it could still have performed a slightly different function in a different environment, such as on a calmer lake or on solid land. Thus there are functional physical precursors that are accessible to Darwinian selection even though they are not functional with regard to the conceptually specified original function. Moreover, if Paley's imagined self-replicating watches existed and operated in a Darwinian fashion whereby random variations occurred in the replication process, then given appropriate selective pressure, these could potentially evolve to solve the longitude problem. If a watch "species" reached that point, it might well be or become irreducibly complex for that function. I could make the same argument pointing to the various intermediate forms of the eye, or other irreducibly complex organs, but that should not be necessary. Indeed, we see now that the "irreducible complexity" of Behe's molecular machines, if that is what they are, makes no argumentative advance over the original argument with macroscopic organs.

Moreover, even if we grant Behe a fuzzy notion of function, there remains another way that a Darwinian mechanism could gradually produce an irreducibly complex system for "the same" function. Biologist Allen Orr explained the process in a review of Behe's argument:

An irreducibly complex system can be built gradually by adding parts that, while initially just advantageous, become—because of later changes—essential. The logic is very simple. Some part (A) initially does some job (and not very well, perhaps). Another part (B) later gets added because it helps A. This new part isn't essential, it merely improves things. But later on, A (or something else) may change in such a way that B now becomes indispensable. This process continues as further parts get folded into the system. And at the end of the day, many parts may all be required.[101]

Nor is this a new scenario that Orr concocted to deal with Behe. Orr notes that this kind of evolutionary possibility had been suggested back in 1918 by the geneticist and Nobel laureate H. J. Muller, and that it was worked out in some detail in 1939.

One further point: Behe had claimed that the problem for Darwinism increases as the number of parts of an irreducibly complex system increases. But it should by now be clear that this is not necessarily true. The reason his mousetrap seemed to be persuasive for his conclusion that irreducible complexity could not have arisen by small steps is that it is not a very complex system. This is what initially made it difficult to see how any changes to its design wouldn't result in loss of its mouse-crushing function, and in a loss of any related function as well. It is the simplicity of that system that gives us the well-known truism about trying to build a better mousetrap. However, once we have been made aware of the unwarranted conceptual constraints that Behe builds in to the example, it becomes easier to think of alternative arrangements of the pieces or slight variations which eliminated or replaced one or another component that would work, perhaps not to catch mice but still to rid one's house of them. My students have had fun thinking up hypothetical functional variations on Behe's irreducibly complex mousetrap. I leave this as an amusing exercise for the reader. Behe had hoped to crush the Darwinian mouse in his trap—but his contraption is not nearly swift enough, and the mouse easily escapes. Let us now return to reconsider Behe's elephant. By now we should have a pretty clear picture of what will happen when the elephant sees the mouse.

Looking at Behe's own assessment of the import of his theory we see why Woodward called up such powerful images to describe him. Behe writes:

The result of these cumulative efforts to investigate the cell . . . is a loud, clear, piercing cry of *"design!"* The result is so unambiguous and so significant that it must be ranked as one of the greatest achievements in the history of science. . . . Why does the scientific community not greedily embrace its startling discovery? Why is the observation of design handled with intellectual gloves? The dilemma is that while one side of the elephant is labeled intelligent design, the other side might be labeled God.[102]

Hyperbolic congratulatory statements aside, however, the grand conclusion that Behe wants to draw does not follow from the data he provides. Behe will no doubt complain that I have not addressed the biochemical details of his real examples, but as we have noted, the evidence is not yet in on those questions. The most that Behe has done here is to point to a number of interesting research problems. One wonders why he, as a biochemist, does not begin the research himself. He is correct that these remain explanatory gaps for science, but he has failed to demonstrate the single point upon which his whole case rests, namely, that irreducibly complex systems, assuming that this is what these are, could not in principle have arisen by Darwinian (or by any other natural) means. It certainly might take the mouse a long time to figure out these molecular mazes, but there is no reason in principle that it could not do so. Of course, Behe has no motivation to pursue the research himself, since he thinks he already knows the answer—biomolecular complexity was produced by the intentional action of an intelligent designer. With intelligent-design theory, Behe has found a way to save himself a lot of hard work.

The elephant of intelligent design is not unlike Paley's elephant. As do other creationist claims, it maintains a stiff-necked rigidity in its opposition to Darwinism, and it is able to compensate for any possible countervailing evidence by virtue of its vague and infinitely flexible hypothesis. If that were not enough, Behe makes explicit what we knew all along, that intelligent-design theory desires much more than to be a "better scientific theory" than Darwinian evolution: it hopes to represent God Himself. With God on their side, IDCs can explain anything, fly right out of mazes, and cross the profoundest of gaps. Behe's pachyderm of intelligent design cannot but remind me of Dumbo, Disney's little circus elephant, who believed he was able to defy gravity not so much because of his big ears but because of the "magic feather" he held in his trunk. One doubts that God would care to be used in this manner. And as the song says, I think I will have seen everything, when I see an elephant fly.

A Day at the Chariot Races

To review and conclude our day at the races, let us now recall Johnson's core claim, echoed by all the other intelligent-design theorists, that the only reason Darwinian evolution is accepted is that the alternative theory of intelligent design is never considered since scientific naturalism bars all supernatural possibilities from consideration. An "open-minded" science that allows theistic interventions would, they say, immediately hear the loud cry of "Design!" We are now in a better position to evaluate this argument.

Science has indeed investigated intelligent design, but so far our only examples of "intelligent designers" are human beings and perhaps some other animals. To speak of talking pulsars and love notes in the sand as though these counted as evidence of a generic intelligence is misleading. In both of these cases, we recognize a message because they are given in what we recognize as a language and, indeed, it is a language we already understand. Of course, we would say that these would be good evidence of an intelligent signaler like ourselves since the observed pattern (e.g., recognizable sentences in Morse code or Roman characters) matches the pattern of our own signals. The issue is whether the information in DNA is relevantly comparable to this sort of situation—and the answer is that it is not. Yes, DNA contains complex information, but not of any sort that is specifically unique to conscious intelligence.

When we look carefully, we see that what IDCs propose as a positive argument for their view is still nothing more than another negative argument, another example of the God of the gaps. Their *Pandas* textbook is structured on the same dual-model argument that creation-science had used. In the introductory chapter, the authors explain their intention for the text:

From . . . six areas of science, we will present interpretations of the data proposed by those today who hold the two alternative concepts: those with a Darwinian frame of reference, as well as those who adhere to intelligent design.[103]

But in speaking of "the two" alternatives, they set up a false dichotomy in the same way creation-scientists did. *Pandas* is also misleading in suggesting that intelligent design is a theory in its own right on a par with evolutionary theory. One must comb through the text very carefully to find anything beyond negative arguments against evolution. Even positive statements of the "theory" of intelligent design appear in only a few paragraphs in the book and amount to no more than saying of one or another biological fact that the designer designed it that way. Of course, this comes as no surprise, given that there has never been any more to the theory than this. A search of the primary literature, using five major scientific and academic indexing databases covering over 6,000 journals, failed to turn up the implied research of "those who adhere to intelligent design." George Gilchrist, the University of Washington research scientist who conducted the literature search, reported that:

Although Davis and Kenyon may claim that intelligent design represents a viable alternative to neo-Darwinian evolution, the scientific literature does not support that claim. Compared with several thousand papers on evolution, the combined searches produced only 37 citations containing the keyword "intelligent design."[104]

Most of these thirty-seven were irrelevant and dealt with computer software or hardware, architectural or engineering design, advertising art, literature, fertilizer manufacture, or welding technology. Only seven had anything to do with biology, and of these, five were discussions of the debate over using *Pandas* by various school boards and two were comments on Behe's book in a Christian magazine. Gilchrist writes: "This search of several hundred thousand scientific reports published over several years failed to discover a single instance of biological research using intelligent-design theory to explain life's diversity"; he also points out that "although Davis and Kenyon are both professional scientists, neither has apparently published anything in the professional literature about their theory."[105] In *Darwin's Black Box,* Michael Behe devotes an entire chapter, "Publish or Perish," to showing that scientists have not published answers to his questions about the evolution of complex molecular systems in any professional journals, and he insists that "[i]f a theory claims to be able to explain some phenomenon but does not generate even an attempt at an explanation, then it should be banished."[106]

Actually, I think Behe's criterion for banishment is a bit harsh, since a perfectly reasonable scientific question might not be answered in the scientific literature only because the tools that would be needed to answer it have yet to be developed, or because there is still insufficient evidence available, or simply because no one has yet gotten around to addressing that particular question. I do believe that intelligent-design theory should be discarded, however, though not simply because neither Behe nor any other intelligent-design theorist has published on it in the professional journals, but rather because the "theory" has no good explanatory resources to offer. Intelligent-design creationism is a one-trick pony (or elephant, if they prefer). Although Behe derides Darwinists for not having given specific explanations for the visual cascade or the bacterial flagellum, his own "explanation" of each complexity he describes is the same—an (unidentified) Intelligence designed it that way. Such an explanation is vacuous. Intelligent-design theorists have no way to identify any specific intended function for biological organisms, and without this the design inference can't get off the ground. Since they have no way to check the postulated Designer's purposes, their hypothesis is compatible with *any* possible state of affairs. In the final analysis, intelligent-design theory has even less of a claim to being a science than does creation-science, for at least young-earth creationists have the courage to take a stand on a few explicit empirical claims, absurd though they may be in light of the evidence. IDCs, on the other hand, can reveal nothing positive about their intentionally nebulous Designer without undermining the key element of the strategy they have adopted.

Raëlians say that we can look to the skies to see the UFOs in which our designers will one day again return. But even those IDCs who are forthright that the Designer they have in mind is God will never predict that He might once again make a spectacular appearance and swing in on a shining chariot like the one described by Ezekiel. According to Heeren, remember, God would decline to appear "in person" because to do so would be to compel belief, and adoration that is not freely chosen is worthless. Indeed, for many Christian believers, one's true faith is only proven when it survives in the face of events that would naturally cause one to doubt God's presence. But this just further illustrates the difference between the religious and the scientific attitude. To hold on to belief come what may is a sign of religious virtue. Contrarily, science takes it to be a virtue that one withholds belief in the truth of a proposition until it is supported by the weight of evidence.

The philosopher Bertrand Russell, an admitted nontheist, was once asked what he would have to say if after death he were ushered into the presence of the Almighty, and God were to ask why he had not been a believer. Russell said he would reply, "Not enough evidence!" Responding in this way, Russell was thinking like a scientist. A theologian might well judge that he simply didn't get the point. Creationists, however, cannot make this argument because they want to portray themselves as scientists and keep the religious nature of their beliefs hidden. Intelligent-design theorists must therefore argue that Russell and scientists like him are blind to the evidence because their naturalist prejudices preclude their even considering the supernatural possibility. However, our comparison of IDC to ET-IDC shows that this is a red herring, considering the conclusion they want to draw. Suppose that upon death, rather than being confronted by God in heaven, Russell found himself facing Raël and the extraterrestrials on the Elohim home planet. The Raëlian Creation hypothesis is entirely naturalistic, positing no preternatural entities or miracles that break the order of natural law. Yet Russell would be right to answer in exactly the same way. Not enough evidence! It is not that creation of the biological world by extraterrestrial intelligences could not possibly be true; it is just that there is no good evidence for it. Like the new intelligent-design creationism, the Raëlian Movement's ET-ID creationism is a "scientific theory" in name only. To believe such a view of creation in the absence of good evidence is a matter not of science, but of faith. At some point, bad science is the same as pseudoscience, and continuing to believe in it is to make it a religion. Whether the supposed "intelligent designer" is alien or divine, the conclusion is the same.

Were the new creationists to hold their views explicitly as matters of faith, though we might still consider such religious views false, there would be no worry of their importing their views into science classrooms. But as we have seen, intelligent-design theory is disingenuous or at least extremely misguided in its claim of religious independence. In this regard it is nothing more than the old creationism now dressed up in designer clothes. Behe insists, without breaking his poker face, that the "intelligent designer" he credits with the complexity of biomolecular systems is neutral with regard to whether the designer is supernatural or natural. This is exactly the same argument made by the creation-science defense in the Arkansas trial, though there the term was "creator":

In this context, the word "creator" and the idea of creation are neither religious nor supernatural; creation science presupposes the existence of a creator no more and no less than evolutionary science presupposes that there is *no* creator. Each is, therefore, equally scientific and equally nonreligious, and the relation to fundamentalism is merely coincidental.[107]

No one is taken in by such disguises. Talking amongst themselves on the Internet lists or in person, creationists dispense with the code words or use them with a broad wink. Introducing intelligent-design creationist Walter Bradley at a special breakfast talk where he witnessed his faith to members of a campus Christian group, the moderator described intelligent-design with a wink as the "politically correct way to refer to God." IDCs are presenting a fake ID to have their day at the races. But isn't it rather unseemly to outfit the Almighty in Groucho glasses with a fake mustache and big plastic nose?

When entering the scientific arena, intelligent-design theorists keep their racing stripes covered and in this way hope to sneak in their supernatural chariot. To date, *Of Pandas and People* has the slickest camouflage, but we should expect further refinements. But we have seen that this intentional ambiguity will not do. So why not just openly admit in the next textbook that it is a supernatural entity whom the authors believe to be the Designer? Why not just press forward honestly with the project of theistic science and show how its methodology is superior to that of ordinary scientific naturalism? The simple answer, of course, is the political one. If intelligent-design theorists were to wear their religious colors openly, they could not hope to gain a foothold in the public school classrooms. But there is a deeper answer that is more philosophically interesting, and that is that a theistic science is neither viable as science nor as a generic theology. We have already seen, in the previous chapter, some of the reasons that science hews to methodological naturalism. Let us now return to that issue and examine why it makes good sense, both scientifically and theologically, to exclude supernatural chariots from the stadium of science.

6

Deus ex Machina

We can define a new term, *creascience*, that allows for the recognition of disconti-
nuities in nature that indicate the intentional, immediate intervention of a first
cause that resembles a person.
—J. P. Moreland

Evolution went underground in the Middle Ages in the witchcraft movement.
—Henry Morris

Time for a Paradigm Shift?

Intelligent-design theorists see themselves as revolutionaries. As I noted
in chapter 4, they call upon Thomas Kuhn and suggest that it is time to
shift the scientific paradigm. However, it is not just evolution that has to
go, in their view. They want to transform science at its most fundamental
methodological level. They want science to reject the methodological
principle of naturalism and to reincorporate appeals to supernatural
entities and powers. Even Alvin Plantinga endorses this notion, recom-
mending a shift to a "theistic science" that allows the incorporation
of theistic hypotheses. Making supernatural interventions a part of scientific theorizing would indeed be revolutionary, but would it make
sense? In this chapter we'll look in detail at the creationist's proposal to
scrap methodological naturalism and to replace it with a supernatural
science.

An odd paradox in human desire drives us to a fascination with the
supernatural. On the one hand, we hunger for understanding. We want to
know how things work, what makes people tick, the real story. Why are

things the way they are? Our natural curiosity demands satisfaction. On the other hand, we are attracted to the mysterious. We want to believe that there are higher planes of reality, that there are forces not commonly seen working toward purposes we cannot understand, that the world is more complex and wonderful than we know or perhaps even can know. Our love of the mysterious is always tempting us towards cover-ups and conspiracy theories that can never be checked. Understanding and mystery: the supernatural can seem to fulfill our desire for both these things. Much of the world does elude our understanding, and many phenomena appear inexplicable. In such cases, appeal to supernatural entities and powers appears to provide a sort of explanation when we can find no other way to understand the event. The astrologer explains the serendipitous opportunities that open and the unpredictable obstacles that block the pathways of our lives by the influence of the planets as they were positioned within the ring of the zodiac relative to the time and place of our birth. The burners of witches thought there was no way to account for a sudden sickening of farm animals or a mysterious souring of milk other than by the malevolent use of demonic powers. The creationist finds an explanation of the origin of the clockwork of the world in general and of human beings in particular through belief in miraculous interventions by an omnipotent divine watchmaker.

However, we should look more closely at supernatural explanations, especially when they are offered as alternatives to naturalistic, scientific explanations. What does it mean to say that some phenomenon has been explained? Are supernatural "explanations" explanatory at all, and if so, are they good explanations? Ought we consider them to be on a par with scientific explanations? These questions are especially important given the essential way that the explanatory virtues of a hypothesis or theory are involved in assessment of its confirmation or disconfirmation. Creationists want to establish the legitimacy of a "theistic science" and to claim that their creation hypothesis of supernatural intervention is the best explanation for the origin of species. Are they right about supernatural explanations? What are the prospects for a theistic science? These are the questions we will consider in this chapter.

A Brief History of Supernatural Explanation

The Ghost of Explanations Past

Perhaps it is a Nietzschian "will to power" that underlies our paradoxical desires for understanding and for the mysterious, and which leads us to belief in the possibility of supernatural explanations. We desire understanding in part because of the control that it could give us. Knowledge is power, as Bacon said, and with it we feel more in charge of our own fates and sometimes the fates of others as well. But what of the mysterious, which by nature seems to be the antithesis of power? For many, the mysterious bears the promise of special, hidden powers. In part this comes from an idea that mysterious powers are greater than ordinary ones; perhaps more important is the seductive notion that if we were but to uncover its source we would thereby possess a unique control that others lacked. Read the stars and foresee the future. Contact the spirit world and be guided to a higher wisdom. Pray with piety and fervor and God will grant eternal life after death, and perhaps special favors before death as well. Even demonic forces, the dark side of the mysterious, can be harnessed to one's special benefit, some believe, by the casting of spells or making an unholy pact with the devil. We find this seductive hope for special, mysterious power exhibited in a variety of ways whenever there are appeals to the supernatural for explanation.

One of the earliest forms of supernatural explanation was animistic religion, which populates the world with gods. According to Japanese Shinto the *kami* reside in each tree, spring, and mountain, and so to ensure a good catch of fish the village holds a festival in honor of the *kami* of the fishing grounds. On the island of Hawaii, Pele is the goddess of the volcanoes; bits of *pahoihoi* or *a-a* lava may sometimes form delicate fibers or tiny smooth droplets, and the native Hawaiians claim that these are Pele's hair and tears. When Kilauea erupted, the explanation was that Pele was angry and must be appeased with sacrifices before she would stop the lava's destructive flow. In other religious forms, the gods have a more independent existence and sometimes more fully developed personalities. The Homeric epics reveal that, for the ancient Greeks, the world was populated by a panoply of competing gods and goddesses who regularly, sometimes kindly and sometimes cruelly, would intervene in the world and in human affairs. Homer explains how the fates of battling armies on the ground were often decided by the favors or jealousies of the Olympian gods above, watching and exercising control for their own purposes.

The switch to monotheism saw no change in this sort of use of supernatural explanation. Yahweh was regularly moved to anger, even toward his chosen people, and in his wrath he would bring forth destruction and pestilence. In the mid-fourteenth century, Christians in much of Europe tried to make wholesale atonement for their sins, which they thought must have led Yahweh to set upon them the plague of Black Death. Prominent Evangelicals offered a similar explanation for a more recent plague in the 1980s when it appeared that the AIDS epidemic was primarily attacking gay men: homosexuals were "reaping the whirlwind" of God's anger for disobeying His supposed commands against homosexual behavior.[1] God's displeasure with gay men was the supposed explanation for the occurrence of this baffling new disease and the horrible death it caused, and the implicit message was that good Christians would be able to avoid that fate.

Of course, it is not only to deities that people have appealed to try to make comprehensible the baffling phenomena of the world, and to possibly bring it under their control. There is a wondrous company of preternatural beings that have figured in supernatural explanations—ghosts and poltergeists, angels and demons, spirit guides and familiars are just some of the more common in the Western tradition. One also finds a similar array of associated occult powers that supposedly can be tapped using prayers and spells, blessings and curses, talismans and potions.

Also, though religions have probably included supernatural explanation most systematically, they are not alone in this predilection. Until just the nineteenth century, even the natural philosophers and scientists who studied the world would often include supernatural elements in their theories. Sir Isaac Newton, for instance, spent as much of his energies delving into alchemy and studying ancient wisdoms as he did investigating mathematics and physics, for which he is now famous. He worked to show, contrary to the view of the Cambridge Platonists, that atomism was not only not atheistic but Christian, for it could be traced back to Mochas, who was identified with Moses. Newton held that atoms and the laws they obeyed were fixed at Creation by God. "All these things being consider'd, it seems probable to me, that God in the Beginning form'd Matter in solid, massy, hard, impenetrable, moveable Particles, of such sizes and Figures, and with such other Properties, and in such Proportion to Space, as most conduced to the End for which he form'd them. . . ."[2] Newton also included forces in his ontology, holding that they too were created by God and superimposed upon matter to serve as causes. As I pointed out earlier, however, in his rules of reasoning Newton endorsed some of the standard elements of methodological naturalism—not to admit unnecessary causes when explaining phenomena, and to regard the conclusions of inductive methods as "accurately or very nearly true"[3] and to eschew contrary hypotheses until new evidence requires them.

Among nineteenth-century British geologists, many of whom were also clergymen, it was mostly taken for granted that the Genesis account of the Noachian Flood was true. Members of the Geological Society spent much of their time searching for evidence of the global flood, and they used the deluge hypothesis to try to explain diluvial gravel deposits, river valleys, and other large-scale geological features. It was only in mid-century that, following a protracted debate between the "uniformitarian" Charles Lyell and the "catastrophist" Rev. Adam Sedgwick that the latter finally admitted that the evidence did not support and indeed went against the biblical supernatural account. In his final address as President of the Geological Society, Sedgwick publicly renounced the supernatural view:

Having been myself a believer, and, to the best of my power, a propagator of what I now regard as a philosophic heresy, and having more than once been quoted for opinions I do not now maintain, I think it right, as one of my last acts before I quit this Chair, thus publicly to read my recantation.[4]

This debate helped spell the end of appeals to supernatural agents in scientific theorizing. Darwin's and Wallace's natural theory of the origin of species followed close behind and showed that no appeal to supernatural powers was necessary to explain how species evolved one from another.

Before Darwin's proposal of a simple mechanism for biological evolution, British naturalists were almost exclusively creationists. They had profound disagreements among themselves about, for example, whether God's organic plan was based upon perfect adaptation of every organism to its environment or upon creating organisms according to ideal archetypes that in many cases were not properly adapted, but they agreed that the biological world was specially created through God's supernatural agency. Nevertheless, Darwin's natural theory was quickly accepted for the origin of most animal species, except for *Homo sapiens*. Even Wallace had difficulty accepting a natural theory of the origin of human beings, and he ultimately fell back on a supernatural account. He allowed that human beings had evolved, but under the direction of "spirit beings," and he spent considerable effort investigating mediums and other spiritualists who claimed to have access to beings in the supernatural realm.[5]

Since then, science has completely abandoned appeal to the supernatural. In large part this is simply the result of the consistent failure of a wide array of specific "supernatural theories" in competition with specific natural alternatives. But there is also a deeper and more generally compelling reason for the abandonment of supernaturalism by science. Though in his early work Johnson failed to distinguish methodological from metaphysical naturalism, in his more recent pronouncements he argues that naturalism is dogmatic even as a method and, unfairly, speaks of it as "methodological atheism." However, science adopts naturalism as a principle of inquiry not because of any hidden atheistic agenda or even any special antagonism to theism. Rather, as we shall see, science eschews theistic explanations for the same reasons that it eschews other supernatural explanations.

"Godless Evolution" and "Methodological Atheism"

It is misleading for creationists to characterize science in general and to define evolution in particular as "godless." Science is godless in the same way that plumbing is godless. Evolutionary biology is no more or less based on a "dogmatic philosophy" of naturalism than are medicine and farming. Why should Johnson and his allies find methodological naturalism so pernicious and threatening in the one context and not the others? Must we really be seriously "open-minded" about supernatural explanations generally? As Bertrand Russell said, it is good to keep an open mind, but not so open that our brains fall out! Surely it is unreasonable to complain of a "priesthood" of plumbers who only consider naturalistic explanations of stopped drains and do not consider the "alternative hypothesis" that the origin of the backed-up toilet was the design of an intervening malicious spirit. Would it not be bizarre to reintroduce theistic explanations in the agricultural sciences and have agronomists tell farmers that their crop failure is simply part of God's curse upon the land because of Adam's disobedience, or suggest that they consider the possibility that the Lord is punishing them for some moral offense and that it might not be fertilizer they need but contrition and repentance?

Johnson and all other creationists insist that such interventions are indeed true. Even though they might be right, we should acknowledge that such spiritual possibilities fall under the purview of the priest and not the scientist. Given the nature and limitations of scientific modes of investigation, the proper role of the scientist is to search for natural causes of such occurrences and not to beg off the investigation by attributing them to supernatural interventions, divine or otherwise. Clearly scientists are not being dogmatic or atheistic in proceeding under the methodological heuristic that such events have natural explanations.

To take one important case, consider the medical sciences. It was once commonplace to attribute the origin of certain illnesses to curses or demonic possession. Indeed, Jesus is said to have performed some miraculous cures by expelling devils from the body of the diseased. If we accept the intelligent-design creationists' diagnosis, medical schools and research physicians are doing a terrible disservice by not teaching students how to perform exorcisms and by not taking seriously the possible supernatural origins of diseases.

Some of the more sophisticated new creationists recognize that evolutionary theory is not "godless" in any dogmatic, ontological sense, but they remain critical of science's naturalistic methodological stance and try to portray the methodology as being essentially equivalent to atheism. For example, they disparagingly characterize science's naturalistic methodology as "methodological atheism." This sort of rhetoric has the effect of making it seem that science has a particular antipathy to theism. Of course it is true that methodological naturalism eschews appeals to theistic interventions, but not because of some special distaste for God. The creationist's rhetoric is misleading in the same way that the following case would be. Suppose someone criticizes a lawyer, saying that she refuses to represent any Jewish person in civil rights cases. This makes the lawyer sound like a bigot. In fact, it turns out that the lawyer specializes in corporate tax law. So, although it is true that she refuses to defend the civil rights of Jewish individuals, this is just because she does not represent *any* individuals in *any* civil rights cases. It is simply not the sort of law that she handles, and so it is unfair to make her appear a bigot by narrowly characterizing her rule of practice. Similarly, science does not have a special rule just to keep out divine interventions, but rather a general rule that it does not handle any supernatural agents or powers since these are taken by definition to be above natural laws. That is what it means to hold methodological naturalism, so it is quite unfair to equate this with methodological atheism.

I hope that such points are obvious so I will not belabor them. What might not be so obvious is the role that explanation plays in science and why introducing miraculous interventions as scientific explanations would be deleterious to scientific method. It is to a more detailed examination of this issue that I now turn.

Scientific Explanation Lost and Found

Creationists want science to reintroduce divine entities and powers into its theories and theorizing. They claim that their supernatural "alternative explanation" should be given equal time with the natural theory of evolution. Literalist creationists hold that biblical revelation should be admitted to explain and justify the supernatural origin of the world, of animal species, of human beings, and more. Johnson and other IDCs want science to incorporate the reality of God and to cite His preternatural divine intelligence as the best explanation of biological complexity. But are these sorts of appeals to supernatural explanation reasonable? In particular, are they reasonable in science?

Many scientists would immediately answer in the negative on the grounds that explanation, supernatural or otherwise, should play no role at all in science. Science, they would claim, does not explain the world, but merely *describes* it and leaves the explanations to philosophers and theologians. Indeed, Arno Penzias made just such a statement in an interview with creationist Fred Heeren about his and Robert Wilson's work that led to the observation of cosmic background radiation and evidence for the Big Bang. Asked why Wilson had been disposed to accept the steady state theory before those 1965 observations, Penzias offered: "[Wilson], like most physicists, would rather attempt to describe the universe in ways which require no explanation; there's the economy of physics. And since science can't *explain* anything—it can only *describe* things—that's perfectly sensible."[6]

Although this is a rather common view among contemporary scientists, it is quite mistaken, and it is important that we see why it is mistaken. First, however, we should note that as a response to the creationist this would be a weak argument. It provides no good reason to rule out supernaturalism. The supernaturalist could easily agree not to call creationism an "explanation," but just a straightforward descriptive hypothesis about the world, what things exist in it and which relations hold among them. They are simply offering an alternative *description* of the world. This gives us a clue about the nature of the scientist's mistake.

The mistake arises in part from ambiguity in the notion of explanation. To see this let us take a moment to examine the concept. At the generic level, an explanation of X is something that "makes X plain." That is, explanation brings understanding where before there was confusion or obscurity. But there are many species of explanations, which can be distinguished by the nature of the phenomenon to be explained as well as the question that is being asked about it. For example, one may ask someone to *explain what* is the temperature of the sun. Here we see without difficulty that the explanation one would offer would be just a description—the surface temperature of the sun averages 6,000° K. The scientist would probably quickly accept this sort of case and then claim that the problematic case is when we ask the person to *explain why* the sun so shines. But here is the source of the ambiguity that leads to the mistake: There are at least two different questions that we might have in mind when we ask "why X?"

The first is a question that inquires after the intended purpose, or the ultimate end, of sunshine. We sometimes express this more explicitly asking it in the form of an *explain what for* question. That is, when we ask "Why does the sun so shine?" we might mean "What is sunshine for?" This is the *teleological* sense of why-explanation. (The term "teleology" comes from the Greek term *telos*, which means "goal" or "end.") In his classification of explanatory types, Aristotle called these "final" explanations. If one considers how to explain, for instance, a sculpture in this sense it would not be appropriate to simply start listing the sculpture's many physical effects (it tips the scales at 2,000 kg., it casts an irregular shadow, it makes many observers shrug and say that their 6-year-old child could have done better, and so on). Rather, the relevant teleological explanation is, say, that the sculpture is meant to express the artist's alienation from contemporary material culture and feelings of irony in having accepted the commission for the sculpture from a major Wall Street trading firm. Teleological explanations are "final" because they refer to ultimate intended goals. It is to this notion of explanation that the scientist is probably objecting. When science investigates the sun it can tell you that sunshine warms the earth and helps make plants grow, but it cannot say that doing these things is what sunshine is for. If that is what you mean by asking for an explanation of sunshine then let the romantic poet give you the answer. Science could also discover that sunshine can burn one's retina or cause skin cancer, but it could never discover, and offer as an explanation, that one or another of these effects was sunshine's ultimate purpose.

However, there is a second notion that someone could have in mind in asking why the sun shines in the manner it does—the interrogator might be inquiring about the physical processes that produce the observed light and temperature. This is the *genetic* sense of why-explanation, and there is no good reason for the scientist to object to this sort of explanation. Indeed, giving accounts of processes that give rise to phenomena is the main thing that scientists do. Furthermore, an explanatory account of this sort is just a special sort of description, and so we see that the distinction between explanation and description that Penzius appealed to was mistaken.

There is a simple reason that many contemporary scientists make this mistake; probably without realizing it they are following a philosophical position that was the received view during the first half of the twentieth century. The position was advocated under a variety of different names and its specific tenets evolved over time, but we may speak of it under the general name of "positivism." Positivism's influence continues to be felt in science, even though philosophers of science themselves have long since abandoned many of its tenets after continued argument and analysis revealed its conceptual flaws. Positivists held that science should not go beyond what is physically observable, and they explicitly rejected explanation in science because they thought that it was necessarily metaphysical. Their maxim, which Penzias echoed above, was that science describes but does not explain. But positivists readmitted explanation into science after Carl Hempel, in a series of important articles beginning in 1948, showed how it could be explicated in a way that was not dangerously metaphysical. The contemporary "positivistic" scientist probably absorbed the antiexplanatory view that dominated until the time of Hempel's work and is simply not aware of the more recent developments in philosophy of science.

Hempel developed several explanatory forms to deal with different sorts of scientific cases, but all fall under what was called the "covering law conception" of scientific explanation.[7] The idea is that we can explain X— the explanandum—by showing that it follows from the empirical law (or laws) governing that sort of phenomena together with background information such as the initial conditions of the variables in the law. Let us take his central deductive-nomological (D-N) model to give an example. Suppose one were to ask for an explanation of why a cannonball takes a given number of seconds (say six and a half) to hit the ground after being dropped from a tower. Here the explanandum, X, is the specific duration of the cannonball's fall. This is explained by giving a logical *deduction* from the gravitation law (*nomological* has the Greek stem *nomos*, which means "law") and the values of its variables for the case at hand. That is, we can derive X from Galileo's gravitation law that governs bodies falling near the earth (or some more general gravitational law) and plug into its equation the figure for the height of the tower from which the cannonball was dropped (the initial condition). We explain the duration of the fall by showing that it follows in this way from the law of gravity.

On Hempel's conception a scientific explanation is thus a special sort of deductively valid argument, namely, one that contains at least one general law in the premises from which one derives the explanandum. Furthermore, Hempel specified that the laws must have empirical content, by which he meant that they had to be testable by observational data. This condition prevented explanation from falling back into metaphysics. Science could indeed explain empirical phenomena by reference to covering laws so long as it was careful to stay within the bounds of empirical testability. Explanation was now acceptable to the positivists, and the antiexplanation tenet was dropped. Actually, positivism itself was abandoned as a unified philosophical view shortly thereafter when sufficiently many of its other central tenets were also rejected for other reasons. Today very few philosophers of science consider themselves positivists, though there are still scientists who are vaguely "positivistic" in the old, outmoded sense without realizing it.

So, science can indeed offer explanations. However, it is unfortunate that we cannot rest with Hempel's precise D-N model of explanation and proceed immediately with our assessment of supernatural explanation by comparing it to the detailed logical structure of his model. Hempel's work was successful in reintroducing explanation to science, and many of his broad conclusions remain in force, but extensive discussion of the technical details of his particular logical models revealed weaknesses that he was unable to overcome. Other philosophers of science took up the task and have made significant conceptual progress since then.

The problem was that in some ways Hempel's conditions were too weak and in other ways they were too strong. For example, the specific logical form of the D-N model was too lenient and thereby allowed in cases which were not truly explanatory. A variety of proposals have been offered as ways to strengthen the requirements. One important version, developed by Philip Kitcher, emphasizes that explanatory understanding can be achieved by unifying our knowledge and thereby reducing the number of "brute facts" we must accept. On this view, for a derivation to count as an explanation it must belong to a restricted set of derivations that optimizes unification by minimizing the number of explanatory patterns needed while maximizing the number of conclusions that can be generated.[8] On the other hand, Hempel's requirement that an explanation be an argument that cites a law is arguably too strict. Most philosophers of science now agree that an explanation need not take the form of an argument at all, and that a description can be sufficient. On the influential account developed by Wesley Salmon, to explain X it might be sufficient to describe the causal process that produced X.[9] Salmon has a detailed theory of causal processes and their interactions that forms the framework of this sort of explanation, but we need not go into its details here. Salmon's main point is that at the most basic level the explanation of something in the world involves something else in the world—the causal processes that led to it. An explanatory account need not include an explicit statement of the causal law, though of course it is understood that the cited causal processes are lawful. Finally, philosophers now agree that Hempel's hope for a theory of explanation that made use of only syntactic and semantic constraints was not possible, because pragmatic considerations must also enter the picture. Bas van Fraassen has developed this point, showing how explanations of X are fixed in relation to a contrast class—some alternative Y that depends upon the question in which we are interested.[10] Thus, explanation-seeking why-questions are too vague if they take the form of "Why X?" and need to be further specified by contrast, such as by asking "Why X, rather than Y?" Philosophers of science have developed other elements of Hempel's view as well, but the three we have mentioned give us enough to proceed with our discussion of supernatural explanations.

Supernatural Explanations

Once we consider the requirements for a scientific explanation we see that creationist theories necessarily fare poorly. But this is not because of any special bias against creationism, for the conclusion holds for any supernatural theory simply because of the characteristics of the notion of the supernatural. Let us review three of the main characteristics of the supernatural to see why this is so.

The first and most basic characteristic of supernatural agents and powers, of course, is that they are above and beyond the natural world and its agents and powers. Indeed, this is the very definition of the term. They are not constrained by natural laws. Indeed, on some views, it is a supernatural Creator who makes the laws in the first place, and those who make the laws have the power to break them. Of course, this is why humans hanker after access to occult powers, since these would supposedly free us from the natural laws that bind us.

If supernatural agents are constrained at all, it may only be by logic. Even God cannot make it so that something is at once P and not-P. When storytellers need a way to save their protagonist from a misguided pact with the devil, they invariably do it by placing a bet that the devil is unable to win by virtue of the rules of logic. But other than bringing about logical impossibilities, there is allegedly nothing a supernatural agent could not do.

The second characteristic of the supernatural, which I have mentioned before and which follows rather directly from the first, is that it is inherently mysterious to us. As natural beings, our knowledge all comes via natural laws and processes. If we could apply natural knowledge to understand supernatural powers, then, by definition, they would not be supernatural. The lawful regularities of our experience do not apply to the supernatural world. If there are other sorts of supernatural "laws" that govern that world, they can be nothing like those that we understand. Occult entities and powers are profoundly mysterious to us.

The same point holds about divine beings—we cannot know what it is that they would or would not do in any given case. God works, as they say, in mysterious ways. We cannot have any privy knowledge of God's will, and those who have tried to claim it are quickly brought back down to earth. When the complex Ptolemaic epicycle theory of the planetary system was explained to Alphonso X, King of Castile, with its equant points, eccentrics, deferents, and epicycles—wheels upon wheels within offset wheels—he is reported to have commented that "if God had consulted him at the creation, the universe should have been on a better and simpler plan."[11] Defending the complexity of his theoretical models from another critic who made the same point, Ptolemy is said to have replied, "You may complain that these models are not simple, but from the point of view of God, who knows what is simple?" And, of course, Ptolemy was right; we cannot say that our notion of simplicity is at all relevant to what God's might be, or even if God values simplicity at all. Scientific models must be judged on natural grounds of evidence, for we have no supernatural ground upon which we can stand since any such ground is necessarily a mystery to us.

A final relevant element of the notion of the supernatural, closely linked to the previous ideas, is that supernatural beings and powers are not controllable by humans. Though our secret desire might be to gain esoteric power through contact with the supernatural, we seem to understand at a deep level that such control would be impossible. Folk tales and literature consistently tell us that those who would steal the fire of the gods are bound to get burned. Information about the future acquired by supernatural foresight did not allow Oedipus to escape his fate and only made the inevitable outcome all the more tragic given that his parents' foolhardy attempts to use that information to outwit fate became the very means that sealed it. The very notion of the "Faustian bargain" carries this warning against the temptation of thinking one can control supernatural powers for one's own benefit; pacts with the devil inevitably turn out for the worse. The best protagonists can do in such tales is to return in the end to their prior state, having learned, one hopes, to be content with their natural estate and powers.

This pattern certainly holds true of our relation to the divine Creator as Christian creationists want us to believe in Him. God controls the world, and though we can control ourselves, we cannot control God. Indeed, part of what it means to accept Christ, on the Evangelical view, is to relinquish even the control we have of ourselves and to turn our lives over to God's will. We need to recognize that the wishful belief in the possibility of human control of divine and occult powers actually contradicts the idea of the supernatural in a profound manner, for by definition the supernatural is beyond the reach of we mere creatures of the natural world. If the supernatural could be controlled by the natural then it would cease to be "super." If we can control the natural world it is only because the world is governed by physical laws that must be "obeyed" even when we are pulling the strings, whereas the very idea of the supernatural is that it stands above natural laws and thus outside the possibility of our control. If God were really under our control in any sense then He could certainly not be said to be omnipotent and probably would not be thought very godly.

These characteristics of the supernatural show why supernatural explanations should never enter into scientific theorizing. Science operates by empirical principles of observational testing; hypotheses must be confirmed or disconfirmed by reference to intersubjectively accessible empirical data. One supports a hypothesis by showing that certain consequences obtain, which would follow if what is hypothesized were to be so in fact. Darwin spent most of *The Origin of Species* applying this procedure, demonstrating how a wide variety of biological phenomena could have been produced by (and thus explained by) the simple causal processes of the theory. But supernatural theories can give us no guidance about what follows or does not follow from their supernatural components.

The appeal to supernatural forces, whether these are taken to be divine or occult, is always available for we can cite no necessary constraints upon the powers of supernatural agents. This is just the picture of God that Johnson presents. He says that God could create out of nothing or use evolution if He wanted;[12] God is "omnipotent."[13] He says God creates in the "furtherance of a purpose,"[14] but that God's purposes are "inscrutable,"[15] and "mysterious."[16] A god that is all-powerful and whose will is inscrutable can be called upon to "explain" *any* event in any situation, and this is one reason for science's methodological prohibition against such appeals. Given this feature, supernatural hypotheses remain immune from disconfirmation. Young-earth creation-science includes supernatural views at its core that are not testable, and it was rightly dismissed as not being scientific because of these in the Arkansas court case. But it at least was candid about a few specific nonsupernatural claims that are open to disconfirmation (and indeed that have been disconfirmed), such as that the earth is less than 10,000 years old and that many geological and paleontological features were caused by a universal flood (the Noachian Deluge). So far, at least, Johnson has declined to offer any specific positive claims of this sort by which his notion of creationism could be tested.

Experimentation requires observation and *control* of the variables. We confirm causal laws by performing controlled experiments in which the hypothesized independent variable is made to vary while all other factors are held constant so that we can observe the effect on the dependent variable. But we have no control over supernatural entities or forces; hence these cannot be scientifically studied.

Finally, if we were to allow science to appeal to supernatural powers even though they could not be tested, then the scientist's task would become just too easy. One would always be able to call upon the gods for quick theoretical assistance in any circumstance. Once such supernatural explanations are permitted they could be used in chemistry and physics as easily as creationists have used them in biology, geology and linguistics. Indeed, all empirical investigation could cease, for scientists would have a ready-made answer for everything. For example, consider intelligent-design creationists Davis and Frair's alternative creationist explanation of the many general similarities among animals (such as common reactions of humans, rats, and monkeys to drugs). These, they say, "can be explained as originating in basic design given by the Creator. Evolution is not needed to account for the similarities."[17] In short the "explanation" does not go beyond claiming that this pattern is so because God designed it so. Clearly, science must reject this kind of one-size-fits-all explanation.

Nor has intelligent-design theory made much progress since adopting its new name and terminology. In his IDC text *Of Pandas and People*, co-authored with Kenyon, Davis substitutes the term "designer" or "master intellect" for "Creator," but the explanation is the same. The intelligent design interpretation of homologies, for instance, *Pandas* gives in terms of the "functional constraints of a particular set of design objectives" of the "designing intelligence" together with the designer's "choices of individual discretion."[18] Neither is it surprising that Michael Behe, who is listed as a Critical Reviewer for *Pandas*, follows suit in *Darwin's Black Box*. After spending several chapters describing how complicated are the molecular systems for blood clotting, cilium motion and so on, Behe's own explanation is simple—an "intelligence" designed them so. He speculates that "the designer made the first cell, already containing all of the irreducibly complex biochemical systems discussed here and many others."[19] Behe then wonders why scientists don't "greedily embrace" this explanation.[20] The reason is that scientists know that when it comes to explanation, as elsewhere, it is wrong to be greedy.

By disqualifying such short-cuts, the naturalistic method also has the virtue of spurring deeper investigation. If one were to find some phenomenon that appeared inexplicable according to some current theory one might be tempted to attribute it to the direct intervention of God, but a methodological principle that rules out appeal to supernatural powers prods one to look further for a natural explanation. And it is not merely because such persistence has proved successful in the past that science should want to encourage this attitude.

The scientists' appeal to supernatural agency in the face of a recalcitrant research problem would be as profoundly unsatisfying as the ancient Greek playwright's reliance upon the *deus ex machina* to extract his hero from a difficult predicament. Sydney Harris, the preeminent scientific humorist, cleverly made the point in a cartoon that appeared in *American Scientist*. He pictures two scientists standing before a chalkboard which the first had covered with an intricate series of symbols and equations, but with one gap in the sequence at which is noted "Then a miracle occurs." The second scientist, gesturing towards this notation, states his considered assessment: "I think you should be more explicit here in step two."

Without the binding assumption of uninterruptible natural law there would be absolute chaos in the scientific worldview. Supernatural explanations undermine the discipline that allows science to make progress. It is not that supernatural agents and powers could not explain in principle, it is rather that they can explain all too easily. As such, we may think of them as the explanation of last resort, since, like the Greek god in the machine, they can always be hauled down to "save the day" if every other explanation fails. They are the poor person's explanations, or rather, the explanations of the intellectually poverty-stricken, since they are available for free. Surely it is not in this sense that the poor in spirit are to be blessed with seeing God.

The Prospects for a Supernatural "Theistic Science"

The abstract considerations of the previous section provide sufficient reason to reject appeals to supernatural explanations in science. Nevertheless, it will be worthwhile to make the ideas more concrete by considering a few specific effects of reintroducing the possibility of supernatural interventions in a practical setting. Rather than addressing the possibility of a theistic science immediately, I propose that we first consider the effects of introducing theism in another area that Phillip Johnson recommends— namely, his special area of expertise, the law.

The Prospects for a Theistic Legal System
In *Reason in the Balance* Johnson claims that naturalism has eaten away at the law in the same way he says it has infected science, and he nostalgically recalls the era in which "lawmakers assumed that authoritative moral guidance was available to them in the Bible."[21] Thus, one result of introducing theism into the law would be that the content of the law would change, with secular rules and notions of justice replaced by ones with a religious foundation in the law of the Bible. He gives the example of adultery, noting that in a naturalist legal setting one could oppose it as a breach of contract but one could not condemn it merely because God forbade it. Johnson tells us that every culture must base its laws in some "creation story"[22] and that the naturalistic evolutionary tale has replaced the traditional story of our Creation. We see what Johnson takes to be the proper basis for laws about

adultery when he writes: "If God really did create us 'male and female' and intended male and female to play different roles in the family, and intended sexual intercourse to be confined to the marital relationship, then the system of traditional family morality makes sense."[23] Johnson recommends this approach in part because he believes that moral duties make no sense except as commandments from a divine authority. I will return to the connection of the Creation story to morality shortly, but first I want to focus on two other significant ways in which the introduction of the creationist variety of theism would be likely to transform our legal system.

The first follows from Johnson's and other IDCs' insistence that science admit the reality of supernatural influences in the daily workings of the world. For the law to take this seriously as well, it would have to be open to both suits and defenses based on a range of possible divine and occult interventions. Imagine the problems that would result if the courts had to accept legal theories of this sort. How would the court rule on whether to commit a purportedly insane person to a mental hospital for self-mutilation who claims that the Lord told her to pluck out her eye because it offended her? How would a judge deal with a defendant, Abe, accused of attempted murder of his son, Ike, who claims that he was only following God's command that he kill Ike to prove his faith? As Johnson says, such divine interventions could indeed occur. But though in our private religious faith we might respect their possible authenticity, surely the law, a public institution, is not being dogmatically biased in discounting them. How, for instance, could the legal system handle torts if it had to consider accusations that a defendant caused the plaintiff's miscarriage by casting an evil eye, or hexed the plaintiff's cow? We need only look to legal history to see the sorry effects of such a system.

The law once did take such accusations of occult interventions seriously. We could pick any number of supernatural possibilities that the law considered, all based in the authority of Scripture, but the case of witchcraft is a good example. The law took the Bible seriously in its descriptions of witches, sorcerers, demons, transvection, and familiars. It also incorporated the scriptural command that one not suffer a witch to live. During the Renaissance the Catholic Church wrestled with the legal implications of this worldview; in 1484, Pope Innocent VIII appointed Heinrich Kramer and Jakob Sprenger as inquisitors and they authored the *Malleus*

Maleficarum, laying out accepted procedures for investigating and prosecuting people accused of witchcraft. There were some safeguards against
too quickly accepting accusations of witchcraft at face value, so in some
cases it might require half a dozen persons to testify (perhaps anony-
mously) that the defendant had bewitched a child or a cow. On the other
hand, since witchcraft was practiced in secret, making it difficult to witness,
and since witches were especially malicious, judges were allowed to be
deceitful to trick them into inadvertent confessions. In the following centuries the procedures were elaborated upon. Jean Bodin's influential *Daemo-
nomania* of 1580 advised that when a woman was reported to be a witch,
there was a strong *prima facie* presumption that she indeed was one and
so could be tortured if there was any other corroborating evidence. Wayne
Shumaker, from whose fascinating history *The Occult Sciences in the
Renaissance* I recount these facts, sadly observed that in Europe from 1484
to 1700 some 200,000 to 300,000 persons were executed as witches, usually by burning at the stake, and that "the situation of any person accused
by malicious neighbors was regularly desperate."[24] Is there any doubt that
we should thank naturalistic science for bringing an end to the need to fear
such "possibilities?"

The Problem of the Devil in the Lettuce

The second significant effect of introducing Johnson's philosophy of theistic realism into the law would be a radical dismissal of ordinary standards
of evidence. The most common evidence upon which someone was found
guilty of witchcraft was simply the accusations of others, against which
the only real defense was to try to have the testimony thrown out on the
grounds that the accuser was one's mortal enemy. Interestingly, there were
a few physical signs that were supposed to count as evidence, such as the
so-called devil's mark. This was an area of skin that seemed to be insensitive
to pain, and was supposed to have been caused by contact with the devil's
claw when the pact was sealed. Confessions under torture were also
accepted, though again defendants were at a disadvantage for it was
thought that *refusing* to confess under torture was also a sign of guilt—
for only a witch insensitive to pain, perhaps with supernatural aid, would
be able to withstand the torture. Judges were warned that they had to be
especially wary because the interventions of demons could cause illusions.

As proof of this power, one author cited the story from St. Gregory's first Dialogue, telling of a woman "who thought she was eating lettuce but instead ate a devil in the form of a lettuce or, possibly, invisible within it."[25] The authority of St. Gregory was supposed to be proof enough of this possible supernatural power. It was apparently inappropriate to ask, as Shumaker suggests a skeptic today might, "How do you know it was the lettuce?"[26]

We here consider posing such a question in the legal setting, but it is the same problem of evidence that is at issue for a theistic science, which Johnson says is supposed to sanction the possibility of supernatural interventions. We have seen this exact problem arise in the creationism debate before, when Gosse and current "mature-earth" creationists suggested that the earth is in fact only six thousand years old but that God gave it the appearance of great age. However, the issue here is not just whether or not God would deceive us in this way, but how we could ever tell that something was produced by a supernatural agent. Some commentators assumed that witchcraft made direct use of supernatural power, while others thought that the mischief was really all done by demons who used secondary causes, that is, by natural though perhaps yet unknown causal laws. In either case we have the same trouble finding out whether or not there is a devil in the lettuce. When the defendant claims that "the devil made me do it," how will we test her defense? How could Gregory have reached his conclusion? We may call the problem of how to determine whether or not a supernatural agent intervened in particular cases *the problem of the demon lettuce.*

Perhaps, as a saint, Gregory was privy to some special lettuce revelation, but I am still taking Johnson at his word that he is not defending the use of revelation in science. And even if one were to allow revelations as evidence how would we test that they were true revelations? Given that the core creationist hypothesis involves special supernatural interventions we should expect some answer to the lettuce problem. The Darwinian view holds that the evolutionary processes are working all the time, and can point to observations thereof. We can observe mutation, recombination, inheritance, natural selection, and the resultant changes in gene frequencies in populations. Can the creationist do as well with the Creation hypothesis? On this point, I now issue another challenge to Johnson to come clean: Are divine interventions occurring today in particular cases? If so, which ones, and how do we tell? If not, why not, and again, how do we check?

Johnson wants us to believe, as he writes in *Reason in the Balance,* that "The possibility that divine intervention may occur . . . emphatically does not imply that all events are the product of an unpredictable divine whimsy."[27] Perhaps not, but we want to see the test for distinguishing the specific cases. On this point we should recall again that in his earlier book, *Darwin on Trial,* Johnson's single example of how the creationist theistic explanation is better than evolution was the case of the elaborately tailed peacocks, which he says an uncaring evolutionary process would never allow to develop, but that they are "just the kind of creatures that a whimsical Creator might favor."[28]

Returning to Johnson's definitions of creationism, we see that the problem of the demon lettuce affects his view in deeper ways than even that of simply identifying the presence or absence of supernatural interventions. Johnson's creationism dismisses the worth of deistic views of Creation and demands ongoing direct control. It thus seems fair to ask how the theistic scientist supposes that control to work. The Darwinian can specify a fair number of the sorts of causal processes that control evolution, fulfilling the basic requirement for a scientific explanation. The third challenge to the creationist is to tell us their alternative divine control process. Creation-scientists, like the Morrises, who keep to Genesis literalism at least are forthright in specifying special Creation from nothing or from dust. Johnson and other intelligent design creationists are silent about the control process. May theistic science appeal to *ex nihilo* miracles or other miraculous control processes? In the case of the development of life-forms, does God create by selecting the variations that will survive? (Young-earth creationists reject this possibility because it makes death God's instrument of Creation, and I suspect that IDCs would not allow it because it could undermine their basic argument from information.) Does God create by causing the variations upon which selection occurs? This control mechanism was proposed by some of Darwin's contemporaries to keep God in direct control of the process. But, then, does God also create fatal and deleterious mutations? How could we tell? The specter of the demon lettuce problem reappears in all these possibilities.

Finally, consider the third, and most important, core element of Johnson's definition of creationism—that God creates for a *purpose*. How is a theistic science to discover God's purposes? Recall the example mentioned in passing of the claim about the purpose of AIDS. How would Johnson's theistic science test the hypothesis that AIDS was created by God for the purpose of punishing homosexuals, drug-users, and others for their sinful lifestyle? Naturalistic science simply proceeds by seeking a natural explanation: it treats AIDS like other diseases and discovers that it is caused by an acquired human immuno-deficiency virus, and nothing in its methodology allows it to test such moral or teleological hypotheses about God's possible purposes. The problem of the demon lettuce is particularly keen here and its implications particularly chilling.

I do not bring up this case of the alleged theistic purpose of AIDS merely to present a provocative hypothetical scenario, but because the questions of the morality of homosexuality and the explanation of AIDS figure significantly in Johnson's work, and because he thinks both are intimately connected to core issues in the creationist debate. Johnson and other creationists see the debate as being as much about the proper moral order as about the proper explanation of biological order, and Johnson brings up homosexuality as one of his standard examples of immorality in *Reason in the Balance* to illustrate this point. Johnson is more subtle than those creationists who note that God "created Adam and Eve, not Adam and Steve," but his point is the same: We are supposed to learn from the Creation story that homosexual behavior breaks the created order and thus the moral order. Johnson's views about the explanation of AIDS are less well known to those who have only followed his antievolutionary writings. The AIDS issue has been his other avocational pursuit. He has written against the current scientific view that AIDS is caused by the human immuno-deficiency virus, and in support of Peter Duesberg's radical contrary view, that AIDS is the result of the lifestyle behaviors of homosexuals and drug-users that cause general ill-health. Biomedical journals gave Duesberg's view a thorough hearing after he first challenged the HIV view in 1987. AIDS researchers have concluded that the available evidence does not support Duesberg's alternative view, but Duesberg and a small coterie of public supporters like Johnson continue to press the point.[29] Duesberg's view meshes rather well with the hypothesis about God's purpose for AIDS (though of course the latter could also apply even if one accepted the standard HIV view). For our purposes it does not so much matter what is the efficient cause of AIDS but how, whatever the efficient cause, a theistic science could test this teleological hypothesis about God's ultimate purpose. Johnson never proposes this or any other specific hypothesis about God's purposes, but he does explicitly tell us that such hypotheses are essential to theistic explanations (i.e., in terms of the purposeful intentions of the divine intelligent Designer) and to his basic notion of a theistic creationist science, so it is fair to ask how his theistic science will check them. This is my fourth challenge to Johnson's proposal for a theistic creationist science. As scientists, I claim, the best we can do is to try to find some natural explanation for AIDS, and though it is likely that science will find there is more to the story than the virus itself, science's naturalistic methodology can never go further to test and discover God's purposes in this or any other case.

Let me just add here that Johnson was in the audience when I presented all these challenges (together with the arguments in the section that follows) at a small academic conference on naturalism and theism for which he was the main plenary speaker. He declined to comment or respond to any of the claims or arguments. This is not surprising, for he consistently has refused to say anything positive about how a theistic science is supposed to work. The reason for that, I contend, is that a true theistic science is unworkable. Johnson cannot say anything positive about the "methods" of creationist "science" because it would be obvious immediately that it has no method but special revelation. Such a supernatural science is unworkable in the same way that the supernatural legal system was in the past and would be if instituted again. Thus, despite Johnson's preaching to scientists to incorporate the "possibility" of supernatural interventions in their profession, it is also not very surprising that he never incorporates these in his own professional texts on criminal law.[30] If in some future edition of his textbook on criminal procedure Johnson adds a chapter on how to prosecute witches and lets trial lawyers know how to evaluate "evidence" for the interventions of other supernatural intelligences, then maybe scientists will begin to take him seriously. I expect, however, that we will have to wait for the hexed cows to come home before that day arrives.

"Super Natural" Explanations

We earlier saw why science has good reason to rule out appeal to supernatural powers and entities as a methodological principle, and in this last section we have seen a few of the unacceptable consequences of abandoning that principle in a comparable context. A "theistic science" that opened the door to supernatural interventions could not function as a science. Let us now turn to a second problem that the proposal for a "theistic science" encounters.

A Science-Fiction Scenario

Having made this argument against appealing to supernatural entities and powers I now need to make sure that it is not misunderstood. The argument for scientific naturalism is an argument based on a methodologically necessary respect for natural laws. To say that science does not deal with "the supernatural" does not mean that everything that we currently think of as supernatural—ghosts and extrasensory perception, for example—really is. Perhaps these are actually natural law-governed phenomena whose causes are yet to be discovered. Philosophers are fond of appealing to scenarios from *Star Trek* to illustrate hypothetical conceptual possibilities like this one, so let me take a case from one episode to develop my point here.

The episode I have in mind involves the people of a world who transported themselves to an asteroid in the belief that their souls would there be set free of their bodies to live on in a blissful afterlife. There are the usual conflicts and misunderstandings to be worked out as the *Star Trek* crew tries to deal with this seemingly absurd practice. In the end, however, they are made to reevaluate their skepticism when their sensors detect unusual energy patterns around the asteroid that exhibit individual coherence and excitations that appear to match the electrical activity patterns of people's brains. We Trekkers know this is all "technobabble" of course, but we willingly suspend our disbelief and, within the fictional universe of the series, we recognize that the peoples' belief in a ghostly afterlife is true after all and has a scientific explanation.

In this science-fiction example it looks as though science has tested and confirmed the existence of ghosts and a "spirit afterlife." In one sense this surely seems right; if such evidence were found then there seems to be no reason that there could not arise a new scientific specialty that would investigate and test hypotheses about this afterlife. In the real world we have not found such evidence (though misguided believers of an earlier century did try to weigh the soul by weighing a body just before and after death), but isn't it possible that we could? If we agree with this then, similarly, why couldn't there be a science that incorporates theistic interventions? This is the creationist's complaint that science is close-minded in ruling out such possibilities. But here is the question: even in the *Star Trek* example, are we really talking about "ghosts" and a "spirit afterlife" in the way we ordinarily conceive of them? The danger of such hypothetical examples is that they mislead us about what sense of "possibility" we are talking about, since one can hide a lot of question-begging assumptions in some well-spun technobabble. In the episode, the departed "souls" turn out to be "coherent energy patterns." They interact causally with other matter and energy, of course, or the sensors would not have picked up their "energy signature." Indeed, if they are energy in the ordinary scientific sense, then it would be possible to exert causal influence upon them in the usual ways. Presumably, we could manipulate or disrupt them as we can other forms of energy. Presumably, such disruption could "kill" them. At this point we should be beginning to feel a little uncomfortable about our earlier conclusion about what was confirmed by the crew. Let us step back now and analyze what has been going on in this science-fiction example.

By discussing the confirmation of "ghosts" or "souls" in this way, we have tacitly taken them out of the supernatural realm and placed them squarely in the natural world. To conceive of ghosts as supernatural entities is to consider them to be outside of the natural realm, outside the law-governed world of cause-and-effect physics. But to say that science could test and confirm their existence, as in our hypothetical case, is to reconceive them as natural entities. Perhaps there really are "coherent energy patterns" as postulated in the story, but such "ghosts" are no longer supernatural—they have been naturalized. The Christian would quite properly object that, whatever these things are, they are surely not departed souls in the religious sense of the term.

Putting God to the Test

So what should this tell us about Johnson's and other creationists' idea for a theistic science? How does God figure in this picture? Will theists really be happy from a religious viewpoint to think of God as a scientific hypothesis as we just considered the hypothesis of a spirit afterlife?

As we saw above, for a hypothesis to be scientific, it must be intersubjectively testable and fit within the framework of law-governed cause-effect relations. This is the core of what it means to be a natural object and to be amenable to scientific investigation. Agreeing to the constraints of this sort of epistemological approach as the means of gathering public knowledge about the empirical world is just what it is to be a methodological naturalist. This is no different than what we tacitly assume for everyday knowledge, and all science does is make careful extensions of our ordinary experience in what is simply a more precise and explicit version of the ordinary way we get such knowledge. In proposing a theistic science, Johnson claims to be expanding science to supernatural possibilities undreamed of in this philosophy, but what he and other Creation scientists are really doing is reducing God to a scientific object, placing God in the scientific box.

Creationists are especially fond of quoting the nineteenth Psalm, but they have a rather odd way of interpreting it.

The heavens are telling of the glory of God; and [the sky] is declaring the work of His hands. Day after day they pour forth speech; night after night they display knowledge. (Psalm 19:1–2)

Instead of appreciating the lovely poetic sensibility of these images—imagining the dome of heaven to be like a clay bowl that praises the potter—creationists talk about this as though it had something to do with the search for extraterrestrial life-forms. They speak of talking pulsars and information-bearing signals from a divine intelligence as though God were a broadcaster beaming out messages for us on a literal carrier wave and waiting for some enthusiastic ham-radio operator to pick up the word of His existence.

Returning to the design argument, we see that it works in just this way, drawing an inference to the nature of God from what is already known and familiar to us in human, natural terms. The intelligent-design theorists have given us a scientifically gussied-up version of Paley's venerable argument: God becomes a big watchmaker in the sky, a divine genetic engineer, or a souped-up "intelligence." But long ago philosophers revealed the flaws in the design argument, and Scripture itself warns against analogizing God to human experience. As Isaiah rhetorically asks, "To whom then will you liken God, or what likeness compare with him?" (Isa. 40:18).

Johnson tells us that scientific naturalism is the root of all evil, but it turns out that he is doing nothing less than naturalizing God. Ironically, Johnson might not be a supernaturalist after all, but a "super naturalist."

To recognize this irony is to suddenly understand what previously seemed to be a rather puzzling feature of the creationist debate, namely, the surprising similarity of outlook between creationists and some atheists. Scientific organizations and religious institutions have issued explicit statements, correctly, that evolutionary biology neither affirms nor denies the existence of God and that Christianity is not threatened by evolution, but both creationists and atheists seem to think that God hangs in the balance over the truth or falsity of evolution. In the present instance we see this in the affinity between Johnson and a few atheistic scientists, Will Provine and Stephen Weinberg in particular, who have been happy to engage him in debate. Provine and Weinberg concur with the essentials of Johnson's framework and the implication that theism is incompatible with evolution. Johnson is an unwitting friend of such atheists, for they agree in naturalizing God and making God an empirical hypothesis, amenable to scientific investigative testing. Adolf Grünbaum, for example, a distinguished philosopher of science who has argued an atheistic position against theists such as William Lane Craig and Philip Quinn (who want to appeal to divine interventions to explain cosmological facts), quite agrees with creationists that science can test the God hypothesis. The difference is that atheists have looked at the world and concluded that from this perspective the balance of evidence weighs heavily against the ordinary conception of an omnibenevolent personal Creator. Naturalize God and put the Creation hypothesis to empirical test, and such atheists claim a knockout.

Because creationists see the dispute in the same terms as Grünbaum, they fear that atheism is the inevitable conclusion if one accepts evolution. This is why they fight so fervently to deny evolution and to try to uphold the scientific status of the biblical account, often to the point of absurdity. To defend the scientific plausibility of Noah's Ark, ICR creation-scientist John Woodmorappe provides a book-length feasibility study[31] and finds himself arguing that Noah solved the problem of animal waste management by training the animals to urinate and defecate upon command as someone held a bucket behind them.

As an intelligent-design creationist, Johnson refuses to state any position about Noah's Ark and focuses on more generic biblical claims, but because he too thinks of God as amenable to scientific test he holds the same incompatibilist view of the relation between Christianity and evolution. He writes:

One might have thought that Provine and I would be bitterly opposed, since I am a Christian who emphatically affirms that the world is the product of a purposeful Creator, not a blind material mechanism. But in fact I think Provine has done a lot to clarify the point at issue, and I agree with him about how to define the question.[32]

Provine says that compatibilists have to check their brains at the church-house door, because he thinks both the Creation hypothesis and the evolutionary hypothesis are on a scientific par and can be evaluated by standard naturalistic methods, and on this basis the former is a clear loser. Johnson holds out hope that evolution might still be proven false, but he frames the conflict in the same way: "Christianity makes sense only if its factual premises are true. . . . The essential factual premise is that God created us for a purpose, and our destiny is a glorious one in eternity."[33] On Johnson's view naturalistic evolution is incompatible with this premise. He seems to think that the only way that God could give human beings purpose is by creating their bodies by directly intervening in the causal processes to produce them. The only option remaining for him, then, is to stand and slug it out on science's home turf.

To enter the field of intellectual argument is to accept the risk that we may lose by being proved wrong. But accepting the risk of being wrong is the inescapable price for making any meaningful statements about the world. The best scientists have never feared to accept that risk. . . .[34]

Having recognized Johnson's inadvertent crypto-naturalism, we should now be sensitized to the significance of such remarks. Whether or not he realizes it, in this statement Johnson is advocating a return to the verifiability criterion of meaning—the view that only statements that can be tested are meaningful. This view of meaning was one of the central tenets of positivism in its most scientistic vein, and its unworkability was one of the reasons that positivism was overturned. Its reappearance in Johnson's conceptual framework shows the extent to which he is unwittingly caught up in an even more pervasively naturalistic outlook than the one he decries. That statements be empirically testable is indeed a requirement for scientific knowledge, but to think that such statements *alone* are meaningful is to buy into the scientistic view that *only* scientific knowledge is meaningful. Christian critics of creationism have pointed out that young-earth creationists, in their insistence upon the need for creation-science, are actually venerating science above God. The critics are right. After all, what is a literal, "scientific" reading of Genesis but exegesis from a naturalistic perspective? Johnson may or may not be a biblical literalist but he exhibits the same inverted perspective, protesting just a bit too much against the supposed naturalist mote in the scientist's eye while remaining unaware of the super naturalist log in his own. This is the reason that he and some atheists see eye-to-eye. Continuing just beyond the above quote, Johnson writes:

If Christian theists can summon the courage to argue that preexisting intelligence really was an essential element in biological creation and to insist that the evidence be evaluated by standards that do not assume the point in dispute, then they will make a great contribution to the search for truth, *whatever the outcome.*[35]

He concludes by acknowledging that this courageous approach allows the possibility that biologists might respond with convincing evidence that shows that direct creation by such intelligence is not necessary. If we are to take him at his word, this means that he would then have to admit that the Creation hypothesis is false. This is just the line of reasoning that the atheists have already followed to its "natural" conclusion.

However, do we really expect that the intelligent-design creationist would accept this conclusion even in such a case? More importantly, is the Christian really forced to accept this scientific conception of God that IDCs and some atheists put forward? The answer is certainly "No" in both cases. As a scientific creationist, Johnson would have us put God on a slide and peer at Him under the microscope of science. But there is a tension in Johnson's conception of God, as we saw, for at other places he expresses the traditional notion of a supernatural, omnipotent God who is mysterious and inscrutable. We would expect IDCs to withdraw to this notion in the same way that other creationists, such as the navel-contemplating Omphalos YECs, have done. Indeed, we already observed Michael Behe's retreat into the shadows in his answer to the problem of biological imperfections, when he argued that it was "virtually impossible" to know the purposes of a designer unless the designer tells you specifically what they are. But that takes us back to special revelation, and intelligent-design theorists have told us that this religious notion has no place in their science. If, on the other hand, we naturalize God and assume that we can investigate God's purposes scientifically from observations of the world on analogy with our own purposes, then we can proceed but we must be willing to accept some rather odd conclusions. When the biologist J. B. S. Haldane was asked following a public lecture what his study of the biological world had taught him about the mind of God, he could only reply, thinking of the relative proportions of all the different species, that it appeared as though God was inordinately fond of beetles.

Christians would be wise not to even start down the dead-end road of creation-science or theistic science, for it is unlikely that they would find a naturalized God to be worthy of worship. Johnson quotes John as the Scriptural basis of his Theistic Science.

In the beginning was the Word, and the Word was with God, and the Word was God. He was in the beginning with God. All things came into being through him, and without him not one thing came into being. (John 1:1–3)

Somehow, Johnson seems to think that this passage tells us that God's creative purposes are open to scientific scrutiny. Somehow he seems to think that the "Word" really just means "information" and that we could pick it apart and study it like "a computer program."[36] Pantheist hackers might find this notion of God appealing, but it is unlikely to win many other converts from the scientific community. Perhaps Christians might better judge this piece of Scripture and the prospects for a theistic science in the light of another New Testament passage from Romans:

O the depth of the riches and wisdom and knowledge of God! How unsearchable are his judgments and how inscrutable his ways! (Rom. 11:33)

God in the Machine?

Tempting though the assistance of the gods may be, calling upon their help in scientific matters would be improper from both a scientific and a religious point of view. I have tried to point to a dilemma in the proposal for a theistic science. If one takes God to be supernatural, then God and the Creation hypothesis have no place in science. On the other hand, if one naturalizes God to make the Creation hypothesis scientific, then we find ourselves faced with a God who is not very godly. Though he himself outlined the argument that IDCs follow, Norman Geisler recognized the danger of this horn of the dilemma and criticized IDCs for not clearly distinguishing themselves from "naturalistic (pantheistic) 'creationists' who see the creator within the universe."[37] Unless one simply assumes with Plantinga a vague, unstated set of revealed truths that Christians supposedly "know," and ignores the widely divergent views of what these are, the only alternative is to try to psychoanalyze God's design intentions on analogy with those of human beings. It is odd that people who take Scripture as seriously as creationists do should fail to remember that it warns explicitly against analogizing God in this way. "For my thoughts are not your thoughts, neither are your ways my ways, says the Lord" (Isa. 55:8). When they try to infer God's "intelligent design" by naturalizing God, IDCs are making God a part of the machine.

7

Burning Science at the Stake

According to the Bible, all the ancient nations developed from the different families radiating out from Babel after the confusion of tongues. Even though their languages were different, they all still retained the same false concept of cosmogony taught them by Nimrod, the great rebel against God. This false religion, with its false cosmogony and its false pantheon of gods . . . thus became the progenitor of all the world's religious systems. . . .

The real author of this vast religious complex—this great world religion of pantheistic, polytheistic, demonistic, astrological, occultistic, humanistic evolutionism—can be none other than the one who is called in the Bible the "god of this world." . . .

Satan invented the evolutionary concept and is using it as his vehicle to deceive the nations and to turn men away from God.

—Henry Morris

The Devil's Chaplain?

One of the great dangers of creationism is the way that it, like other extremist views, tends to demonize the enemy. I do not like to put the matter so bluntly, but the conclusion is inescapable. I first became aware of this pattern when I reached the final set of exhibits at ICR's Museum of Creation and Earth History. It began with small comments, such as the guide's telling us that Darwin stole his theory from others without giving them credit, and then quickly escalated to blaming evolutionism for Hitler and Nazism. As I read and talked more with creationists, I found that these sorts of accusations and insinuations are not confined to the YECs.[1] Eventually I realized that for creationists this is not just a debate about the evidence for or against a scientific conclusion, but a classic clash of good and evil. As creationists see the issue it is a battle in defense of the one true form of

Christianity against the philosophy of the devil. Behind all the arguments against evolutionary theory is the real war against "Evolutionism." In their view, creationists wear the halos and evolutionists the horns. I wish this were hyperbole, but it is not. Henry Morris makes the point explicitly, taking us back one more time to the Tower of Babel.

Morris tells us that almost all of the worlds religions are based upon the theory of evolution. He mentions Buddhism, Confucianism, Taoism, Hinduism, Shintoism, and Lamaism by name, as well as atheism and humanism, which he also considers to be religions.[2] Though they vary in the language they use, all religions, except those that accept the Genesis cosmogony, he claims are varieties of evolutionism. Why is this so? It is because all the ancient nations had their origin at the confusion at Babel where they all had been taught "this false religion, with its false cosmogony and its false pantheon of gods" by Nimrod, ruler of Babel and the builder of the great tower there.[3] What was the Tower of Babel but the expression of a rebellion against and a challenge to God? According to Morris, the religion that drove this rebellion was Nimrod's "pantheistic, polytheistic, demonistic, astrological, occultistic, humanistic evolutionism."[4] And who was the original author of evolutionism? Apparently, it was Satan, the devil himself. Morris writes:

Modern humanistic evolutionists, of course, scoff at such notions. They believe in neither God nor Satan, worshipping only themselves. So the idea that Satan invented the evolutionary concept and is using it as his vehicle to deceive the nations and to turn men away from God, is to them naive foolishness. Our purpose here, however, is not to court the humanists, but to show Christians the great dangers in compromising with evolution.[5]

Morris is no maverick in linking evolution to the devil. In her study of Fundamentalism, Ammerman found this creationist belief to be common, and she quotes a tellingly representative viewpoint: "Satan attacks the word of God anyway . . . particularly Genesis and Revelation. In Genesis his downfall is prophesied, and in Revelation it is completed; and he doesn't like either one. So he tries to throw out the beginning by bringing in evolution, and throw out the ending by saying that man does not owe anything to God. And then, there you come up with the atheist."[6] The new creationists may be more subtle in the way they draw these links (or at least more circumspect about how they express them), but we shall see that they view the debate in the same stark light.

Back in 1856, as Darwin contemplated what he had learned of the struggle for existence in nature, seeing it in quite a different light than he had when still under the influence of Paley's rosy descriptions of "happy" nature, he commented in a letter to botanist Joseph Hooker: "What a book a Devil's Chaplain might write on the clumsy, wasteful, blundering low & horridly cruel works of nature!"[7] It was actually the Rev. Robert Taylor who had earned the sobriquet of the Devil's Chaplain, having been jailed for a year for blasphemy because of his anti-Christianity proselytizing, but Darwin's biographers Desmond and Morris suggest that Darwin worried that he would be the next to earn the title. It now appears that Darwin's fears were well justified.

By demonizing evolution, creationists are implicitly demonizing evolutionists. By rejecting scientific naturalism and calling for a return to "theistic science" that recognizes supernatural entities and occult powers, creationists are directly charging scientists with atheism. Of course, creationists certainly would not say that we should return to burning heretics at the stake, as Giordano Bruno was burned; if they did scientists would really be in deep trouble given what we saw in the previous chapter about execution of witches on the basis of supernatural evidence. But what they are saying, in effect, is that we need to burn evolutionary theory and with it much of the rest of science, not simply because it is false (they claim), but because it is morally evil. Creationists believe that moral value itself is at stake. Understanding that belief and trying to assuage the fear that accompanies it is the purpose of this chapter.

Creationism's Crisis of Meaning

Listen carefully to creationists for long enough and you will realize that they are not so much worried about evolution as they are worried about meaninglessness. Most people are probably initially both amused and puzzled by creationism. We think that denying evolution is now on a par with the view that the earth is flat. It perplexes us how someone can ignore the clear evidence for evolution. We all know the ways that science has conflicted with the Bible in the past and how Christians have worked through these conflicts, and we wonder why creationists get so worked up about this particular scientific conclusion. I think we can better understand

the situation if we look beyond the scientific and pseudo-scientific arguments and recognize that what underlies and fuels the issue is a deep philosophical and theological concern about loss of purpose. The creationism
controversy is not just about trying to avoid being descended from apes,
it is about trying to avoid an existential crisis.

Although there are theistic existentialists, it was the atheistic French
existentialists, especially Jean-Paul Sartre and Albert Camus, who most
keenly dramatized the bleak worldview that leads to a crisis of meaning.
They saw the universe as amoral and supremely indifferent to us, and they
argued that human life had no essential value. Creationists do not want to
believe that life is meaningless, but they seem to fear that it would be so if
evolution were true. Often, one must read between the lines to find this
usually unarticulated fear, but many creationist writers have identified and
expressed it quite directly.

For example, John Ankerberg, who supports young-earth creationism
in his widely broadcast television show, writes in one of his pamphlets:
"In the evolutionary or materialistic worldview, man has no unique status
other than which he may choose to give himself."[8] Hugh Ross, whose
Reasons to Believe ministry defends an old-earth creationism, puts the
point this way: "[I]f the universe is not created or is in some manner accidental, then it has no objective meaning, and consequently, life, including
human life, has no meaning."[9] Finally, Phillip Johnson, the Berkeley law
professor who defends the minimal "intelligent-design" creationism,
attacks evolution by trying to undermine its naturalistic basis. He makes
the worry most starkly explicit: "Scientific naturalism is a story that
reduces reality to physical particles and impersonal laws, [and] portrays
life as a meaningless competition among organisms that exist only to survive and reproduce. . . ."[10]

Almost all of the debate between creationists and scientists has dealt
with the scientific evidence for evolution, but this obscures the underlying
concerns that have little to do with how we got here and everything to do
with the philosophical and theological consequences of thinking that we
got here one way rather than the other. However, looking at such statements as a philosopher, one sees the expression of existential angst—a
gnawing anxiety that life is without value or meaning. These next sections
will explore the nature of the creationists' existential worries. I will refer

to a variety of creationists to demonstrate the scope of this anxiety, but I will take Phillip Johnson as my main case because he exhibits the concern most thoroughly. In particular, I will focus upon his book *Reason in the Balance* in which he tries to express what he thinks is wrong with the naturalistic philosophy that he claims props up the evolutionary picture of the world and tries to show what is really at stake in the creationism debate. If we can understand these beliefs and how they lead to a fear of meaninglessness perhaps we may be able find a way to assuage the crisis of values that drives the creationist crusade.

The "Evil Fruits" of Evolution

Phillip Johnson opens *Reason in the Balance* with the following question and statement of intent:

Is God the true creator of everything that exists, or is God a product of the human imagination, real only in the minds of those who believe? This book is about how people answer that question, and the consequences of answering it one way or another.[11]

The mention of God simply as the true creator tacitly signals an important difference between Johnson and other creationists, for he refuses to reveal his positive view about the details of Creation and officially advocates only the generic notion that God creates purposefully. As we have seen, Johnson's strategy is to define evolution in opposition to this minimal notion of creationism and then to support the latter by negative argument against the former. His innovation is the argument that evolution wins only because science rules out the possibility of divine creation dogmatically by what he calls its speculative metaphysical doctrine of naturalism.

Johnson does not maintain a consistent notion of naturalism, but the central idea he has in mind is that it rules out by fiat any intervention by God in the physical world of cause and effects. In his main discussion of naturalism in *Reason in the Balance,* Johnson explains his notion of the concept as follows:

"Naturalism" is similar to "materialism," the doctrine that all reality has a material base. . . . The essential point is that nature is understood by both naturalists and materialists to be "all there is" and to be fundamentally mindless and purposeless.[12]

Johnson's central argument against evolution is that it rests upon a dogmatic speculative metaphysics, and unfairly rules out the better theory—
the creationist view that God designed the world for a purpose—because
that metaphysics is essentially atheistic. For example, he writes:

Naturalism is a *metaphysical* doctrine, which means simply that it states a particular view of what is ultimately real and unreal. According to naturalism, what is
ultimately real is nature, which consists of the fundamental particles that make up
what we call matter and energy, together with the natural laws that govern how
those particles behave. Nature itself is ultimately all there is, at least as far as we
are concerned. To put it another way, nature is a permanently closed system of
material causes and effects that can never be influenced by anything outside of
itself—by God, for example. To speak of something as "supernatural" is therefore
to imply that it is imaginary, and belief in powerful imaginary entities is known as
superstition.[13]

We have already seen the problems with Johnson's argument from naturalism so here let us just focus upon what worries lie behind it. The rest of
the first quote points us toward the answer. Johnson sees the creationist
controversy as a weighing of two incompatible metaphysical/moral
worldviews. In our right hand is the generic creationist view in which the
world and human beings were designed by God with a purpose in mind,
and in our left hand is the view of evolutionary naturalism in which we
are the result of "a purposeless material process"[14] and God is merely an
idea in our minds. Weigh these two worldviews carefully, warns Johnson,
for the consequences of choosing incorrectly are dire indeed.

Johnson is not alone in seeing the creationism debate as fundamentally
about the consequences of two opposing moral world views. A recent
posting on the *Christian Answers* page on the World Wide Web describing
what is at stake in the Creation/evolution debate mentions several supposed consequences of Darwinism for society and ethics:

[E]volution has given a scientific justification for rejecting the absolutes of God's
Word in Western society. . . . Increasing lawlessness (no one owns me, there are
no rules); abortion (we're all just animals anyway); marriage breakdown (Jesus'
teaching on marriage and divorce always went back to the historical basis in Genesis) and ever more open homosexual practices are just some of the fruits being
reaped by a society rearing its young in this anti-biblical world view.[15]

The writer goes on to highlight the atrocities of Nazi Germany as the
"starkest example" of "openly admitted application of consistent evolu-

tionary thinking." Such outrageous statements about the supposed consequences of believing in evolution are not uncommon, as I discovered during
my visit to ICR's Museum of Creation and Earth History. The exhibit,
which attacks "Evolutionism" and promotes the young-earth creationist
view, concludes with two panels that depict the "fruits" of the two world
views. On one panel we find a list of the good fruits of creationism:

True Christology	True Faith	True Family Life
True Evangelism	True Morality	True Education
True Missions	True Hope	True History
True Fellowship	True Americanism	True Science
True Gospel	True Government	

On the other panel we see the purported evil fruits of evolutionism:

Communism	Racism	Abortion
Nazism	Pantheism	Euthanasia
Imperialism	Behaviorism	Chauvinism
Monopolism	Materialism	Infanticide
Humanism	Promiscuity	Homosexuality
Atheism	Pornography	Child Abuse
Scientism	Genocide	Bestiality
Slavery	Drug Culture	

The viewer is asked to choose between these two worldviews.

Although Johnson is subtler and more indirect, he too implicates the
worldview of evolutionary naturalism and its handmaiden, liberal rationalism, for many of these same items including, for example, abortion,[16]
homosexuality,[17] pornography,[18] divorce,[19] genocide,[20] and even bestiality.[21]
As we shall see, he blames the worldview of this unholy duo—the combination he terms "modernist naturalism"[22]—for other evils besides, including
"the irrationalist and tribalist reaction that is so visible all around us."[23]

Could it really be that Darwin begot all this? It would take more space
than we have here, but it would be fairly easy to go through these lists and
refute each charge in turn. However, one example should suffice. The evil
of slavery was perpetrated long before the theory of evolution ever arose.
Certainly evolutionary theory itself does not justify slavery, and Darwin
himself was an adamant abolitionist who wrote "How weak are the argu-

ments of those who maintain that slavery is a tolerable evil!"[24] Sadly, among the most common proslavery arguments were those made by Christians who quoted Leviticus (25:44–46), I Timothy (6:1), and a variety of other scriptural passages to show that slavery was endorsed by the Bible. Indeed, creationists should know that it was once common to cite Genesis (9:27), in which the righteous Noah curses his son Ham and his descendants to be slaves, as the "creation story" of slavery.

However, we should not let the offensiveness and absurdity of the implications distract us. The important point is to recognize that creationists believe Darwinism and its worldview lead to such consequences. Now we can go on to try to discover why.

Johnson on the Modernist Naturalist Worldview

For Phillip Johnson, the worldview that he opposes includes not only an array of social evils, but a bevy of philosophical views, such as Rortian neopragmatism,[25] hermeneutics,[26] ethical relativism,[27] liberal rationalism,[28] postmodernism,[29] and reductive materialism,[30] which he thinks give rise to them. Biologists who have evaluated Johnson's critique of evolutionary theory often express irritation with Johnson in that it is so clear that he lacks a biologist's understanding of the material.[31] Similarly, a philosopher finds it frustrating to read Johnson's work, not only because he ignores distinctions we find critical, thereby conflating quite distinct conceptual positions, but also because he regularly uses terms in nonstandard and inconsistent ways that lead him to make unjustified conceptual leaps. He also has the lawyer's habit of making every issue into a winner-take-all choice between two positions.

In *Darwin on Trial*, the parties in the case were specially defined versions of creationism and evolution, but in *Reason in the Balance* we see that the disputants each have several aliases. Johnson now identifies his client as theistic realism—the view that "God is objectively real"[32]—and on its behalf he brings suit against modernist naturalism. These are the supposedly contradictory viewpoints in the metaphysical dispute that Johnson had announced in his opening question. Again, one of the greatest dangers of creationism is the way that it draws a line in the dust, asks "Are you with us or against us? Be ye friend or be ye foe?" If we are to defuse tensions in the creationist controversy, we must try to get past such posturing. One way to do this will be for us to call attention to the ways in which Johnson and other creationists construct this and other false dichotomies and draw misleading inferences from them, and to reject them forthrightly and explicitly.

Ethical Relativism, Sex, and Authority

Johnson locates the roots of the modernist worldview in the "liberal rationalism" of seventeenth- and eighteenth-century philosophers such as Thomas Hobbes, John Locke, David Hume, Adam Smith and John Stuart Mill. As Johnson uses the term, "Its essence lies in a respect for the autonomy of the individual." He goes on to explain that, although the founders of liberalism were theists, contemporary liberalism "incorporates the naturalistic doctrine that God is unreal, a product of the human imagination."[33] On this combined modernist-naturalist view, there are natural rights that derive from our status as autonomous beings, but, says Johnson, there are typically no natural obligations, particularly none from God, since God is an illusion. The only legitimate obligations are those that individuals consent to be bound by, in the same way that the only legitimate government is one that has the consent of the governed. Johnson concludes that because of this basis in individual rights and autonomy, modernist morality becomes "permissive" and "relativistic."[34] Elsewhere he makes the point this way:

From a modernist standpoint, morality is subjective. Some people may have the opinion that certain conduct is immoral, but others have an equal right to disagree.[35]

We must challenge such misstatements and non sequiturs.

First of all, the modernist philosophers Johnson cites as the progenitors of liberalism certainly did not think of morality as relativistic or subjective, nor do the vast majority of philosophers today. Relativism certainly does not follow from respect for individuals' autonomy, for that simply holds that individuals are responsible for their own moral choices and actions, and this view is equally at the heart of any religious morality that appeals to a person's conscience. Indeed, the notion of autonomy is most closely associated with the philosophy of Immanuel Kant, who is a theistic liberal rationalist on Johnson's definition, but who has as radically antisubjectivist and antirelativist an ethical system as one can imagine. According to Kantian ethics, rational autonomous agents are able to discover the moral rules that fix their ethical duties by careful application of reason, and these moral imperatives hold universally and absolutely (no person is exempt from the rules and the rules have no exceptions).

If anything, even without the theistic element, respect for individual moral autonomy more clearly generates rather than removes obligations. Similarly, to say that people have a right to disagree about their moral opinions is not to say that morality is subjective. The defense of the right of individuals to hold and express different moral opinions is not a declaration that those opinions are all of equal merit, but rather a defense of a political right that is itself rooted in moral principles. As for whether liberal rationalism is "permissive," that judgment depends upon how much one thinks should be forbidden or restricted.

Johnson seems to think that modernist naturalism does not forbid nearly enough and that that is the cause of the social evils we see. Continuing from the previous quote, he writes:

> The term *morality* in modernist usage is usually associated with sexual morality, and therefore with traditional prohibitions of sodomy, fornication, adultery and abortion. With respect to such matters involving personal taste and one's own body, consenting adults must be free to do as they like.[36]

Given that Johnson opposes modernism, the clear implication here and elsewhere is that people should not be free to choose in such matters.

This may seem to be taking us rather far afield from the issue of creationism, but Johnson tells us that in debates with evolutionists the topic invariably turns to sexual politics. I can only think this is because creationists like Johnson regularly argue that such matters are at stake. Even without an evolutionist to bring up the matter, in *Reason in the Balance,* Johnson introduces issues of sexual morality in every chapter. The majority of his examples of the purported conflict between theistic and naturalistic worldviews involve sex: chastity versus premarital sex; faithful, stable marriages versus adultery and divorce; heterosexual versus homosexual relationships; flexible gender roles versus "proper" roles of male and female. He *twice* tells us that modernism has led to sex-education classes in which girls practice unrolling condoms over cucumbers.[37]

It is probably true that the person on the street associates morality with prohibitions regarding sexual behavior, because these are likely to have been foremost among the moral rules that one is taught as an adolescent. What is surprising and quite incorrect is Johnson's statement that modernist philosophy usually makes this association. Issues of sexual morality form a small subset of the topics that ethicists consider. To take as examples a few of the modern philosophers Johnson mentioned, Hobbes's main concerns involved political rights and obligations and the justification for government; Smith wrote mostly about economic liberty; and Mill's main concern was how a government could form ethical public policies that avoided the irresolvable deadlock of differing religious viewpoints, each of which appealed to revealed Truth.

On the other hand, Johnson's emphasis upon sexual morality is not surprising since this is among the main moral concerns of the religious right and it is here that we find most creationists. What is the connection? It is their idea that knowledge of morality and of the world has its source in the authority of the Bible, rather than in some other way, and that each depends upon the other. Here again we see the sharp delineation of the worldviews:

> The rationality of any moral code . . . is linked to a picture of reality that contains both fact and value elements. . . . The Christian story is one of human beings who are created by God, but who are separated from God by their own sin and must be saved from that sin to become what they were meant to be. The Enlightenment rationalist story is one of human beings who escape from superstition by mastering scientific knowledge and eventually realize that their ancestors created God rather than the other way around.[38]

According to Johnson, modern society has wrongly abandoned belief in the authority of the Bible. He complains that modernists hold that God is dead and thereby reject the Christian's appeal to the authority of the Creator as being irrational. Johnson wants to resurrect the legitimacy of appeal to biblical authority. Later, we will see what is wrong with Johnson's analysis of modernist naturalism and its relation to God and morality; for the moment let us briefly examine how he applies his analysis to the cases of education and, his specialty, the law.

Johnson devotes a long section of *Reason in the Balance* to criticizing Yale law school professor Bruce Ackerman's defense of liberal rationalism in *Social Justice in the Liberal State*. He especially decries Ackerman's application of this anti-authoritarian philosophy in education, in which "[t]he goal is to produce self-defining adults who choose their own values and lifestyles from among a host of alternatives, rather than obedient children who follow a particular course laid down for them by their elders."[39] Johnson wants a return to an authoritarian education that teaches children to be what their parents wish them to be, whether that be a priest or businessman for Jack or a mother and homemaker for Jill.[40] What are the effects of abandoning the authority model in education? A society in which Johnny cannot tell right from wrong. And what if Bible-believing parents in some school district want to have their children learn arguments against evolution or to be taught to save sex for marriage? Johnson says that modernists in the media will ridicule them as irrational religious extremists and that the educational bureaucracies will see that "rationality" prevails. Furthermore, the court system will back this up.

For much of the past in the West, "lawmakers assumed that authoritative moral guidance was available to them in the Bible,"[41] but the legal system, too, is now infected with modernist naturalism. According to the standards of naturalist rationality that the system now assumes, says Johnson, adultery may be wrong because it is a breach of contract or because of the damage it causes to human relationships, but since the Creator God of the Bible is as unreal as Zeus, "to condemn adultery merely because God forbids it would be, in modernist terms, irrational."[42] Johnson's view is the opposite, that adultery and other sexual immoralities are immoral merely because God forbids them, and that it is rational to believe so and proper to set public law upon this base.

And what are the "fact elements" of the picture of reality mentioned in the previous long quote that supports the rationality of this moral code? They are the traditional creationist theses that human beings are not the product of evolution by natural selection but rather were specially designed and created by God for some divine purpose. For the creationist, the fact and value elements of the Creation story are inseparably linked. For instance, that Genesis reveals that God purposely created us "male and female" is critical for understanding that men and women should have different moral roles. Creationists hold that the Bible is authoritative on all such issues. Johnson recommends that the law reflect these fundamental truths, and he suggests that the Supreme Court erred in its rulings against teaching creationism in the public schools.

If a high-school curriculum incorporates the subject of biological origins, and if supernatural creation is a rational alternative to naturalistic evolution within that subject, then it is bad educational policy as well as viewpoint discrimination to try to keep students ignorant of an alternative that may be true.[43]

The last part of this quote, to the effect that divine Creation may be *true*, takes us to the next major element of their concern.

The Desire for Absolute Truth

As we saw in ICR's list of the good fruits of creationism, all of them are true. Johnson thinks that defenders of evolution try to avoid talking about truth. He thinks creationists know the truth about biological origins and that evolutionists set up "rules of science" that exclude creationism specifically to suppress that truth. Evolution appears to win only because the scientific priesthood stands guard at the door to defend their dogma. Johnson devotes a full chapter to a discussion of truth, taking the work of two major philosophers, Richard Rorty and John Searle, as exemplars, respectively, of the pragmatist and "traditionalist" views of truth, and tries to show that both are infected by naturalism. He wants creationism to be considered as a candidate for absolute truth, and claims that this concept does not even make sense in the non-theistic framework of naturalism.

This emphasis on truth as an absolute seems almost quaint in an era dominated by naturalism and hence by pragmatism. Pragmatism is less concerned with what is absolutely true than with what is useful for some specific professional agenda (the scientific outlook) or for some worthy social program (empowering the victims, saving the environment). The very idea of absolute truth, independent of and superior to the consensus of opinion among the most educated people, is fundamentally a theistic concept that makes little sense in terms of modernist metaphysics.[44]

This remarkable set of statements is the most important passage in Johnson's discussion of truth, and one would have to take several pages to enumerate and explain the half-dozen ways in which it is seriously mistaken or misleading. Because in this chapter I am only interested in these issues as they bear upon the problem of meaning and value, I will just briefly mention a few of them here without going into detail.

First of all, it is not clear what Johnson means by "absolute truth." One standard meaning of "absolute" in this context is to express a contrast with *relativism*. Johnson does claim that relativism is one of the bad consequences of modernism, but if that is his intended meaning, then opposition to relativism is certainly not antithetical to science, and can be found among many naturalists and even among those who hold some form of pragmatism. Another possible, though less common meaning of "absolute truth" is that it refers to *degree of certainty*, so to claim absolute truth is to claim *absolute certainty*. If this is Johnson's intended meaning, then it *is* fair to say the idea is "quaint," for at least in science one is rarely in a position to claim complete certainty in the truth of scientific conclusions. In science, one typically makes only some probabilistic assessment of the degree of warranted belief (perhaps expressed in terms of a statistical significance level) in the truth of a conclusion, and even such assessments are taken to be revisable in principle should countervailing evidence be found. Finally, one might take "absolute truth" to indicate *absolute precision*. If this is Johnson's intended meaning, again he is correct, because scientific claims of truth are almost always to be understood in terms of approximation. That is, when scientists claim that a law is true what they mean is not that it is perfectly precise, but that it is accurate to within some stated limits of tolerance. However, being careful and explicit about margins of error and not claiming absolute precision is one of the strengths of science. Thus, if Johnson demands absolute truth in the first sense (i.e., of being contrasted with relativism), then it is not "quaint," let alone incompatible with naturalism. If he means it in the second or third sense, then absolute truth *is* now disavowed by science and, I think we should agree, quite properly so.

Second, Johnson's characterization of pragmatism that specifies utility in terms of "professional agenda" or "worthy social program" is incomplete and seriously misleading. A much more common pragmatic notion of utility might be couched in terms of usefulness for "getting around safely in the world," and this hardly has the ideological baggage that Johnson's examples suggest.

Third, it is not true that pragmatism is a logical consequence of naturalism; one can find almost every type of theory of truth among naturalists. It is also wrong to describe pragmatism in terms of "the consensus of opinion among the most educated people," which falsely suggests not only that the pragmatic notion of truth simply deals with combining opinions but also that it involves a kind of academic elitism.

The last two claims, that absolute truth is "fundamentally a theistic concept" and that it "makes little sense in terms of modernist metaphysics," are also wide of the mark. In many of these cases, perhaps because he takes Rorty as his example, Johnson winds up attributing to modernism points of view that are closer to postmodernism. As the name suggests, postmodernism is a more recent philosophical view that is a reaction against modernism. It typically rejects the modernist ideas of rationality, objective knowledge and mind-independent reality, holding instead that reality is "constructed" rather than rationally discovered and that what we call "knowledge" is nothing more than the specific constructed fiction that those in power are able to enforce. On the postmodernist view, science has no special privilege to interpret the world, and its high status derives from the cultural-political power of its practitioners. Given that postmodernism is almost as opposed to the scientific worldview as Johnson is, it is particularly unjustifiable that he conflates them as he does.

Although Rorty is certainly the most important representative of one influential form of the new pragmatism, his is hardly the only version of that theory of truth. Furthermore, pragmatism is not necessarily the enemy of theism. Indeed, the generic notion of pragmatism—interpreting truth in terms of the test of what "works" (that is, in terms of outcomes)—allows a range of possible specific alternatives, many of which are compatible with or even support some forms of theism. For instance, there is an element of pragmatism in the most common Christian test of truth: "By their fruits ye shall know them." (Remember it was this very test that evolutionism supposedly fails.) In this advice Jesus was explaining that one can distinguish true Christians and true Christian doctrines from false ones by seeing whether or not they work to produce, say, a loving community, a heavenly kingdom on earth. Because Johnson wants to blame naturalism for all that is bad, he paints a greatly oversimplified picture of the philosophical landscape.

We see this again in his discussion of Searle, taken to represent the "traditionalists," who holds to a correspondence view of truth. On this view a statement is true if it reflects what is so in the world. Although Johnson does note that this notion of truth includes the ideas of objectivity and reality that he likes, he immediately attacks Searle's view, which he characterizes as being essentially tied to reductive materialism. It appears that Johnson finds reductionism unacceptable because it makes the human mind into nothing more than a form of the physical and rules out the possibility of dualism and vitalism. These views say, respectively, that mental activity and life-processes involve special substances that are ontologically distinct from physical matter and energy, and thereby stand outside the laws of physics and chemistry.

Johnson's conflation of Searle's "traditional" notion of truth to reductionism is another case of his oversimplified philosophical discussion, for Searle is perhaps best known for his work opposing simple reductionist accounts of the mental.[45] As in other cases, Johnson ignores a wide range of intermediate viewpoints and presents only a stark dichotomy. Explaining the reductionist, evolutionary, naturalist viewpoint to his fellow creationists he writes:

You may think that humans are created in the image of an omniscient God, magnificently endowed with minds that fall far short of their potential because they are flawed by sin, but what we really are is baboons with surplus neurons that caused us to imagine God before science gave us knowledge.[46]

Setting the simplistic dichotomy aside, we can agree that Johnson is correct to say that science thinks of dualism and vitalism, at least in their naive forms, in much the same way that it thinks of creationism. However, this should help make it clear why his general charge that science is dogmatically atheistic is unjustified. Science does not impose a special rule to prohibit using God as an explanatory hypothesis specifically to exclude creationism or because of a general prejudice against religion or theism. Rather, science rejects all special ontological substances that are supernatural, and it does so without prejudice, be they mental or vital or divine.

These points about the nature of truth and of the mind deserve greater attention, but they are relevant to the main topic of this chapter only insofar as they play a role in the central problems of value and meaningfulness and the way these figure in the creationists' concerns, so I set them aside here and return to the main discussion.

Is Evolution to Blame for the Loss of Values?

To say that life is meaningful is, in part, to say that our lives are valuable. The fear of meaninglessness arises out of a fear of loss of values. Although we might disagree about some of them, we have seen that Johnson and other creationists do believe that the values they hold dear are at risk.

But what does this have to do with evolution? Creationists believe that evolution is to blame. Johnson thinks that most other people deceive themselves into believing otherwise or else just try to ignore the issue, but that everyone who has thought about it understands, for example, the conflict between evolution and morality. Discussing the influential 1897 lecture "The Path of the Law" by Supreme Court Justice Oliver Wendell Holmes, Johnson writes: "As a convinced Darwinist who profoundly understood the philosophical implications of Darwinism, Holmes found it difficult to take morality seriously."[47] The philosopher of biology winces to read such glib statements about the relation of evolution and ethics.

I cannot say whether or why Holmes drew this faulty inference, but it is easy to locate Johnson's error. Johnson holds that the basis for law, education, philosophy and morality lies in a culture's creation story: "If we want to know how we ought to lead our lives and relate to our fellow creatures, the place to begin is with knowledge about how and why we came into existence."[48] Johnson does not say enough about how knowledge of our origin is supposed to help us know how we ought to live our lives for me to evaluate his position fully but, on the surface at least, it appears that his view commits either the genetic fallacy or the naturalistic fallacy or both.

Darwinian evolution, Johnson says, matters only secondarily as science; its primary importance is that it has become the West's dominant creation story. On this view of the ground of values, it is no wonder that law, education, philosophy, and morality are in crisis, if they are based upon an inherently meaningless creation myth. Johnson reiterates this point: "[The] naturalistic creation story implies that knowledge of the Creator's mind and purpose is inherently illusory; the true creator—evolution—has no mind or purpose."[49] This takes us full circle to the problem with which we began.

The critical issue for the creationist is not really about the truth or falsity of evolution as a descriptive and explanatory scientific theory, or even about the validity of "creation-science" as a scientific alternative, but rather about their relative viability and worth as value-grounding creation stories. Creationism tells of a world that God planned with us in mind in which we have special roles to play that, properly understood and followed, will fill our lives with meaning. "Evolutionism" tells of a godless, material world in which we are the accidental result of meaningless mechanical processes that no more had us in mind than aphids or fly larvae. Creationists fear that if evolution is true, then the only basis for value, the only source of purpose, the only foundation for meaningfulness would be lost.

We can see this expressed by other creationists as well, such as Fred Heeren:

If our universe came about by some strange fluke and there is nothing outside of it, no purposeful Creator beyond its time and space to value it or give it meaning, then it must *remain* without meaning. The universe can't generate its own meaning or value any more than a rare rock sitting on an uninhabited planet can ever be valuable sitting there all by itself.[50]

Hugh Ross, the old-earth creationist, says much the same thing of the evolutionary worldview:

A mechanical chain of events determines everything. Morality and religion may be temporarily useful but are ultimately irrelevant. . . . On the other hand, if the universe is created, then there must be reality beyond the confines of the universe. The creator is that ultimate reality and wields authority over all else. The creator is the source of life and establishes its meaning and purpose. The Creator's character defines morality.[51]

This notion that the Creator's authority alone grounds morality—indeed, that this is the only possible ground for meaning—is the important assumption that underlies and drives the creationist existential worry. We have already seen how it runs through Johnson's book: Students should follow the authority of their teachers who will mold and shape them. Children should obey the authority of their parents who brought them into the world. We, as children and creations of God, should follow God's authority. Being moral involves recognizing and obeying our Lord's commandments. According to this view, morality, value, and meaning are necessarily based on the authority of the Creator who brought us into being *with a purpose in mind*.

For Johnson, this idea is so central that it becomes part of his very definition of what it is to be a theist. He writes: "[T]heists . . . believe . . . that we were created by God, a supernatural being who cares about what we do and has a purpose for our lives which is to be fulfilled in eternity."[52] Given this understanding of the source of purpose, it is easier to see why creationists believe that evolution threatens meaningfulness. The deep connection between God and value comes out clearly in another passage in which Johnson says, "In fact, one way to define theism is that it is a story about the universe that proclaims the reality of the true, the good and the beautiful."[53]

If believing in evolution meant rejecting truth, goodness, and beauty then who would not want to be on the side of the angels? However, this idiosyncratic notion of theism is just another example of Johnson's creative redefinitions, and, in any case, neither evolution nor scientific naturalism is in any way contrary to truth, goodness, or beauty. More to the point, neither are they incompatible with the existence of God, though Johnson, in his effort to paint the world in the black-and-white terms of his two worldviews, would have us believe otherwise.

As do other creationists, Johnson dismisses one after another alternative or compromise position and chides Christians who defend evolution or who believe that religion can be reconciled with scientific naturalism. He calls such theologians "theistic naturalists."[54] Since, on Johnson's definition, naturalism denies God, the notion of a theistic naturalism is a contradiction in terms and Johnson's implication is that such Christians hold a logically absurd view and delude themselves in believing otherwise. Johnson rejects such attempts to chart out a course of reconciliation as "accommodation" to scientific naturalism.[55]

In his earlier works Johnson tried to appear to defend only the ecumenical notion of creationism as the view that God creates for a purpose, but that apparent tolerance quickly evaporates whenever he needs to draw the line more clearly between the two worldviews. Values and meaning rest upon matters of fact, in particular the truth-status of specific claims about Creation. He puts the matter most starkly in two passages:

If . . . the universe was created by God for a purpose, the truth claims of Jesus Christ may well be credible and meaningful. Those claims are not even conceivably credible or meaningful if the universe is a meaningless chain of material causes.[56]

Christianity makes sense only if its factual premises are true and if it is providing meaningful answers to questions that people ought to be asking. The essential factual premise is that God created us for a purpose, and our destiny is a glorious one in eternity.[57]

We have already seen how Johnson claims that scientific naturalism leads to atheism and ethical chaos, erodes family values and the dignity of the human mind, and undercuts the basis of law and moral education. In the two passages above, he raises the stakes to even more incredible heights. Is accepting scientific naturalism really tantamount to calling Jesus a liar? Must Christianity itself topple if evolution is true? Does the hope of eternal life disappear with Darwinism?

Calming the Creationist's Fears

The purpose of this chapter is not to respond in turn to the many specific allegations that Johnson makes, but rather to address the general malaise that underlies them. Johnson is only standing in here to represent the worries of a wide range of creationists because he has articulated those worries most systematically. Our general goal has been to see what creationists believe to be at stake in the debate and why, and then to try to alleviate the dread that their view engenders. Having identified the creationists' existential crisis of meaning and traced it to its source, we can now better understand why they are so fearful of evolution, and how their leaders' divisive rhetoric is likely to play upon these anxieties in others. Let us now focus upon how we might relieve such anxieties. There are several ways we could try to do this.

We can offer reassurance indirectly by continuing to identify errors in the creationist leaders' arguments about the relationships among evolution, naturalism, morality and meaning. In this chapter I have highlighted several faulty assumptions, imprecise concepts, ambiguous definitions, and fallacious inferences in Johnson's recent writing, but much more could be done to show why his two-worldview dilemma is vastly over-simplified and that theists need not accept the either-or predicament he tries to set up. However, we should not resort to a merely negative campaign but try to offer some positive reassurance as well. Here I recommend two approaches. First, we should point out that the possibility of moral values does not depend upon the authority of God, so a meaningful, value-filled life would be possible even if evolutionary naturalism did imply the "death" of God. Second, and most important to reassure the troubled theist, we can show that evolution does not imply that God does not exist, and that scientific naturalism is not equivalent to atheism.

Meaning Independent of God

Starting with the first point, let us reconsider the worry that if there is no God then life would be meaningless. As we have seen, creationists believe that value, morality, purpose, and meaning necessarily depend solely upon God's authority and commandments. But why should we assume this?

Beginning with a few simple psychological observations, we should note that the world is full of people who do not believe in God, and yet find their lives to be meaningful. The creationists' notion of a personal God as described in Genesis who created us for a specific purpose is a well-known but a minority spiritual view. It would seem very odd to claim that people who do not share the creationists' notion are simply wrong when they report that their lives are valuable and fulfilling. That a creationist would find life meaningless if the God of Genesis is undermined by evolution tells us more about creationists than it does about meaningfulness.

If we ask people what is most valuable in their lives, what gives their lives meaning, we get a wide range of answers. Certainly, some people will cite their faith in God (though for most of these their faith does not depend upon whether or not the Genesis account is literally true). Many more will mention the pride and joy they feel for their children, the tenderness they feel for their lovers and friends, the sense of accomplishment they derive from their work, the pleasure they receive from music and art, or the deep satisfaction they feel in the struggle to build a better tomorrow. People find value in a well-crafted novel and in a well-cooked meal, in vigorous athletic activity and in quiet moments of reflective contemplation. They find purpose in the building of a home, the furtherance of social justice, and the pursuit of scientific knowledge. How easy it is to extend such a list! Thus, the creationists' fear that life would be devoid of meaning without belief in God seems ill founded—value and purpose, at least in the straightforward, felt psychological sense, can be found at every turn.

But what if we expect more? Suppose that one wants not just the feeling of value, but values that are justified. A creationist could rightly complain that psychology cannot supply more than a subjective, individual notion of value and meaning. On this point the philosopher and the creationist can agree—by itself the simple identification of individual psychological value does nothing to justify those values. We would not want to fall into a form of subjective ethical relativism, which is antithetical to the most basic meaning of morality. But now the creationist and the philosopher once again part ways, for the former holds, as we saw, that only God's authority could ground morality and value. There are two parts to the philosopher's disagreement here.

The first is to note that rejecting relativism is only the first step in ethics, usually accomplished in the first week of one's freshman philosophy class, and that it is at this point that the positive work of moral philosophy begins. We need not leave the values mentioned above behind as merely psychological, for there is a whole history of ethical thought that is available to justify their worth. I will not attempt to present even a partial list of the wealth of substantive work that has been done in ethical theory that supports objective moral values, but will just briefly note one solution to the existential crisis, that offered by existentialism itself. I mention this not because I think it is the best approach, but because the existentialist accepts the creationists' worst fears about purposelessness and the subjectivity of values and yet finds a way to continue forward.

According to the existentialist, we are right to feel worried about meaninglessness because the world really is meaningless. We are moral beings in an amoral world, so it is quite understandable that, thrown into such an absurd situation, we might wonder whether life is worth living. Nevertheless, let us not give in to despair, counsels the existentialist, for as moral beings we have the freedom to interpret the world as we will, and thereby to impart meaning to life. If we are not given a purpose, we can generate our own purposes. We can thumb our noses at meaninglessness and rise above the amoral contingencies of the world, creating value as we go, by the choices we make and the actions we take. This is a philosophy that challenges us to be masters of our own fate and to carry on in the face of hopelessness. Think of the Man of La Mancha as the existentialist hero who conjures a world of knightly responsibility and honor. That he tilts only at windmills, rather than true giants, does not take away from the valor of his deed or his courage of character. That his Dulcinea is no true lady does not diminish her value to him or the true chaste love he feels for her in his heart.

The second point of disagreement is stronger: God's authority cannot serve to justify morality and value. This might seem paradoxical or even arrogantly irreverent at first but it is in fact neither, and is a widely acknowledged point among both philosophers and theologians. The classical version of the argument comes from Plato, who makes the point in the *Euthyphro*[58] with a simple but profound question, put here in contemporary language: Is something good because God commands it so, or does God command it because it is indeed good?

Let us suppose the first to be the case—Johnson's and the creationists' view—that moral value comes only from God's authoritative word, that moral value is *by definition* that which God commands. If so, then if God commands us to love one another, then loving one another is morally good, by definition. However, it is equally true on this view that if God were instead to command us to hate and enslave all those who are of a different race then, still by definition, the hate-filled slave-holder would be morally good and praiseworthy. Similarly, if God were to have created us such that our purpose was to kill each other for fun, then the peacemaker would be a demon and the serial murderer would be a moral saint, again by definition. Surely such conclusions about morality are absurd. Indeed, the creationist will likely say that such ideas move beyond irreverence into blasphemy and that it is impossible to think that God would ever command such immoralities. However, notice that such a reply would contradict itself in the mouth of someone who says that morality is merely that which God commands. Johnson and the other creationists cannot truly be serious that morality is simply what God commands or that our purpose is just whatever God chose for us in Creation. Given their other beliefs, it must be simply a mistake that they proclaim this view. We can put the point another way. If we hold the creationists' view consistently, then such claims as "God is good" or "God is omnibenevolent" would be reduced to vacuous trivialities instead of being important statements with content about the character of the Creator. Again, we know creationists cannot seriously believe this. Plato's point is that this view—that God's authority as the origin of value—is fundamentally flawed. It is rather the second view that makes more sense, namely that God commands something because it is indeed good. That means, therefore, that goodness must have a basis that is independent of God. The lesson for us here is that the creationist version of the existentialist fear is ill-founded—the possibility of value, purpose, and meaning are not lost even if God does not exist.

Evolution Does Not Imply Atheism

Plato's *Euthyphro* argument showed us that the authority of God per se does not provide a ground for morality—loving one's neighbor is not good simply because God commands it; rather, God commands it because it is good. Morality could thus have an objective ground whether or not there is a God. Therefore, even if evolution implies the death of God it would not imply the end of objective values. An atheist who thought that atheism obviates moral responsibility would be making the same mistake the creationist makes. However, let us now consider as a worst-case scenario the creationist who is not satisfied with objective values but holds that God's existence itself is essential for a meaningful life in some other way, perhaps to guarantee an eternal afterlife. Can we assuage this fear? We can, because evolution does not preclude the existence of God.

Creationist activists often note that Darwin lost his belief in God after he came to the evolutionary viewpoint. They speak of Joseph Stalin who had been a theology student but who, they say, upon reading Darwin became an atheist and led the Communist purges, slaughtering thousands of Russians. They mention prominent contemporary biologists who they say are atheists. Darwin's biographers do not think that his loss of faith is so simply attributable to his discovery of evolution,[59] and I doubt that we can blame the theory of evolution for all of Stalin's atrocities, but creationists are no doubt right that some biologists are atheists, and it is probably true that some have become atheists because they thought that Darwinism led to atheism. However, all this is irrelevant. The only relevant question is whether Darwinism actually does imply that there is no God. Just as we should not condemn Christian morality in general simply because some people have committed horrible crimes in the name of Christ, so we should not condemn evolutionary theory simply because some people have illegitimately leapt to conclusions from it about the death of God.

What does evolution imply about the existence or nonexistence of God? To answer this question, let us quickly review the central elements of evolutionary theory. The first is the basic general fact that populations of organisms change over time in such a manner that new species arise from modifications of their ancestors. A second element is that the pace of this change is more or less gradual; though the rate might not be regular. A third element involves the reconstruction of the pathway of evolution, the ancestor-descendant relationships among species that form the tree of life. A fourth element is the mechanism (or more properly, mechanisms) of evolution, especially the Darwinian mechanism that involves the nonrandom natural selection of randomly generated heritable variations. Nowhere in evolutionary theory does it say that God does not exist, for the simple reason that, like cell theory and relativity theory and quantum theory and every other scientific theory, it says nothing at all about God. But to say nothing about God is not to say that God is nothing.

Evolutionary theory is naturalistic in just the same way that all scientific theories are, in that they proceed without any appeal to any supernatural entities or powers. Given that this is true of science generally, why should evolution be any special worry to the theist? If it is science's naturalistic methodology that is inherently problematic, then Johnson should be equally worried about chemistry and meteorology and electrical engineering. He should also be concerned about automobile mechanics, for this field too proceeds under the naturalistic assumption that God does not intervene in the workings of the motor. But surely no one thinks that these naturalistic sciences imply that God does not exist.

One conclusion about God that we *can* draw from evolutionary theory is that God is not necessary to explain the modification of species one into another or the adaptedness of organisms to their environment. This might be thought to be a threat to theism in that it undermines one positive argument for the existence of God. One version of what philosophers call the *teleological argument* tries to prove the existence of God by saying God is necessary to explain the apparently designed character of creatures and the fit of organisms to their environments. The most famous example of this sort of argument, which Darwin had read and appreciated as a student, is William Paley's classic argument that as we infer from the complexity of a pocketwatch to the existence of a watchmaker, so we may infer from the even more intricate complexity of the world to the existence of a divine world-maker.

Philosophers had found the faults in this version of the teleological argument well before Darwin's theory arrived on the scene, but evolutionary theory put the final nail in its coffin by showing in a clear and simple way that God is not needed to explain the biological world since the evidence points to a natural process that accomplishes the same results. Johnson focuses on just this point, claiming that the whole issue rides on whether the Darwinian mechanism really can function, in Richard Dawkins' words, as a "blind watchmaker." He gives the impression that if the evolutionary mechanism is the true explanation of biological phenomena—what he calls "the blind watchmaker thesis"—then all is lost. Now it is true that, if someone thought that the biological version of the teleological argument was the *only* reason to believe in the existence of God, then evolution would indeed be likely to lead that person to atheism. However, this is a long way from saying that evolution implies atheism. Perhaps for some people, perhaps even some creationists, biological complexity is the sole support for their belief in God. If so, we could understand why they would defend it against evolution with such vigor. Most theists, however, have a more robust faith, and it is important that they not allow creationists to mislead them, with a simplistic two-worldview dilemma, into believing that they must choose between evolution and God.

Defenders of evolution should be clear about this as well. We tend to allow creationists to set the agenda. Biologists who accept a challenge to debate from the Institute of Creation Research easily agree to their conditions that religion not be mentioned and that the debate be limited to the contest between evolution and creation-science. ICR boasts that Duane Gish has won every one of the hundreds of debates he has had.[60] How can this be, given the weaknesses of the creationist arguments? Clearly, just presenting the scientific evidence for evolution in a debate with a creationist will be of little use if the audience feels that truth, beauty, morality, and their Christian faith are at stake and riding on the outcome. Even if one is not able to make explicit the other arguments we have discussed, at the very least, defenders of evolution need to reassure their listeners that evolution is not synonymous with atheism, since that fear is at the root of all the other worries. How can we make this point clear?

First of all, there is a simple matter of logic: "not necessary that *X*" does not imply "necessarily not *X*" or even "not *X*." It is easy to get confused about such modal locutions and one can understand that someone could mistake the weak claim that evolution shows that God's existence is not necessary to explain biological phenomena for the stronger claim that evolution shows God's non-existence or the necessity of God's non-existence.

I find very few evolutionists who make this simple sort of mistaken argument, but I can certainly support Johnson and other creationists' complaint when they find someone who does.

Second, losing one positive argument for the existence of God, even an important one, does not undermine other positive arguments. The germ theory of disease gave a mechanistic explanation of the origin and transmission of infectious diseases like the plague, obviating the "need" to postulate that such diseases were a divine punishment for moral degeneracy or disobedience of God's laws. The germ theory of disease was naturalistic in the same way that evolutionary theory is, but we do not think that showing that God is not necessary to explain plagues implies that there is no God, since undermining that positive argument does not impugn other ones. Of course evolutionary theory is powerful in that it undermines a whole class of such arguments, but theists historically have offered a wide variety of reasons for belief in God, including other versions of the teleological argument, which evolution does not touch.

Finally, no matter how strong the scientific evidence for any empirical conclusion, from a merely logically point of view it can never completely negate the possibility of the existence of God, for the theistic realist always has a way to bring God back in at any point. As an omnipotent, supernatural being, God could intervene at any or all points in the process, either to create organisms wholesale or to guide evolutionary development, and to do so in a way that we could never discover scientifically. An extreme example of this is the "appearance of age" creationist view, which holds that the world was created by God 6,000 years ago as the Bible says, but in such a way that it appears to be much older. More subtle versions, such as one extensively discussed on one of the creationist electronic discussion groups, suggest that God creates the mutant variations upon which natural selection works, but in so nuanced a manner that they appear random to us. Evolutionary theory can never disprove such views for the same reason that science cannot disprove philosophical skepticism, so even Genesis literalism will always remain a logical, if not a reasonable, possibility. But it is exactly because such creationist views are radically untestable that creationism falls outside the realm of science. Science excludes appeal to supernatural entities as a point of method, and thus it is improper to draw directly the atheistic conclusion that God is ontologically unreal from evolution or any other scientific conclusion. Such questions are not scientific and must be left to the theologian and the philosopher.

In most of his writings and talks Johnson quotes Richard Dawkins's comment that Darwin's theory finally made it possible to be "an intellectually fulfilled atheist," suggesting that upon the truth or falsity of evolution rests the entire issue of atheism versus theism. But we are now in a position to understand the force and limits of Dawkins's claim. He is not saying that evolution makes one an atheist, but that evolution allows someone who is already an atheist to feel intellectually satisfied. Why? Because, prior to Darwin, the atheist had no good natural explanation for biological phenomena. Evolutionary theory filled the large explanatory gap that had made the atheist feel ill at ease. Perhaps Dawkins is one of those for whom this was personally the last significant gap to be filled, and, if so, it could be that he thus thinks that evolution does lead to atheism. But we should now be clear that this is not a foregone conclusion. Evolution does not necessarily lead to atheism, and if defenders of evolution regularly made this clear it might open the fearful hearts of their audience, which is the first step to opening their minds to the evidence.

Johnson does occasionally admit, usually in a footnote or appendix, that evolution is not necessarily atheistic. For example, he writes:

The blind watchmaker thesis makes it *possible* to be an intellectually fulfilled atheist by supplying the necessary creation story. It does not make it *obligatory* to be an atheist, because one can imagine a Creator who works through natural selection.[61]

Elsewhere he acknowledges that: "Scientific naturalists do not claim to have proved that God does not exist."[62] However, as we have seen, Johnson forgets this critical fact in places where it counts most. He forgets it in his discussion of the basis of natural law in which he says that "From a naturalistic standpoint . . . the Creator God of the Bible is every bit as unreal as the gods of Olympus, and the commands of an unreal deity are in reality only the commands of an ancient priesthood."[63] He forgets it in his condemnation of contemporary liberal rationalism, which he says "incorporates the naturalistic doctrine that God is unreal, a product of the human imagination."[64] He forgets it in his original definition of naturalism: "*Naturalists* . . . assume that God exists only as an idea in the minds of religious believers."[65] I could easily cite a dozen or more similar statements or implications throughout *Reason in the Balance* and many more in Johnson's earlier works, in which his central claim is that evolution rests upon a dogmatic naturalistic metaphysics that says the world is a closed system of material causes and effects and that "that's all there is."

We have been considering three main views one could take regarding the logical relation between biology and God: that the biological facts make the existence of God (1) impossible, (2) possible, or (3) necessary. We have seen that evolution does not imply that God's existence is impossible, only that God's intervention is unnecessary to explain the biological facts. Evolutionary science, like all other sciences, is neither theistic nor atheistic in the ontological sense, but is agnostic, leaving God as a possibility that is outside the boundary of its methods of investigation. But Johnson is not satisfied with the compromise view that scientific naturalism allows God as a possibility. Immediately following each of the above brief admissions that evolution is not atheistic Johnson quickly dismisses the left-open possibility of God as unsatisfactory. It is not surprising that Johnson wants his readers to reject or ignore the intermediate position and its many variants, since it shatters the two-worldview dilemma that he wants to set up. Defenders of evolution would help their case immeasurably if they would explicitly reject the creationists' contention that evolution is atheistic, and reassure their audience that morality, purpose, and meaning are not lost by accepting the truth of evolution.

Can We Seek Meaning Together?

I have argued that creationists are not primarily worried about the status of evolution as scientific theory, nor are they simply repelled by the idea of being descended from apes; rather, they are motivated by a crisis of meaning. Given their theological view that the only source of morality and value is in the divine authority, they fear that life would have no meaning if we were formed by the purposeless processes of evolution rather than by the direct purposeful creative will of a divine intelligence. We have seen how creationists in general, and Phillip Johnson in particular, see the debate about evolution as a holy battle between two incompatible worldviews, one—theistic realism—that upholds truth, goodness, and beauty, and another—evolutionary naturalism—that undermines the same, leading to moral decay in every aspect of society, not to mention the loss of hope in an eternal afterlife.

With so much seemingly at stake, it is no wonder that creationists are unable to judge the evidence for evolution with an open mind. I have recommended that defenders of evolution attempt to calm these existential fears that creationist leaders engender in the way they frame the dilemma, by explicitly challenging the two-world dichotomy not just on the empirical level, but also on the level of values, reassuring their audiences that evolution does not imply atheism or moral nihilism. Meaningfulness can be found everywhere we turn, no matter how it was that we came to be. The meaning we can find in the beauty of nature, in the love of our families, in the expression of our hopes and ideals, or in the discovery of an elegant equation that expresses a natural law depends not at all on whether our capacity for appreciation arose in an evolutionary process or in a special act of creation. The way to assuage the existentialist fear will vary depending upon the specific form that the worry takes, but let me conclude by showing how one might address the fear in the most extreme sort of case.

Let us consider someone who is unmoved by arguments that morality and values are grounded independently of God and who, following Johnson, insists that it is by divine authority alone that our lives could come to have purpose and meaning. Let us also be forthright in stating the truth of evolution as best science understands it given the evidence at hand: we are descended from earlier primate forms that also gave rise to our closest evolutionary cousins, the pygmy and the common chimpanzees and the gorillas, though many of the details of our evolutionary pathway are still unknown, and that the mechanism that produced us involved natural selection of heritable random variations. Finally, for the sake of argument, let us even go beyond these empirical conclusions and hypothetically accept the unscientific metaphysical claim that Johnson unfairly attributes to evolutionary science, namely the dogmatic assertion that God in fact did not directly intervene physically in any stage of this creative process. Is such a person's life thereby necessarily rendered purposeless and meaningless? Certainly not. We are still the same beings with the same capacities and longings.

Whatever the purposes we might be expected to fulfill on Johnson's or some other creationist's view of origins, we are no less capable of fulfilling them by having arrived at these capacities by an evolutionary process. Ask creationists what they believe our purpose to be and then ask why we should fear that our evolutionary origin precludes our fulfilling it. Surely, nothing of value is necessarily lost by acknowledging the truth of evolution. Indeed, most Christians will probably answer that our purpose in life is to praise God, to accept Jesus as our Lord and Savior, to ask for spiritual grace and, as Johnson says, to ultimately fulfill our purpose "in eternity."[66] John Morris says that "our purpose is to have fellowship with [God] and to bring Glory to Him."[67] What is it about evolution that in any way rules out the pursuit of such notions of purpose and meaning? Even the confirmed atheist would have to admit that evolutionary theory by itself does not exclude such a possibility. If there is a God, then God could give us spiritual purpose no matter how we came to be.

When Johnson or another creationist says that we must choose between evolution and morality, or evolution and purpose, or evolution and meaningfulness, let us reject the dilemma and try to calm the fears that the purported choice engenders, if only with a rhetorical query: "Are you saying, Professor Johnson, that an omnipotent God, omnipresent in space and time, could not have allowed humans to arise through an evolutionary process while still giving us a moral code to uphold, a purpose to fulfill, and a life filled with meaning? Surely this would not be impossible for God, so certainly the choice is not so stark as you portray it."

Let us not polarize the debate and return to a view that holds religion to be necessarily at war with science. Christianity was eventually able to recognize the truth of Galileo's disturbing findings and to adapt to their implications, and most Christian denominations have absorbed Darwin's findings in a similar manner. Creationists no doubt find meaning in their picture of the world, but meaningfulness cannot be long sustained upon a falsehood. Evolutionary theorists are no less concerned about morality and about the meaning of life and should be forthright that these are not incompatible with evolution. If we are able to calm the divisive fears that evolution is the root of an atheistic philosophy that leads to purposelessness and immorality and reassure the creationist that evolution does not bar the roads to meaningfulness, then, and perhaps only then, will the creationist controversy be put behind us so that we can travel those roads together.

Committing Darwinism to the Flames

In a 1991 interview in the *Sacramento Bee,* Johnson makes what biologist Thomas Jukes rightly calls a "vile accusation," namely, that "scientists have long known that Darwinism is false. They have adhered to the myth out of self-interest and a zealous desire to put down God."[68] Although we now have a better appreciation of what Johnson takes to be at stake that leads to such a view, it will not be possible to seek meaningfulness in unity if he and other creationists continue to fan the flames in this manner, making preparations to burn Darwinism at that stake. We need to remember that there are other values at stake as well, and we must be aware that to toss evolutionary theory and scientific naturalism onto the pyre would be to commit much of the rest of science to the flames as well.

Often one hears creationists claim that evolutionary theory does not really do very much and that we could do very well without it. Darwin developed evolutionary theory only a century and a half ago, they say, and science operated just fine, thank you, without it before then. The impression creationists give is that evolutionary theory could simply be snipped out of science and discarded without much fuss or bother. I hope by now it is clear that this could not be farther from the truth. As Theodosius Dobzhansky pointed out, "Nothing in biology makes sense without evolution." Evolutionary theory is the unifying explanatory framework that provides us with an understanding of the complexities of the biological world. However, more is at stake for science than even the huge explanatory losses within biology that would result from burning Darwinism, for science is an interlocking network of theories and evolution is central and very well connected. The myriad empirical threads that are the evidential warp and woof of evolution, tethering it to the world with thousands of supporting facts, also bind the theory to every other part of the fabric of science. As we have seen, when creationists try to poke holes in evolution, they necessarily cut deeply into areas as far ranging as cosmology and geophysics. The sixteen chapters of Henry Morris's magnum opus, *The Biblical Basis of Modern Science,* reflect this in their subtitles, which include not only "Biblical Biology," but also everything from "Biblical Astronomy" to "Biblical Chemistry and Physics," and "Biblical Ethnology." We see the same pattern in the rejection of radiometric dating, continental drift, language evolution, and on and on. Equally serious would be the profound methodological losses that would result from the torching of scientific naturalism. Is it wise to burn the notions of scientific truth and objectivity? Do we really want to see this all go up in smoke at once?

Much is at stake for science and I hope that people will recognize that its value is worth defending against creationists fears and false accusations. But there is more at stake even than this, for science and scientific values play an important role in society. In burning science together with evolution, creationism also puts at risk other values that are important for American society generally. In the final chapter we turn to this most important issue.

8

Babel in the Schools

[T]o the extent that it was true in an earlier period . . . that evolution was not mentioned or was just mentioned very briefly and vaguely in the curriculum, I think the reason for that very largely was that the educators thought they ought to stay away from the issue of creation and not deal with the issue at all. Now what we're getting today is that they're dealing with the issue but only with one side of it. You see, they're saying that . . . since there is a separation of church and state we should indoctrinate all the children in naturalism. Now that isn't separating the church and the state, that's making naturalism into essentially an official religion of the state and so I'll say they can either stay away from it or they can deal with the issue honestly. But to say that the Constitution requires that we tell a loaded story is just not, in my opinion, correct.

—Phillip Johnson

Creationism in the Science Classroom

In January, 1995, the school board of Merrimack, New Hampshire took up a proposal of a local Baptist pastor, the Reverend Paul Norwalt, to give creationism equal time to evolution in science classes. Rev. Norwalt had recently held meetings in the high school with Duane Gish, co-founder of the Institute for Creation Research, who has been a tireless campaigner for creationism for over two decades. Norwalt also had already gained the support of two board members (one a member of his church, the other a Roman Catholic), who had been elected in May in campaigns that emphasized a return to education of the "three R's" and opposition to the school's health curriculum, which included distribution of condoms. Two other members of the board opposed the proposal and the fifth did not reveal his view. In September the board had voted in a proposal that instituted a daily silent prayer in Merrimack schools. The new proposal would replace current science textbooks with ones that taught that both creationism and evolution were "assumptions," and would also use the Bible as a text. A mid-February article in the *New York Times* described the resulting division among the townspeople over the proposal as well as the supportive attitudes of several officials in the state House Education Committee.[1]

In March 1996, the Tennessee state legislature considered a creationist bill introduced by State Senator Tommy Burks, whose home district is just 45 miles northwest of Dayton, where the Scopes trial took place 71 years earlier. His legislation would fire any teacher who taught that evolution was factual. The state's attorney general gave his opinion that the bill was unconstitutional, but the bill was nevertheless being supported by the conservative Eagle Forum and was expected to pass.[2] State Senator David Fowler took the floor in support of the bill, arguing: "If evolution is true, then it has nothing to fear from some other theory being taught; the truth will prevail. But if intelligent design is the truth, then God forbid we should not teach it to our children."[3]

These are not isolated cases; one finds, for instance, similar recent cases in California and Texas, the two states that set the model for most textbook publishers. In 1996 the school board of Sultan, Washington considered adopting the intelligent-design creationism textbook *Of Pandas and People* as had school boards in Alabama, California, Idaho, Texas, and elsewhere. In June 1997, the Texas State Board of Education held public hearings on its proposed curriculum standards, the Texas Essential Knowledge and Skills (TEKS), and one after another creationist speaker rose to testify against the inclusion of evolution in the science section, arguing that it is not a very important theory, or even really scientific, and thus should be omitted. Creationists suffered major legal defeats in 1982 and 1987 when state and U.S. Supreme Courts struck down creationist-sponsored state legislation, but they have regrouped in the last few years and have begun to pursue a new local strategy. Creationist "stealth candidates" run for school boards and reveal their agenda only once elected. They advocate breaking holes in the wall of separation between church and state wide enough to allow religion back into public education. Creationist speakers argue that the Constitution should not be interpreted to disallow people's free expression of their important religious views about divine creation of the geological and biological world when the topic of their origin is discussed in science classes. They claim that evolution and creationism are on a par as theories and that schoolchildren should be allowed to hear both and make up their own minds about which to believe.

This chapter examines some of the conflicting religious, political, and epistemic values that are at play in the creationism controversy. In particular, it focuses upon the distinction between the public nature of scientific knowledge and the inherently private nature of religious faith and questions the wisdom of setting them in opposition to one another in the schools.

What do creationists want? The various legislative and curricular proposals give a clear indication of at least their immediate goals. They want theism to be taken seriously in science classes. They want equal time; to the extent that biology teachers teach about the scientific view they should to the same extent teach the supernatural view of the origin of life. They want textbooks that present both evolution and special divine creation as "assumptions." They want to break through the wall separating church and state so that their religious views may receive governmental support in the public schools. They want to bring the Bible back into the classroom, and they want it to be used as an authoritative text on matters of geology and biology.

The purposes of our system of public education are rooted in political values, specifically, those of the American democratic ideal. So, when creationists demand that creationism be included in public school science classes, they are necessarily making a claim about how we should understand a tripartite relationship of values—they are asserting an equivalency of the religious values of their form of Christianity and the epistemic values of science vis-a-vis the political values that underpin education. Here then is the central question we must address: Are the creationist demands to teach their religious beliefs in science classes in accord with the values and purposes of public education? I will argue that they are not.

Educational and Political Values

Our public school system serves several important purposes. One major function is to serve the individual by compensating for the contingencies of birth. Wealthy parents have the money to give their children the time and resources to develop their potentials (though some nevertheless neglect their parental responsibility to do so). The poor, on the other hand, typically do not have the ready ability to pay to give their children a good education (though, of course, many sacrifice and find creative ways to do so despite their circumstances). But being born to irresponsible or poverty-stricken parents is no fault of the child. No child in a democratic society should be deprived of the opportunity to acquire basic knowledge and necessary skills because of such accidents of birth. Though the school system does not succeed as well as it should, free public schooling aims to make sure that no child is denied a basic education. Of course, some people may want more than a basic education. They may desire specialized skills or esoteric knowledge. They may want to be schooled in some exclusive system of thought. There are all sorts of extra educational goals that someone may want to achieve, and for such people private schooling at various levels is available. But the first task of the public schools is to assure that no one is denied access to the basics of public knowledge. As we shall see, public knowledge is exactly what science provides, which is what makes it of central importance in public education.

A second major function of public education, more pertinent to the question at hand, is to serve society. Individuals in a democratic society do not simply have a right to acquire a basic education; ideally they have a responsibility to do so. They must be able to read, write, and compute. They must learn the basic facts about the world in which we live and learn how to acquire other facts about the world that may be needed. They must learn their political duties and the civic values that will allow them to live in a pluralistic democracy. A form of government that invests power in the people can only work when those people have an understanding of their civic duties and have the requisite abilities to acquire and process the information needed to make informed decisions as those duties require. This is why basic schooling is not just available, but compulsory—individuals must be educated in the body of knowledge and skills requisite to their role as citizens. Citizenship is an inherently public notion. It is a relational concept. One could not be a citizen in private, independent of society. An individual standing in the role of citizen necessarily takes on a mantle of rights and responsibilities vis-a-vis all other individual citizens in the political system. Citizenship requires that one be mindful of political duties to other citizens. As a citizen, the individual must subordinate allegiance to private values and agendas to public values and principles, and make decisions based upon unbiased knowledge.

Let me be clear that in isolating these two functions of public education and connecting them to the notion of citizenship in a democracy, I am not claiming that these are the historical bases for public education. Compulsory schooling was instituted in all the states only in the 1930s. In part, this significant shift in policy was made for educational reasons, but another part of the rationale was to overcome the abuses of child labor and to protect children from being made to work in the mines and sweat shops, and it is also fair to say that the practical need to reduce unemployment among adults during the Great Depression played a significant role. Nor am I speaking here of the legal aspects of these concepts. Legally, of course, one may be a citizen without being a good citizen in the sense I have in mind. My argument is philosophical, and what I am doing is identifying what I take to be some of the common values that underlie public education and that I am willing to defend philosophically as a worthy ethical ideal.

We have already seen one of the central values of the American democratic ideal exhibited in the rationale for public education—the value of the individual person irrespective of the accidents of birth. This too is an ethical ideal, one that holds individuals ultimately responsible for their actions, including the action of self-governance. It is an ideal of political equality. We reject a distinction between royalty and commoners. We reject special privileges of inheritance or rank. It is true that our nation has not always and may not have yet fully achieved the inclusive ideal even in the most basic sense of the right to vote—earlier in its history the United States denied the franchise to blacks and to women—but the overall trend has been, and we recognize that it ought to be, toward rather than away from the egalitarian ideal. However, we must not think that the democratic ideal starts and stops with the right to vote. Nor does it mean that all matters may be decided by resorting to the ballot box. We cannot vote, for example, to determine the truth or falsity of matters of empirical fact. Nor may we vote away certain values that we assume as fundamental in our political system, such as the value of including the widest possible range of freedom for individuals to hold, express, and act upon their conceptions of what constitutes a worthy and valuable life. We take for granted the rights of life, liberty, and the pursuit of happiness. Furthermore, we agree not to legislate a specific view of what constitutes happiness, but only the prerequisites for the pursuit thereof. Ours is a political philosophy that recognizes the importance of individuals' search for a meaningful life informed by their ethical values. Government shall not institute or promote a particular conception of the good life; instead, it upholds a general conception of the political good, seeking to protect the liberty of individuals to seek the good life after their own lights.

For many people, one of the most important sources of meaningfulness is their religious faith. The Constitution takes special pains to protect this particular form of personal value by explicitly guaranteeing the freedom of religion. Many of the European groups that colonized the New World emigrated specifically to flee religious oppression from their home states, so the framers of the Constitution were keenly aware of the dangers of state-supported religion. The historical record makes it clear that the Founding Fathers intended to keep the state out of the business of the church, and vice versa. They tried to ensure that the Republic they created was not and could not turn into a theocracy. For example, George Washington saw to it that there was no reference to Christianity or God in the Constitution. In answer to a direct question from a Muslim potentate in Tripoli, Washington assured him that "the Government of the United States of America is not in any sense founded on the Christian religion."[4] Of course, for our purposes it is less important to see that the Framers of the Constitution had in fact that intention than to recognize that it was a wise course of action. Among other positive effects, this policy allows private religious beliefs to thrive by protecting them from governmental intervention.

When creationists recommend reinterpreting the Constitution to break down the separation of church and state they are not advocating greater tolerance and religious freedom; in reality they are advocating that the United States moves toward becoming a "Christian nation." Some creationists advocate this explicitly, while others like Johnson do so in a mitigated form, as we saw, in arguing for a return to the view that the law should be based on the authority of the Bible. I contend that it is unwise to follow this path. Religious beliefs are profoundly individual and private and to have the government promote a specific set would undermine the public values that allow for unified democratic action and that protect our freedom to practice those private beliefs.

Religious Values

The history of Christianity is a history of doctrinal schisms, and the fragmentation continues unabated. In 1990, David Barrett, an Anglican missions researcher who served as consultant to the Vatican and Southern Baptist Convention, counted 23,000 separate and distinct Christian denominations. The variety of beliefs and devotional practices within Christianity is astounding. Of course, the differences become further exacerbated when one broadens the scope of inquiry to include those of other, non-Christian religions. This disparity of private beliefs is echoed in the issue of Creation. As we saw in the first chapter, even among Christian creationists who agree that creationism should be taught in science classes, one hears rancorous debate about whether the Genesis account of the six days of Creation describes events that occurred over six literal days, or over six "ages" or "God days," or some other interpretation. Was the Noachian Flood global or local? Which of the two conflicting Genesis Creation stories or the many attempts to reconcile them should we teach? In keeping with the calls for balanced treatment, perhaps the fair thing would be to give equal time to all sides. Of course, we have to recognize that there are dozens, if not hundreds, of other religious stories of creation that would have to be included. Let us look at just a couple of other religious accounts. Like the Genesis account, which centers Creation in the land of the Jews—"the chosen people"—other Creation stories typically bear a local, self-centered point of view.

The Creation tale of the Modoc tribe describes how the Chief of the Sky Spirits carved a hole in the sky and pushed down ice and snow to form the first mountain—Mount Shasta. Walking upon the new land the Sky Spirit touches his finger to the ground and in each spot a tree grows. When the leaves fall he blows them, turning them into birds. He breaks off pieces of his walking staff, using the smaller pieces to make beavers and fish and the larger ones to make other animals, including the grizzly bears. The Chief brings his family down from the Above World to live on the earth, but one day his youngest daughter is blown off the mountain and is found by a grizzly bear who rears the girl with her cubs. When the girl grows up she marries a grizzly bear and they have many children, who are "not as hairy as the grizzlies, yet did not look exactly like their spirit mother, either."[5] These creatures, part animal and part spirit being, are the ancestors of all Indian tribes and they scatter over the earth.

Japanese Shinto describes how the gods Izanagi and Izanami thrust down the jewel-spear of Heaven into a briny ocean which coagulates at the point to form the first land. The two gods then unite as man and wife, and the female deity subsequently gives birth to other lands which form the "Great Eight-island Country" of Nihon.[6] They then produce many children, including Amaterasu no Okami, who is the sun goddess and lord of the universe, and from whom the Japanese imperial line of emperors stems in an unbroken descent.

Or consider the Creation story of the Pima tribe of the Pueblo Indians, which describes how the Magician Man-Maker baked his creations in an oven but was tricked by the Coyote into taking the first batch out too early so they were too light, and the second batch out too late so they were burned too dark. He places these slightly mistaken people across the oceans, and the final time watches carefully so that the beings are neither underdone nor overdone. "These are exactly right," he says. "These really belong here; these I will use. They are beautiful." These are the Pueblo people.[7]

One could cite dozens of similar examples from Africa, South America, or any other area in the world. The cultural anthropologist looks at the local pattern and notes that creation myths help solidify tribal loyalty and provide the group's members with a sense of special identity, purpose, and privilege. Many people find it reasonable to go a step further and conclude that the creation stories are not divine revelations of actual events, but human inventions; ancient peoples cannot explain their own origin or that of their environment so they appeal to supernatural agents, preferably ones that are favorably disposed to help them in their struggles with nature and their enemies.

It is sometimes difficult for Christians to take these unfamiliar stories seriously. It may seem disgusting to think that humans are the offspring of a mating with a grizzly bear. How could one really think that the Japanese emperor is a direct descendant of the Sun Goddess? Perhaps the story of the baking creation of the races of people sound childishly naïve. Of course, from without, the Genesis account sounds equally silly. Did God really shape Adam from the dust and then make Eve starting with one of Adam's rib bones? Did a snake really talk Eve into eating a fruit that gave her knowledge of good and evil, and was it then cursed by God to crawl, limbless, on its belly? And so on. Generally, the stories of other faiths can seem alien and even bizarre, while those of one's own religious tradition resonate with meaning. This is another sense in which religious faith is a private matter: to the faithful such esoteric beliefs are accepted and embraced, sometimes in defiant opposition to reason and good sense, and they serve to establish religious identity, distinguishing members from outsiders.

The fact that the faithful are willing to stake their religious identity upon such beliefs is an indication of their spiritual value to them. One can certainly appreciate the importance that Christian creationists place upon the details of their beliefs about Creation. For them, to deny the literal truth of Genesis would be nothing else than to reject an essential part of what it means to be a Christian. From their point of view, to deny in this way that the Bible is the revealed word of God is tantamount to rejecting membership in the fellowship of Christ. Someone who denies these articles of faith stands outside the fold and forsakes the guidance of Jesus the shepherd. With that would go the moral compass, the hope of salvation, and the promise of eternal reward in heaven. Given the religious commitments of the creationists, their desire to spread the Genesis story is understandable, for they think that those who do not accept the creed are not really Christian, and so will be damned to Hell. Creationists undoubtedly have a sincere concern for the souls of those whom they believe are not yet saved, and this is really what motivates the issue for most of them. Their set of religious beliefs defines their Christian identity, and since one can achieve salvation only through Christ, they are moved to try to help those who they believe are not yet Christian. They may even be correct about the prerequisites for salvation—perhaps all those who fail to accept the literal Genesis story of Creation do risk their souls—but it is also possible that one of the many alternative, conflicting religious views is correct or that all are false. Unfortunately, there is no way to discover which view is correct until after death, so we are forced to take a personal leap of faith.

A commonly held religious value is that one's faith must be freely chosen. Religious belief that is coerced or compelled is worthless from a moral point of view. This provides an answer to the question of why God does not provide public evidence of Himself. It is said that we could not help but believe in God if He were to publicly reveal Himself and, as IDC Fred Heeren argues, in that case such belief would not be freely chosen. Faith is more valuable to the extent that it is chosen in spite of the lack of evidence. As we saw, some creationists hold that we should believe in the Scripture-based 10,000-year-old age of the universe and the contradictory account of Creation in Genesis in the face of the vast amounts of opposing empirical evidence precisely for that reason—the countervailing empirical evidence actually functions as a test of one's faith.

The central importance of such freely chosen, personal leaps beyond or even against the evidence is the most critical sense in which religious belief is private. The contrast with science could not be clearer. It is thus surprising that many creationists try to argue that science itself is a religion. We find a simple version of this view that science is a form of faith expressed indirectly in the Merrimack proposal's claim that both creationism and evolution are simply "assumptions." In 1994, in the *Peloza v. Capistrano School District* case, the Ninth Circuit Court of Appeals upheld a district court finding that a teacher's First Amendment right to free exercise of religion is not violated by a school district's requirement that evolution be taught in biology classes, rejecting the plaintiff's argument that "evolutionism" was a religion. Sometimes the more general charge against science is made via complaints that science is a form of the so-called religion of secular humanism. Phillip Johnson gives a more developed version of the charge, arguing that evolution wins against creationism only because science dogmatically presumes a naturalistic metaphysics. Such views either misunderstand or misrepresent the nature of science.

Science's Epistemic Values

Religion is typically private and exclusive. Religions are like clubs to which one must become a member before one qualifies for the requisite rights and privileges. In extreme cases only the elect may gain membership, and nothing one does can unlock the door to the clubhouse. In other cases one need only ask for membership to be joyously given it. But whether the door is locked or stands open, one may not receive the blessings of the house without first entering. Science, on the other hand, is inherently public and inclusive. Everyone may share in the benefits of science irrespective of whether or not they "believe in" science. You need some mathematical skill, but you need not have faith in physics to be able to use its equations to predict an eclipse. No pious incantations or fervent prayers need be recited to make a chemical reaction work. And even the most vehement anti-evolutionist may still find his or her life saved by a drug discovered using test-tube evolution.

More important, the scientific method of investigation is itself inherently public. Science is not private dogma. Scientific knowledge is not validated by appeals to authority, but rather by appeal to evidence. With the success of relativity theory, Einstein acquired as much personal prestige and authority among scientists as any has ever done, but when he drew a scientific conclusion it was evaluated on the basis of the publicly accessible data, not on the basis of his personal say-so. The debate over quantum mechanics is nicely illustrative of this point. Einstein thought that quantum theory was incorrect because it included indeterministic processes. He continued to hold to the standard view that physical processes were deterministic, a view that he colorfully expressed with a religious-sounding claim that God does not play dice with the universe. But private beliefs about what one believes God would or would not do are irrelevant in the scientific method. Indeed, how could we possibly discover God's view on this matter? Of course, Einstein never intended that the issue should be decided by appeal to such beliefs, and his arguments against quantum theory dealt with the empirical data and with the logic of the theory itself. Scientists who supported quantum theory responded with evidence and arguments of the same kind, and in the end the theory was vindicated.

The epistemic values that are emphasized in scientific research all reflect this basic requirement that hypotheses be evaluated on the basis of publicly accessible evidence. The simplest expression of this is seen in the expectation that scientists publish reports to not only their conclusions, but also the data upon which the conclusions are based and the methods used to uncover the data. One is to accept a scientific theory not as an "assumption" but as a conclusion based upon evidence. (The basic elements of Darwin's theory of descent with modification can be stated in a couple of paragraphs, and *The Origin of Species* could have been a much shorter book if we were just supposed to take the theory as an assumption. In fact, the *Origin* is essentially an extended argument and presentation of some of the vast quantity of evidence for the theory that Darwin had collected over more than a twenty-year period.) Furthermore, evidential support must be open for all to see. Personal revelation cannot count as scientific evidence because it is not a public, intersubjective observation. Indeed, the scientific requirement that evidence be publicly available is stricter still, for even intersubjectivity may not be sufficient if the observation is not repeatable. The claim of Pons and Fleishman to have discovered cold fusion is a case in point—despite the enormous potential importance of the result, other scientists had to dismiss it when they were unable to replicate the findings.

This highlights another major difference between science and religion: scientists are supposed to reject a theory when the evidence is against it even if they desperately want to believe it, and to accept a theory when the evidence supports it even if they would rather not do so. Scientific conclusions are to be drawn without the bias of personal ideology, subjective opinion, or wishful thinking. Of course, scientists are subject to the same weaknesses of character as anyone else, so the scientific virtues might not be fully realized in particular instances. Historians and sociologists of science have shown that scientists do not always live up to the ideal of objectivity. However, the fact remains that scientists may not simply assume or believe what they want but must subject their theories to tests that can be publicly checked. Scientific knowledge is public knowledge.

This thesis has been discussed in detail by several recent philosophers of science; here I briefly mention just two, Philip Kitcher and Helen Longino. These philosophers approach science from quite different perspectives—Kitcher is rooted in the tradition of logical empiricism and Longino in feminism—but both still find agreement in the view that science produces objective public knowledge even when the criticisms of the historicists, sociologists and postmodernists are taken into account.

In *The Advancement of Science* (1993), Kitcher considers the notion of scientific progress in the light of Kuhn's influential "incommensurability critique" of the positivist notion of progress as a series of ever-broader theories that incorporate what is correct in earlier theories as a special case of a more general account. It now appears that science does not always progress in this manner, nor is there a single, simple method that drives it. Though some broad methodological generalizations may be drawn, there remains considerable variation of technique among specific scientific disciplines. Kitcher also takes into account the studies of historians and sociologists of science that show how external factors, such as scientists' desire for personal prestige and their need to compete for limited government funds, may influence research. Though in individual cases such external factors can lead to biased results, Kitcher argues that the overall effect of this sort of variation and competitiveness is to weed out errors caused by personal biases. In effect, Kitcher is saying that scientific practices function as a sort of evolutionary epistemology, whereby increasing truthfulness of scientific theories emerges in a process akin to natural selection. Science is able to make progress in discovering knowledge of the world in large part because it is this sort of *communal* project.

Helen Longino takes even more seriously the charge that scientists may be swayed by ideologies, political or otherwise, that oppose science's constitutive epistemic values. For example, she highlights research on biological differences between men and women that might have been tainted by unrecognized sexist, androcentric presuppositions. Yet she holds that science is able to move past such limitations by broadening membership in the scientific enterprise and including other perspectives which can serve a critical function. She defends a view she calls "contextual empiricism" that takes into account both the constitutive epistemic values and the external contextual ones, without devolving into relativism. Scientific knowledge is *not* constructed (privately) by an individual, but rather through interactions among individuals constrained by empirical evidence, and thus is inherently social. This is the major theme of her book, expressed in its title *Science as Social Knowledge*. Longino also clearly recognizes the connection of the scientific view to the ideals of democracy. She writes,

That theory which is the product of the most inclusive scientific community is better, other things being equal, than that which is the product of the most exclusive. It is better not as measured against some independently accessible reality but better as measured against the cognitive needs of a genuinely democratic community.[8]

We do not need to agree with all the details of Kitcher's and Longino's full positions to accept their important consensus about the point at issue. Kitcher shows how scientific knowledge advances because it is a special sort of community project, and Longino shows how the scientific process produces objective social knowledge. Both illustrate in different ways the thesis that I have put forward—that science is an inclusive public enterprise in both its methods and its results. The epistemic values that underlie scientific methodology are in keeping with democratic political values in that they deny special privilege. Politically we reject claims of authority in government that are based upon a supposed divine right of kings, and scientifically we reject claims of authority about knowledge based upon purportedly divine revelation. Scientific knowledge is public knowledge because scientific conclusions are tested and warranted in a manner that eschews personal bias and requires publicly accessible evidence.

Values in Harmony or in Conflict?

The creationists put forward their proposal to introduce their view into the public school science classroom as though it were an innocuous request that could be supported simply in the name of open-mindedness and academic freedom, but it should now be clear that their proposal is incompatible with the values that underlie public education. In fact, their proposal would put the three sets of values we have discussed—political, religious, and epistemic—into conflict in a manner that would be to the detriment of them all.

As we have seen, the public schools serve the needs of citizens cooperating in a democratic form of government, and they do this in part by teaching basic skills, providing unbiased knowledge, and inculcating the civic virtues that are prerequisite to that cooperation. We govern ourselves based upon shared political values and we explicitly protect our disparate private religious values, by making sure that the government cannot support one over another. However, when creationists propose that the Bible be used as a text in the public schools, as in the Merrimack case, they are proposing that the government use public funds to teach one religious doctrine from among hundreds. When they demand that creation-science or intelligent-design or some other variation of creationism be taught, they are claiming a special privilege for their interpretation of Scripture over those of others. By urging that the wall of separation between church and state be breached to let through their religious beliefs they are weakening the shared political values that provide the common foundation for our pluralistic democracy.

The deleterious nature of the creationists' demand becomes especially pronounced in the effect it would have on science. As we've seen, the epistemic values that underlie the scientific mode of investigation emphasize objective tests of hypotheses based upon publicly accessible evidence. When creationists ask that evolution be taught as an "assumption" they turn upside down the evidential basis of scientific reasoning that allows it to function impartially to establish public knowledge; evolution is not an assumption accepted on faith, but a conclusion supported by a vast array of empirical data. When creationists ask that creationism be given "equal time" with evolution they are asking that private religious beliefs based upon supernatural revelation be given equal weight to conclusions that are supported by publicly accessible data. The "equal time" model might sound democratic, but actually it is antidemocratic in this context for, as we noted, it is inappropriate to vote upon the truth of matters of empirical fact or to decide them based upon personal ideology.

Creationists will, of course, claim that I am dismissing their religious values as inferior to those of science, and that I am simply defending the latter at the expense of the former. In a sense, I suppose this is true in that I do hold that their "method" is inappropriate as a way to discover empirical truths about the natural world. But I do not mean to dismiss religious values generally. Indeed, we must rebuff the creationists' demands in large part to protect religious values. Religious people need to recognize that it is in their own interest to reject the changes the creationists want to make, for when creationists undermine the unifying values that allow for democratic action they undermine the basis of religious freedom. Rejecting a scientific methodology that validates knowledge based upon data available to all in favor of a supernatural view based upon private revelation and faith is preferable only to those who think that *their* faith will be affirmed by the government and *their* private revelations will be the ones that will be promoted. Moving away from an inclusive democracy and toward an exclusive theocracy is preferable only to those who believe *theirs* will be the religion in control. But, as we saw, even in relatively simple matters there is an astounding variety of religious beliefs, so if one view gains power the vast majority necessarily loses. If we wish to maintain the freedom to hold our private religious values then we must not allow creationists to dilute the public epistemic and political values, both epitomized by science, that keep that freedom in place.

Of Ideologues and Intellectual Honesty

Philosopher Edwin J. Delattre, dean of Boston University's School of Education, argues that educators and educational institutions bear an obligation that is grounded in the public trust—a trust in its power to improve their lot and that of their children—and that requires of them a special devotion to common purposes and common values. He argues that it is important that in the schools we teach those values that are concomitant with the scholarly and scientific disciplines, but he warns of the danger of ideologues, whether religious or political, who subvert the common goals of education by trying to manipulate the young to accept their private values. Ideologues who masquerade as teachers might have good intentions, but ultimately they betray the most basic value upon which good learning relies, namely, intellectual honesty. Delattre illustrates this danger using the example of creationism, quoting from a creationist textbook, *Science*, which is used in the eighth grade in some schools. It presents the theory of evolution as being both false and evil, and provides the following explanation of why people nevertheless believe it:

"You see," responded the judge, "people really believe what they want to believe. Because you want to be right with God, you find it easy to believe and accept the facts. A person who is not right with God must find reason, or justification, for not believing. So he readily accepts an indefensible theory like evolution—even if it will not hold water. That is his academic justification for unbelief. In fact, that is what all the many theories of evolution are—a mental justification for unbelief."[9]

Given that, as we have seen, evolutionary theory is well confirmed by a vast array of evidence, and that its truth is recognized generally among scientists, it is intellectually dishonest to tell students that evolution is false and indefensible; moreover, even if evolutionary theory were false, it is a fallacy to argue against it by attacking its proponents as evil and self-deluded. Delattre acknowledges that many school textbooks are much better than *Science*, and that many teachers are neither religious nor political ideologues, but insists that "there is no room in schools, colleges, or universities for manipulation or proselytizing that portray themselves as objective and impartial."[10]

Intelligent-design theory fits this mold precisely, and *Of Pandas and People* is, if anything, worse than the textbook Delattre cites, for at least *Science* is forthright in stating its religious viewpoint, whereas *Pandas* disguises its proselytizing and is systematically misleading. It purports to present a legitimate biological viewpoint, and says that it will compare the approach of those who hold an evolutionary theory to the approach of those who adhere to intelligent-design theory, as though the latter were truly a competitor on the scientific stage. It leads students to believe that they are similarly well developed as scientific theories, and that scientists are actively engaged in research in both. It claims to do "what most textbooks do not do," namely, to impartially present the "evidence" for both, and to allow students to judge for themselves which to believe. But, as we saw, there is nothing to intelligent-design theory beyond the familiar negative creationist arguments against evolution, and its positive "explanatory theory" is limited to the simple claim that the world was created by an amorphous "designing intelligence" and that things are the way they are because the designer purposely made them that way. Furthermore, as Gilchrist's review showed, the only references to it that appear in indexes of the literature are a handful of book reviews and a couple of articles, most of which had to do with controversy over *Pandas* itself.[11] The way in which *Science* and *Pandas* misrepresent science for their religious ends is intellectually dishonest and manipulative, and as Delattre rightly observes, "[A]ll manipulation for the sake of conversion, whether political or religious is tyrannical. It is rooted in contempt for human beings and their capacity to discover truth."[12]

This is just the sort of attitude that IDCs exhibit, and which is especially evident in Johnson's writings. "Truth (with a capital T) is truth as God knows it," he declares. "When God is no longer in the picture there can be no Truth, only conflicting human opinions."[13] Though he might express the view with more subtlety, passages of a kind with the one Delattre quoted appear throughout Johnson's works. Consider the following representative example:

Evolutionary biology is a field whose cultural importance far outstrips its modest intellectual and scientific content. Its sacred trust is to preserve the central, indispensable part of the modernist creation story, which is the explanation of how such things as life, complex organ systems and human minds could exist without a Creator to design and make them. We might say that the point of Darwinism is to refute the otherwise compelling teaching of Romans 1:20, which is that God's eternal power and deity have always been evident from the things that were created.[14]

Here we have yet another example of Johnson's rhetorical attempt to portray evolution as the foundation of a religion of modernism. He tells us that its very purpose is to refute Christianity, suggesting, just as the textbook *Science* does, that scientists promote it for this reason, all the while knowing that it is not a viable theory. In an interview with Yves Barbero, Johnson claimed that what is going on in the classroom is not good teaching but something more akin to a conspiracy of intentional "indoctrination" in "the official philosophy of the educational elite and really of the government these days" by "educators, the rulers of science of education, the organizations that control it."[15]

Johnson continues this attack upon science teachers in *Defeating Darwinism*, telling us that "When students ask intelligent questions [about evolution] like 'Is this stuff really true?' teachers are encouraged or required not to take the questions seriously. Instead they put the students off with public-relations jargon about how the scientific enterprise is reliable and self-correcting."[16] Such statements are incredible. I have never heard from any teacher about school policies for science classes that have such requirements or recommendations. I have, however, heard many biology teachers say that they tread very lightly when they teach about evolution in order to be sensitive to their students whose parents and pastors have told them that evolution is false and unchristian. Many science teachers feel that, in order not to upset such students, they need to tell their classes explicitly that, although all students have to study evolution and geology, they don't have to believe it. If public school science teachers are hesitant to talk about the truth of evolutionary theory and the geology of the ancient earth, it is because creationists have made it such a touchy issue. Good teachers should, of course, always try to be respectful of their students' religious and cultural backgrounds, but when teachers bend over backwards, as they do, to avoid offending religious sensibilities with regard to evolution, they may be doing their students a serious educational and intellectual disservice.

The proper answer to such a question from a student about whether "this stuff" is true would be: "Yes, indeed, and let's take a look again at the evidence that supports it so you can begin to see for yourself." Going on to explain about the reliability and self-correcting nature of the scientific enterprise is also in order, for they are among the properties of scientific method that make it reasonable for us to put our trust in its results. To suggest that biology teachers who explain the scientific basis of evolution in this way are simply spouting "public relations jargon" as part of their intentional effort to refute Scripture and indoctrinate children in an anti-Christian religion is as insulting as it is false. It is equally insulting to suggest that, perhaps, they know not what they do because they suffer from "massive hypocrisy and self-deception"[17] due to the "materialistic philosophy" that blinds them to the Truth. If science teachers are supposed to teach Truth, and if, as Johnson says, Truth is truth as God knows it, then they really should be telling students to use the Bible as a science textbook, for science certainly cannot provide God's-eye biology. Scientists must content themselves with the more humble, natural truths that we can discover with our human eyes and the light of our own intelligence. Delattre is right that creationists are ideologues and he is right that, for this reason, they do not belong in the classroom.

Ideologues cannot bear the trust of education because they do not themselves trust independence of mind. Teachers work in the faith that human intelligence has a future, even a bright future, and they will bring illumination to classroom inquiry in the expectation that students are thereby acquiring lights of their own.[18]

The Creationist "Viewpoint"

Delattre cautions us, however, that "It is not always easy to identify ideologues in education because their effectiveness consists in appearing not to be ideologues."[19] Sure enough, as we have seen, intelligent-design theorists try to turn the table on scientists and charge that it is not creationism but evolution that is ideological, and that it is creationists who are the rational skeptics and evolutionists the dogmatists. Appropriating the terminology of skeptics, Johnson tells us that he would build a high-school curriculum in evolution around principles of critical thinking, and have students learn to train what astronomer Carl Sagan called their "baloney detectors" upon evolutionary theory.

In *Defeating Darwinism*, Johnson goes through Sagan's baloney detection list of fallacious appeals to authority, selective use of evidence, begging the question, *ad hominem* arguments, and so on, but he illustrates these with ways that he claims evolutionary biologists are dishing out the baloney. For example, he says students should be taught to watch for evolutionists' bait-and-switch strategy of starting with what they call "the fact" of evolution and then surreptitiously inflating it to include the mechanism as well. Continuing to intimate that scientists are conspiring to protect their power, Johnson claims that this important distinction between product and process is "just a debating gimmick"[20] to hide problems with the Darwinian mechanism. Gould and some other evolutionary biologists, as I discussed in chapter 3, speak of common descent with modification as "the fact" of evolution to distinguish that from "the theory" of the mechanism(s) by which it occurred. In Johnson's section on curriculum he misleadingly defines and dismisses the former as being just the uncontroversial point that "organisms have certain similarities like the DNA genetic code, and are grouped in patterns,"[21] but he later uses it in Gould's sense to refer to common descent when he rejects that thesis.[22] Johnson tells high-schoolers that they need to "learn to use terms precisely and consistently,"[23] and that their baloney detectors should register "Snow Job Alert!"[24] when they see someone being slippery about terminology. As we have seen throughout his writings, Johnson's own use of terminology—not only for biological concepts, but also especially for the philosophical ones upon which his argument turns—does not exemplify the virtues he rightly praises. As for the point at hand, the theory of evolution includes both the common descent thesis and the Darwinian mechanism (and more besides), and both are well confirmed by the evidence.

We should applaud Johnson's call for teaching critical thinking, but the way his program purports to apply this for a biology curriculum is no model of fair, careful skepticism, and it again betrays the creationist ideology—his seven point program is nothing more than the familiar dual-model strategy of giving negative arguments against evolution and simply saying that we should be open-minded about the alternative divine view. To quote Delattre once more, "it is not intellectually honest for a teacher to imply that all positions are equally logical, factual, and reliable. Open-mindedness cannot entail indifference to quality and evidence; quite the opposite. Teachers should help students to learn to be dubious when there is reasonable ground for doubt, and to be decisive when evidence warrants resolution."[25] As we saw, the dual-model argument is logically fallacious, and it is false to suggest that there is little scientific evidence for common descent or for the efficacy of the Darwinian mechanism. Only an ideologue would think that the proper way to teach evolutionary biology and all the other sciences connected to it is to tell students that the subject is little more than "philosophical dogma" and "orthodoxy," and that scientists who support it are "hypocrites" who intentionally "bluff" and "dodge the hard questions" and who should be "viewed with suspicion," all of which are *ad hominem* claims that Johnson makes in various forms in *Defeating Darwinism* and throughout his other writings.

Even parents who are creationists and would like to see this critical approach to evolution in the schools may be less than pleased to hear that Johnson also recommends that students learn in biology class to turn their baloney detectors upon their own religious beliefs. He argues that to believe in God simply on faith rather than reason is either a "mistake" or a "rational defensive strategy born of desperation,"[26] and that students must confront the theological problems that result from taking on evolution. He admits that some Christians like Darwinism because they think it gets God off the hook of the problem of suffering: "If (for some reason) the divine plan involved creating by means of scientific laws, then God couldn't intervene to prevent suffering without spoiling his own grand scheme."[27] But because he holds that God can and does intervene to break natural laws Johnson knows he can't make this argument, and he says that he would have to tell students that "none of the usual answers to the problem of suffering is entirely satisfactory."[28] These are all interesting questions in philosophical theology, but, believe it or not, such theological matters also come under one or other of the seven points Johnson would include in his biology curriculum.

Are Johnson's curricular recommendations just part of a rhetorical strategy to appear not to be an ideologue, or does he really think that such a framework that mixes theology and science is a serious possibility? It seems as though he may indeed be serious about his recommendation, if we consider his important discussion in *Reason in the Balance* in which he appears to outline a new legal strategy to get around earlier Supreme Court decisions regarding creationism. Johnson's academic specialty is the law, so on this question he is speaking within his area of expertise, and so we need to pay close attention to his interpretation of how the U.S. Constitution should apply to the issue. It is also worth noticing how in his discussion of this he once again links the question about evolution to issues of morality. He makes his argument in a comparison of two Supreme Court cases, the 1987 *Edwards v. Aguillard* case which had overturned Louisiana's creationist legislation, and the 1993 *Lamb's Chapel* decision.

American law has excluded religious speech from public schools on the constitutional grounds that government may not establish or support one religious view over another, but Johnson says that this "official explanation" of neutrality is "wearing thin" now that schools are "actively promoting the progressive viewpoint on sexual behavior to students from traditionalist homes and actively promoting a naturalistic understanding of creation in science classes."[29] How can this policy be genuinely neutral, he asks, when it allows educators to promote an ideological agenda and ". . . patently serves the interests of one side to a major cultural controversy"?[30] Again we return to the two opposing worldviews, both of which, according to Johnson, are religious. How so? In another example of Johnson's non-standard use of terms he defines religion as "a way of thinking about ultimate questions."[31] This definition omits many essential features of religion; although it is true that religion often does include this characteristic, so do many philosophical views that are not religious. However, by using this vague definition of religion together with his ambiguous notion of naturalism (that conflates, as we saw, a metaphysical with a methodological view), Johnson is able to conclude that naturalism is the "established religion" of the United States.[32] With that "conclusion" in place he is able to pose a dilemma:

The decisive question for First Amendment religious law, then, is one of metaphysics rather than legal doctrine. Is the Constitution genuinely neutral between scientific naturalism and theism? In that case both positions should be admitted to public discussion, in the schools and elsewhere, and protected from "viewpoint discrimination." Or is naturalism the established constitutional philosophy? In that case naturalism will have a monopoly in the public arena, while theistic dissent will be restricted to private life. If the latter alternative is taken, then the Supreme Court will in effect have established a national religion in the name of First Amendment freedoms.[33]

The point about "viewpoint discrimination" is the key to Johnson's strategy, for that is the point on which the *Lamb's Chapel* decision turned. The case involved an application by an evangelical minister of Lamb's Chapel in Center Moriches, New York, to use public school facilities to show a film series on parent-child relationships from a conservative Christian perspective. The series featured Dr. James Dobson, Christian broadcaster of the popular *Focus on the Family* radio shows. Johnson describes Dobson's view as follows:

The values he upholds center on a man and a woman who marry for life and who play distinct paternal and maternal roles. In this value system, abortion is equivalent to homicide, and sexual activity outside of marriage is a manifestation of sin.[34]

As we've seen, Johnson and other creationists believe that these moral values hold absolutely because God created the world as recounted in Scripture, giving each being a proper function. On the other hand, the progressive view of the New York schools, Johnson says, takes a relativistic approach to sexual morality.

From the progressive standpoint, to advocate heterosexual marriage as morally superior to homosexual relationships, or premarital chastity as morally superior to sexual experimentation, is . . . a form of viewpoint discrimination.[35]

The school, backed by the attorney general, denied the minister's application to show the series. Supreme Court decisions regarding the First Amendment had banned viewpoint discrimination, which meant that if schools allowed discussion of a topic on school property at all it could not discriminately allow some views and disallow others. New York State believed that it adhered to this because its policy was to ban *all* religious activities from school properties. When the minister sued, both the federal court and the State court of appeals accepted that reasoning and held that the school had not violated the Constitution in rejecting the film series on that basis. The Supreme Court, however, unanimously rejected that argument on the grounds that the appropriate category for the case at hand was not religion but rather family values and relationships, a topic that was part of the school's curriculum. Viewed under that category the school could not reject the film simply because it discussed the topic from a religious perspective.

Having given the above account of the court's reasoning in the *Lamb's Chapel* case, Johnson shows how he believes it should also apply to creationism cases. He says that the *Edwards* case was clouded by the fact that it involved young-earth creationism, which diverted attention to the idea of incorporating details of Genesis literalism to the science curriculum and away from what was really at stake: the general concept of divine creation. Johnson asks the reader to imagine the case replayed, using the *Lamb's Chapel* reasoning, with the details of the timing of Creation set aside and the creationist claim pared down to merely that an intelligent designer was necessary for biological creation. Surely this is a rational, and possibly true, alternative to Darwinian evolution, so if the topic is biological origins it should not be rejected simply because it is religious.

Put most simply, the underlying philosophical question in both *Lamb's Chapel* and *Edwards* was whether the authorities were dealing with separate subjects or with conflicting opinions about the same subject. If a high-school curriculum incorporates the subject of biological origins, and if supernatural creation is a rational alternative to naturalistic evolution within that subject, then it is bad educational policy as well as viewpoint discrimination to try to keep students ignorant of an alternative that may be true.[36]

What intelligent-design creationists mean when they speak of truth is "Truth with a capital T"—God's Truth. Johnson seems to believe that he has found a crack in the legal armor of the guardians of evolutionary naturalism, and a way to bring that divine truth into the public schools. *Defeating Darwinism* is a manifesto to galvanize and inspire creationists in the project of wedging open that crack. In the book, Johnson tells us that he always smiles when he sees the familiar quotation "YOU SHALL KNOW THE TRUTH, AND THE TRUTH SHALL MAKE YOU FREE" emblazoned in bold capital letters on the facade of Berkeley's public high school, for he knows that these are Jesus' words in the Gospel of John, and they remind him that Jesus "warned against the scoffers who build their house on a foundation of sand. The Truth he referred to was himself. . . ."[37]

The complete title of Johnson's manifesto is *An Easy-To-Understand Guide for Defeating Darwinism by Opening Minds*. Johnson asks what we are to do when materialist philosophy is unable to fill the evidential gaps in evolutionary theory, and his answer is that we should scrap scientific methodology and turn to a theistic science. However, as Albert Szent-Gyorgyi said, "For every complex problem, there is a simple, easy to understand, incorrect answer." It is all very well to talk about truth, but one can judge claims of truth only by means of some rational method, and the sole "method" theistic science has to offer is revelation and the design argument. However, we have seen in detail why these do not serve as evidential methods. It is simple indeed to explain the complexities of biomolecules by reference to design, especially when this is one of the things "that we know as Christians." Whether or not we think that supernatural design is true, it is ideological to begin with the answer that one desires.

IDCs claim that they are not ideologues but are merely open-minded about the possibility of design. But IDC equivocates at its core by not being consistent and forthright about the divine nature of its hypothesized designer—*Panda* speaks vaguely of a "designing intelligence" or a "master intellect." In making their arguments, intelligent-design theorists always try to justify their design inference by giving examples that involve human design—the faces of Washington and other American presidents carved on Mount Rushmore, or a message in English scrawled in the sand. But no scientific naturalist would dispute that we can sometimes conclude legitimately that something was created by design in *natural* cases such as these. It is their tacit slide to the supernatural that is question-begging. Paul Nelson challenged me at a conference with an example in which one could infer the design intentions of a taxicab driver by means of a variety of observational data. Taxicab operators, particularly in New York City, occasionally do seem to exhibit superhuman abilities, but never supernatural ones, and I fail to see how we can check whether cab drivers are in any way similar to a divine Creator. Moreover, as we saw, if we were to consider the Creation hypothesis without ambiguity as a natural hypothesis, we could do so in principle, as in the extraterrestrial intelligent-design hypothesis I discussed in chapter 6, but in practice we find no good evidence that supports ET-ID. Not only should supernatural views be excluded from science classrooms, so also should fully natural ones that have not passed the test of evidence. If it is viewpoint discrimination to keep such alternative "ways of thinking about biological origins" out of science classes, then we should not only be using *Pandas*, but also textbooks that advocate Claude Vorilhon's Raëlian model, the Swami Prabhupade's cyclical Hindu model, and so on. As they say, this way lies madness.

Dealing Fairly with Bones of Contention

That complex human beings are the evolutionary descendants of primitive simple protozoa is not, as Johnson asserts, "naturalistic doctrine," but rather a reasonable scientific inference drawn from all that is known of the tree of life, the common chemistry of living things, and the well-confirmed general processes of evolutionary development. School biology and geology teachers are not teaching atheistic, materialist metaphysics when they discuss the evolutionary roots of the tree of life or the ancient depositional origin of sedimentary rock formations any more than English-language teachers are when they explain the Latin or Sanskrit roots of the words students are to learn.

Indeed, many science teachers these days are barely even teaching the basics of evolutionary theory at all. When I ask students in my university classes about what they learned about evolution in high school, I am always dismayed at the large number who did not even have a cursory exposure to it. My experience seems to be typical; in a survey of close to a thousand first-year college students who either had nonscience or undeclared majors, one professor found that more than 25 percent said they believed that God created the earth within the last 10,000 years and formed human beings exactly as described in the Bible (compared to 47 percent of the general population who had that view, according to a 1991 Gallop poll). Another 50 percent said they were undecided as to whether evolution is a valid scientific theory or a hoax.[38] Even teachers who want to give their students a good education in evolution often choose to avoid the topic in local districts where creationist activists have made it a contentious issue for fear of complaints from creationist parents. Eugenie Scott, Director of the National Center for Science Education, writes that "in the face of parental pressure, principals and superintendents frequently fail to support teachers, even when the curriculum mandates the teaching of evolution."[39] When this happens it puts science teachers in a difficult position that seriously impairs their ability to fulfill their teaching duties. Sadly, one even hears of a few individual creationist teachers who place their religious ideology above their professional responsibilities and who quietly use their classrooms to proselytize their creationist views. Scott mentions a case in Stanwood, Washington where a teacher invited a creation-scientist into the class to lecture on the "latest scientific findings" about how humans and dinosaurs lived contemporaneously, and another case in which a teacher was advocating the view that the earth was very young. A student of mine told of one of her high school teachers who was quite explicit and adamant about promoting creationism and denouncing evolution, even bringing into class and discussing the ICR poster I described in the preface that depicts the battle being waged between the pirates in the Tower of evolution and the clerics in the Tower of Creation.

The failure in the secondary schools to adequately teach the key biological and geological findings and central explanatory theories such as evolution leaves students ignorant of basic public knowledge and unprepared for higher education. Moreover, students who enter university who have learned science from the Bible or creationist materials often have serious misconceptions about a whole range of facts related to evolution. They regularly misunderstand the mechanism of evolution, thinking that the theory says that organisms evolved entirely by chance. Some think that the theory says that humans evolved from the monkeys they see in the zoo, leading them to ask why there are still monkeys around. The Bible is not a science textbook and the mistakes that result from thinking that it is are sometimes even more absurd than this. Robert Root-Bernstein, a professor of physiology at a major Midwestern university, wrote of his shock the first time he taught a course on his subject when a student announced that she knew, even without looking, what the difference was between the male and female human skeletons that he had displayed for a class exercise. Males, she said, had one fewer pair of ribs than females. Her belief, of course, had come from the Genesis account of the creation of Eve from Adam's rib. Nor was she alone in her belief; five more students in the same class expressed the same view, and Root-Bernstein says that he has come to expect that at least 10 percent of students will tell him that males and females differ in rib count.[40]

If people insist upon interpreting the Bible as a science textbook, then such absurdities are unavoidable and we cannot expect that complete harmony of scientific and religious beliefs will ever be possible. Johnson is right that scientists cannot contend that science and religion are so separate that the former does not have any effect on the latter. Even teaching the simple fact that men and women have the same number of ribs bears upon and threatens some students' religious beliefs. For those who interpret Genesis in the way that 10 percent of Root-Bernstein's students do, then there is indeed a direct conflict between biology and their Christianity. But such conflicts are not confined to evolutionary biology. In an earlier era, scientific findings had contradicted Christian beliefs that the earth was fixed in the center of the universe, and, as we saw, creationists today also take issue with findings of geology, paleontology, physics, linguistics, and so on. Specific scientific findings also can threaten the beliefs of others besides creationist Christians.

Kennewick man, a fossil hominid skull with apparently Caucasoid features that was recently uncovered in Washington, conflicts with the religion of local tribes whose Creation story says that God created their people there in the Pacific Northwest before all other races. The tribe demanded the right to ceremonially bury it. However, its ownership was contested by a local group of self-proclaimed pagans who contend that their ancestors inhabited the region first, and who want to bury the skeletal remains their own way. As I write this, the fate of the skeleton remains in legal limbo. Paleontologists are interested in the skull for its scientific value, for it could reveal information about recent human evolution and patterns of migration into America, and it might turn out that a study of the skull would yield evidence to resolve the issue of ancestry and ownership. Clearly, the religious groups at odds here have different viewpoints on the origin of the bones, and both stand to have their contrary beliefs affected by the outcome of scientific investigation of them (which may be why there is opposition to scientific analysis of the skeleton by some of these believers). However, every religious group has skeletons in its closet, and it is not reasonable to expect science to ignore any fact that might offend, or science educators to give "balanced treatment" to any topic in the curriculum that happens to conflict with private, religious beliefs in this way.

What about the complaint, though, that science is philosophy? In a simple sense, creationists are right about this too. The use of the term "scientist" is actually a relatively recent development in the evolution of the English language,[41] and people we think of as the seventeenth-century scientists who ushered in the scientific revolution would have called themselves "natural philosophers" and would have said that they were engaged in "natural philosophy." It is also true that science, then and now, has philosophical underpinnings (if it didn't, as a philosopher of science, I'd be out of a job), but so does every other human practice. If this were all there was to the creationists' challenge, then we should have shrugged with indifference and admitted defeat right at the beginning.

However, as we have seen, creationists are making much stronger accusations, claiming that science itself is nothing but an ideology that "simply assumes" the truth of evolution and denies the existence of God "by fiat." Scientific naturalism, they say, is an arbitrary, metaphysical dogma, and a relativistic, immoral, anti-Christian, atheistic dogma at that. They impugn the intellectual and ethical integrity of any scientist who accepts and teaches evolution, asserting that it is a false, evidentially unfounded theory, and that science educators know this but promote it nevertheless as part of their intentional campaign to defend their cultural power and to attack the Christian religion. We cannot afford to let such accusations go without a reply, for they fly in the face of the truth. I have tried in this book to show why the creationists' arguments are philosophically unjustified. I would also submit that their insinuations about the supposedly anti-Christian motivations of scientists and science educators are similarly unfounded. No doubt there are a few atheistic science teachers who have used their classrooms to attack religion, but I have never met one. In my experience, Root-Bernstein's attitude is representative of that of the professionals who teach science, and it seems to be the only fair and reasonable way to deal with religious bones of contention:

I believe just as firmly in religious freedom as I do in the scientific search for understanding. Thus, while I adhere rigorously to teaching the best science and showing how scientists recognize it as the best, I never insist that students believe scientific results. On the contrary, I encourage them to be skeptical—as long as their skepticism is based on logic and evidence.[42]

Root-Bernstein lets his students discover a bit of the evidence for evolution on their own, having them examine skeletons and X-rays directly. He notes that some of the students counted the ribs and then reported that they had verified their preconceived notion, and he said he had to stand beside them and have them repeat the procedure two or three times before they agreed that the male and female skeletons have the same number. He saves for last the clincher that he himself does have one fewer pair of ribs than his mother, not because he has had a pair removed—he has the normal 12— but because his mother has 13. And he mentions the 5,300-year-old man found frozen in the Alps several years ago, who had just 11. Chance genetic variations do indeed occasionally happen, just as evolutionary theory says, and such anatomic differences are what drive evolution, he explains to his students. He has them find homologies between human and chimpanzee skeletons that are evidence of their common ancestry, examine the underside of the bony back-plate of a tortoise skeleton to see how it could have formed from the broadening and fusing of the ribs of a reptile ancestor, and look at casts of the hooves of species leading from the four-toed *Hyrancotherium* to the one-toed modern horse to see some of the evidence for transitional forms. He also has them compare the relative size of human brains and female pelvis width to that of chimps, to figure out one of the reasons that sexual dimorphism is more pronounced in the former— "Bigger brains require bigger hips." But he tells them not to just trust his word, but to check the skeletons for themselves: "Take nothing for granted," he counsels his students, "that is what makes a scientist."[43]

The best education in science involves learning not just the scientific facts, but also beginning to understand how scientists reason and how they test hypotheses against the empirical evidence, which is to say, learning something of science's epistemic values. Root-Bernstein provides a wonderful model of how to get students to understand evolution and the evidence for it. This is not metaphysical indoctrination or materialist propaganda. It is simply good teaching. As Delattre writes,

[L]earning and being taught to reason well—to be objective and impartial—inculcate specific ideals. These are the ideals of reason itself, and any student who does not learn them is forever in thrall to his or her own ignorance. It is no arbitrarily imposed standard to teach a student to identify valid and fallacious formal and informal arguments; to measure probability; to apply relevant criteria; to verify and falsify propositions; and to confirm or disconfirm hypotheses. In fact, it is irresponsible to teach less.[44]

A Final Look at Creationism's Tower of Babel

I began this book with the observation that, responding to selective pressures from both within and without, creationism seems to be evolving, and that a confusing array of varieties are competing for dominance in the creationist Tower. Since the 1980s, all creationists have tried to adapt to what, for them, is a harsh legal and intellectual environment. Though it has not yet and is not likely to ever displace the older forms, intelligent-design theory has been by far the most successful of the new varieties of creationism; its memes are sweeping through the population. In part this is because intelligent-design creationists, by explicitly discussing only a relatively minimal set of commitments, have so far managed to avoid major direct conflicts with the other creationist factions and also has been able to better camouflage its religious basis when in the public environment. IDCs have also evolved more complex ways of expressing the creationist view than the simple formulations of the Institute for Creation Research leaders and their ilk. Furthermore, to their credit, IDCs have recognized the errors in many of the old creationist arguments against evolution and no longer claim, for instance, that evolution is refuted by the Paluxy "man-prints," the depth of moon dust, or the second law of thermodynamics. These arguments, and others as well, are weak and in the end not rationally viable in the face of the scientific evidence. That the new creationism has changed in this way is one evolutionary development that is to be welcomed.

But let me not extend the Darwinian analogy any further, for, of course, there is one crucial difference—the new varieties of creationism do not arise at random but, as one would expect in this case, by design. Despite their serious disagreements, creationists have one common goal, which is to defeat the evolutionary account of the origin of species and replace it with one or another supernatural, divine account of special Creation. Because they all hold on to their desired religious end above all else, the evolution of creationism is teleological—goal directed—so the change is non-Darwinian. It is therefore not surprising that, for all the recent modifications of their coloring, creationists insist upon remaining "true to type."

At first, IDCs spoke only of their opposition to the Darwinian mechanism, but as more details of their position come into view it is beginning to look more like the classic form of creationism, as, for instance, when Johnson recently began to also deny the basic evolutionary thesis of common descent. IDC is one with all other forms of creationism in its primary goal of reintroducing supernatural explanations into science; in this critical sense, Plantinga's call for a theistic "Augustinian" science, Moreland's "creascience," Johnson's program of theistic realism, and so on, are of a kind with ICR's creation-science. Nor is Johnson's linking of evolution to naturalism and modernism new; opposition to modernism is one of the defining characteristics of Fundamentalism and goes hand in hand with its anti-evolutionism (as Ammerman pointed out), and anti-Darwinians such as Charles Hodge[45] had criticized evolution as naturalism as soon as Darwin published his theory. IDC also exhibits the old fundamentalist fear that evolution is the basis of ethical relativism and immorality. Most IDCs (Norman Geisler is one exception) never say publicly that evolution is the tool of Satan, as ICR and many other old-style creationists do, but they do voice the same charge that evolution is to blame for what they take to be a range of social ills and immoralities, from abortion and divorce to feminism, homosexuality, and more. As we saw in chapter 7, Johnson's list of the various evils spawned by evolutionary naturalism is very nearly identical to that of ICR. As I have highlighted, IDCs see the evolution/ Creation controversy as the center of a culture war, and the rhetoric of the new creationists, like that of their predecessors, is heavy with military metaphors. Even ICR's image of the battle between the two towers reappears, though in a less cartoonish and more articulate form.

Rallying his troops to storm the naturalist castle, Johnson declares that "Darwinism is in serious trouble" and "the proud tower of modernism is resting on air."[46] Marx and Freud have fallen, he cries, and "Darwin is next on the block."[47] With regard to the creationist Tower, Johnson knows that the construction and defense of his form of creationism is theologically risky, and he tells his followers that "we had better count the cost before we start to build the tower."[48] The new challenge to materialism might fizzle again, he acknowledges, as it did at the Scopes trial. He writes that "accommodationists in the Christian academic world" and even some fundamentalists, have advised him that "it is futile and dangerous to challenge the truth claims of modernism on secular territory."[49] However, he reassures these doubters of little faith that the Darwinian materialists are overconfident, and likens evolutionary theory to the Soviet Union in the days before its fall, proclaiming that "a cultural tower built on a materialist foundation can look extremely powerful one day and yet collapse in ruins the next."[50] Which tower will remain standing at the end of the battle? Johnson answers that "there is no way to find out for sure without going into the courtroom."[51]

As a lawyer, Johnson wants the final battle to be decided on his home turf. My own view is that the legal arena is not the proper setting to deal with these issues at all. Just as the debate format is not a reasonable way to weigh the merits of a scientific hypothesis, neither is the courtroom format. Scientific hypotheses are to be tested against the evidence in the field, in the laboratory, and held out for criticism at professional meetings and in peer-reviewed journals. As for the philosophical and theological issues, I would prefer that we work to resolve these out of court as well. But we should not simply settle out of court, for it is not intellectually honest to just trade truths over a bargaining table. Johnson argues that because, as he claims, it rests on the institutionalized dogma of naturalism, "evolution has become the focus of a culture war instead of a subject that can be discussed constructively in educational institutions or in the political realm of negotiation and compromise."[52] In his suggestion that empirical truths should be open to "negotiation and compromise" we again see elements of Johnson's postmodernist leanings. But the truth of factual matters is not the sort of thing that is properly resolved by political bargaining or conceptual gerrymandering.

It comes down to the following: If it is correct that evolution and other factual matters about the natural world are things about which there are simply different viewpoints, as the court found was so for the category of family values and relationships, then the conclusion of the *Lamb's Chapel* case would seem to apply to creationism, just as Johnson argues. Indeed, it should then apply not just to Johnson's preferred form of intelligent-design, but to young-earth creationism as well, and to Raëlianism, and to every other creationist viewpoint. At the start of this chapter, I quoted an argument for teaching intelligent-design that a Tennessee state senator gave when he spoke in support of legislation that would fire teachers who taught evolution as a fact. It is not out of fear of the truth, as he claimed, but out of respect for the truth and how we corporately come to know it that we should not teach such "theories."

As I hope has become clear over the course of this book, intelligent-design creationists are wrong to say that evolution is just a "loaded story" or an assumed point of view; rather, it is as well confirmed by the scientific evidence as any of the other great explanatory theories. More important, they are wrong to say that scientific naturalism is metaphysical dogma; rather, it is a methodology that is rationally justified and that is accessible to all. As Albert Einstein reflected in *Out of My Later Years*: "The whole of science is nothing more than a refinement of everyday thinking." In everyday life we take it for granted that the lawful processes of cause and effect are not broken by miraculous interventions. When we are at the grocer's we squeeze the vegetables to check for freshness, but we don't, and can't, check for devils in the lettuce. It is this quotidian, mundane, and yes, entirely *natural* reasoning (in both senses of the term) that makes the knowledge we gather by such a method *public* knowledge. Science only makes this process more precise. Creationists could not be more wrong that this is mere dogma.

Let me close by mentioning one point of agreement with the creationist who has been my main adversary in this book. Phillip Johnson says he "regards the idea of a Christian political party with a combination of horror and amusement," given that "Christian denominations are themselves so confused and internally divided."[53] To his sentiment, I want to say "Amen!" But why, I wonder, does he think that a Christian theistic science would be immune from this fractious factionalism? The confusion he identifies is just the sort of disorder that we observed within the creationist Tower and that we have to work to see that it does not infect our public dealings, especially when it involves our children. Would it really be wise to inject this ancient and ongoing conflict among private religious beliefs into science and the science classroom? Although we should be respectful of individuals' right to express and live their lives in accord with their religious values, we must not compromise the common public values, especially those exemplified in the ideals of the scientific epistemic virtues, that allow us to act in concert.

Agnes Meyer wrote in *Education for a New Morality* that "From the nineteenth-century view of science as a god, the twentieth century has begun to see it as a devil. It behooves us now to understand that science is neither one nor the other."[54] Meyer had a different set of issues in mind, but her good advice is relevant and perhaps even more apropos to the controversy that has been the topic of this book. Science is neither God nor devil, but profoundly human. It is not infallible. It cannot answer every question. It reveals nothing of possible supernatural realms. It is simply the best method that we evolved, natural creatures have yet discovered for finding our way around this natural world.

Notes

Preface

1. (Pennock 1996b)
2. (Pennock 1996c)
3. (Pennock 1996a)
4. (Pennock in press)
5. (Pennock 1998)
6. (Chick 1992)

Chapter 1

1. I adopt this fruitful idea from Richard Dawkins's book *The Selfish Gene* (Dawkins 1976, ch. 11). Dawkins suggests that there is a unit of cultural transmission which is the counterpart of the gene, and he coins for it the term "meme." (It is derived from the Greek word *mimeme*—imitation—and the French word *même*—memory. It should be pronounced to rhyme with "cream.") Memes are any sort of idea, and they may be thought of as competing for survival in the cultural soup just as genes do in the gene pool.
2. (Scopes and Presley 1967)
3. (Anonymous 1982)
4. (Overton 1982, p. 318)
5. See (La Follette 1983) for a series of useful commentaries on the legal and some of the philosophical aspects of the Arkansas case.
6. (Davis and Kenyon 1993)
7. See (Ammerman 1987) for a fascinating, sympathetic study of Fundamentalism by an Evangelical sociologist.
8. In John E. Lothers's study of the views of biology teachers at Christian institu-

tions, he found that at least 75.3 percent held the former view, while 15.1 percent held the latter (Lothers Jr. 1995, p. 185).

9. (Spong 1991, p. 226)

10. (La Follette 1983, p. 16)

11. (Brewster 1927, p. 109)

12. Tom McIver's forthcoming book *Creationism: Origins, Evolution , and Diversity* is a valuable resource on the histories and factional disputes among a wide varieties of creationist organizations. His earlier book, *Anti-Evolution: An Annotated Bibliography,* is the best reference work covering creationist books (McIver 1992 (1988)).

13. <http://www.indirect.com/www/wbrown/>

14. (Brown 1996 (1986)). Brown also publishes a special edition of the book, for possible use in schools, that omits the biblical references.

15. (Frair and Davis 1983, p. 74)

16. (Heeren 1995, p. 156)

17. (Price 1902)

18. (Scofield 1917, p. 3)

19. (Numbers 1992, p. 99)

20. (Morton 1995)

21. (Luther 1958, p. 6)

22. (Archer 1982, p. 60)

23. (Ross 1994, p. 51)

24. (Van Bebber and Taylor 1995, p. 84)

25. (Van Bebber and Taylor 1995, pp. 81–82)

26. (Van Bebber and Taylor 1995, p. 9)

27. (Van Bebber and Taylor 1995, p. 9)

28. (Van Bebber and Taylor 1995, p. 22)

29. (Van Bebber and Taylor 1995, p. 64)

30. (Van Bebber and Taylor 1995, p. 119)

31. (Ammerman 1987, p. 55)

32. Quoted in (Numbers 1992, p. 87)

33. See especially (Peacocke 1979; Peacocke 1993; Peacocke 1997).

34. (van Till 1986, p. vii)

35. (van Till 1986, p. 198)

36. (van Till 1986, p. 188)

37. (Campbell 1997a)

38. Byron Nelson's most important works were his 1927 *After Its Kind* (1952

(1926)) which attacked evolution and his 1931 *The Deluge Story in Stone: A History of the Flood Theory of Geology* (1968 (1931)), which promoted young-earth Flood Geology. His *Before Abraham: Prehistoric Man in Biblical Light* (1948) attacked the fossil evidence for human evolution.

39. A few other anti-Darwinians, such as David Berlinsky and Hubert Yockey, are closely associated with the group, but I do not (currently) include them because it seems (so far) that they would not agree with the central creationist claim of intelligent-design.

40. (MacBeth 1971)

41. (Bird 1991 (1987))

42. (Dembski 1995, p. 3)

43. (Ankerberg and Weldon 1993)

44. (Moreland 1994)

45. (Ankerberg and Weldon 1994)

46. These are quoted from a recent ICR catalog.

47. (John Morris, June 1997 fundraising letter)

48. (Nelson 1997)

49. (Duvall 1995)

50. (Buchanan, Dec. 1995, Interview in *GQ*)

51. (Buchanan, Feb. 18, 1996, Interview on ABC-TV program *This Week with David Brinkley*)

52. (Nelson 1997)

53. (Ammerman 1987, p. 55)

54. (Gilkey 1985, p. 104)

55. (Johnson 1997, p. 92)

Chapter 2

1. (Sibley and Ahlquist 1987; Sibley, Comstock, and Ahlquist 1990)

2. (Morris and Morris 1996b, p. 89)

3. (Morris and Rajca 1995)

4. (Morris and Rajca 1995)

5. (Ammerman 1987, p. 53)

6. (Ammerman 1987, p. 54)

7. (Salmon 1966)

8. There are subtle but significant differences among these that I am loathe to gloss over, but I do just that here since they are not of immediate concern to our subject. For a discussion of some of the differences see (Pennock 1995a).

9. (Numbers 1992, p. 132)

10. See (Sober 1993) for a philosophical explanation of these and other evolutionary concepts.

11. (Desmond and Moore 1991, p. 49)

12. (Paley 1805, p. 296)

13. (Cronin 1991, pp. 12–15)

14. (Darwin 1969)

15. (Desmond and Moore 1991, pp. 160–161)

16. (Desmond and Moore 1991, p. 152)

17. Quoted in (Desmond and Moore 1991, p. 155)

18. (Desmond and Moore 1991, p. 159)

19. Quoted in (Desmond and Moore 1991, p. 128)

20. Quoted in (Desmond and Moore 1991, p. 209)

21. (Darwin 1969)

22. See (Hall 1994) for an overview of the state of the art.

23. (Darwin 1859 & 1871)

24. (Darwin 1969, p. 119–120)

25. Quoted in (Desmond and Moore 1991, p. 116)

26. Quoted in (Cronin 1991, p. 18)

27. (Darwin 1964 (1859), p. 47)

28. (Sedgwick 1988 (1860))

29. (Darwin, Letter to a German youth, 1879)

30. Quoted in (Desmond and Moore 1991, p. 636)

31. (Morris and Morris 1996a, p. 288)

32. (Morris and Rajca 1995)

33. (Lyell 1881, p. 249)

34. (Lyell 1881, p. 256)

35. (Sedgwick 1831, p. 313)

36. (Numbers 1992, pp. 73–101)

37. (Numbers 1992, p. 125)

38. (Numbers 1992, p. 126)

39. (Morris and Morris 1996a, p. 307)

40. (Morris and Morris 1996a, p. 334)

41. (Morris and Morris 1996a, p. 334)

42. (Schrödinger 1967 (1944))

43. (Wicken 1987, p. 5)

44. (Weber, Depew, and Smith 1988, p. 8)

45. (Kauffman 1995, p. 21)

46. (Morris and Morris 1996a, p. 334)

47. (ICR 1997)

48. (Vardiman 1997, p. iv)

49. (Vardiman 1997, p. iv)

50. (Vardiman 1997, p. iv)

51. (Vardiman 1997, p. iv)

52. (Vardiman 1997, p. iv)

53. (Mayr and Provine 1980)

54. (Pennock 1991)

55. (Crick 1968)

56. (Osawa 1995)

57. (Fox 1987, p. 67)

58. (McMasters 1989, p. 168)

59. (de Bono 1970)

60. (Adams 1974)

61. (Koberg 1974)

62. (Osborn 1953)

63. This indicates another problem with Dembski's explanatory filter. It takes the form of a flow chart that considers law and chance and design in order of priority and, by assuming that these are mutually exclusive and taking them in turn, overlooks the possibility that they may work in tandem. Evolution does not occur simply by chance nor does it work simply by deterministic laws; rather it is the special combination of causal mechanisms that does the trick.

64. (Darwin 1964 (1859), p. 29)

65. (Morris 1985, p. 6)

66. (Morris 1985, p. viii)

67. (Morris and Morris 1996a, p. 34)

68. (Morris and Morris 1996a, p. 35)

69. (Johnson 1993 (1991), p. 22)

70. (Johnson 1993 (1991), p. 21)

71. (Popper 1957, p. 106)

72. (Popper 1968 (1962), p. 340)

73. (Popper 1978, p. 344)

74. (Popper 1978, p. 346)

75. Protesting just a bit too much, Johnson dismisses the finch-beak studies in *Reason in the Balance* (chapter 4), several times in *Defeating Darwinism* (e.g., pp. 11, 52, 59), and in almost every one of his articles and speeches. Having been forced to give up their rejection of even microevolution, creationists must constantly insist on a sharp distinction between micro- and macroevolution, and to deny the possibility that the Darwinian mechanism could account for the latter. Johnson says, "Don't let anyone tell you that the mechanism is a mere detail; it is what the controversy is mainly about" (1997, p. 59). Creationists have to rely upon negative arguments against the Darwinian mechanism, for they know that have no mechanism of their own to even offer, let alone observe and quantify. I'll discuss the issue of micro- vs. macroevolution in detail in chapter 3, Johnson's use of negative argumentation in chapter 4, and the emptiness of his alternative "theory" of miraculous intelligent-design in chapters 5 and 6.

76. <http://aero.stanford.edu>

77. (Flam 1994b; Travis 1991)

78. (Flam 1994a)

79. Quoted in (Safire 1993, p. 38)

80. (Darwin 1859 & 1871, p. 397)

81. Quoted in (Desmond and Moore, 1991, p. 230)

82. (Goodall 1986, p. 366)

83. (Goodall 1986, p. 386)

84. (Darwin 1989 (1877), pp. 125–126)

85. (Darwin 1989 (1877), p. 126)

86. (Pennock 1995b)

Chapter 3

1. (Morris 1985, p. 185)

2. (Morris 1984, p. 429)

3. (Morris 1984, p. 431)

4. (Morris 1984, pp. 433–434)

5. (Morris 1984, p. 432)

6. Their argument runs as follows:

Though we are surely not omniscient, nor omnipresent, nor omnipotent, by the powers released in us through the gift of language we are undeniably able to entertain such concepts, and in doing so we give as clear a proof as ought to be required that our capacity for language cannot have originated within the narrow confines of any finite duration of experience. (Oller Jr. and Omdahl 1994, p. 266)

One problem with the argument is that it confuses words and concepts. That we can express something in words does not imply that it is conceivable—you can say "square circle", but can you really entertain the concept of a square circle? Nor is there a good reason to think that even if we could conceptually "step outside the bounds of time and space" (p. 266) that this "necessarily" implies that some "intervening intelligence" beyond space and time created that capacity in us.

7. (Darwin 1859 & 1871, p. 465)

8. (Chomsky 1975, p. 123)

9. (Landsberg 1988, p. vii)

10. (Schleicher 1983 (1863), p. 14)

11. (Schleicher 1983 (1863), p. 30)

12. (Schleicher 1983 (1863), p. 32)

13. (Atwood 1975)

14. (Fisher and Bornstein 1974, p. 282)

15. (Schleicher 1983 (1863), p. 32)

16. (Fisher and Bornstein 1974, pp. 52–55)

17. Quoted in (Desmond and Moore, 1991, p. 215)

18. (Schleicher 1983 (1863), p. 33)

19. Jones, *Works* 3:34–35; quoted in (Cannon 1991, p. 31)

20. (Cannon 1991, p. 45)

21. (Cannon 1991, p. 25)

22. (Mengham 1993, p. 57)

23. In fairness, one can find places where Schleicher recognizes blending. He writes of the Romance languages and the "unmistakable offspring" of Latin, which emerge "partly through the process of ramification and partly through foreign influence, which you, gentlemen, would call crossing ..." (Schleicher 1983 (1863)). It may be that the criticism made of Schleicher was simply because the emphasis put on his branching tree model obscured caveats and subtleties.

24. (Atwood 1975, pp. 108–109)

25. (Darwin 1989 (1877), p. 59)

26. (Schleicher 1983 (1863), p. 44)

27. (Cavalli-Sforza and Cavalli-Sforza 1995, pp. 112–123)

28. (Cavalli-Sforza, Minch, and Mountain 1992; Cavalli-Sforza et al. 1988)

29. (Cavalli-Sforza and Cavalli-Sforza 1995, p. 202)

30. (Schleicher 1983 (1863), pp. 65–66)

31. (Pei 1965, p. 421)

32. (Schleicher 1983 (1863), p. 21)

33. Quoted in (Numbers 1992, p. 132).

34. (Roose and Gottlieb 1976; Soltis and Soltis 1989)

35. (Ankerberg and Weldon 1993, p. 9)

36. (Greenberg 1991, p. 127)

37. (Schleicher 1983 (1863), p. 43)

38. (Gish 1995)

39. (Wray, Levinton, and Shapiro 1996)

40. Quoted in (Numbers 1992, p. 114)

41. See, for example, (Futuyma 1986, p. 553)

42. (Raff 1996)

43. (Johnson 1993 (1991), p. 69)

44. (Morris 1984, p. 431)

45. (Oller Jr. and Omdahl 1994, p. 256)

46. (Oller Jr. and Omdahl 1994, p. 257)

47. (Pinker 1994)

48. I introduced Bradley in the first chapter as an old-earth creationist; intelligent-design creationists are usually silent about the age of the earth in order to provide an umbrella for both OECs and YECs, but Bradley is one IDC who is forthright in his public talks about his stand on this issue. He is co-author with Charles Thaxton and Roger Olsen of *The Mystery of Life's Origin* (Thaxton, Bradley, and Olsen 1984), which argues that the origin of life by purely naturalistic processes is implausible and advocates a distinct "origin science" that incorporates supernatural interventions. Behe also explicitly endorses an old earth in his *Darwin's Black Box*. I will discuss his views later in this chapter.

49. The Foundation for Thought and Ethics, based in Richardson, Texas, describes itself as a "Christian think-tank" dedicated to proclaiming the Gospel and publishing the biblical perspective on contemporary social and educational issues. Charles Thaxton serves as the Foundation's director of curriculum research, and was the editor of *Pandas*.

50. (Behe 1996, p. xi)

51. Darwin, quoted in (Behe 1996, p. 18)

52. (Linhart 1997, p. 388)

53. (Frair and Davis, 1983, pp. 92–93)

54. (Davis and Kenyon 1993, p. 58)

55. (Frair and Davis 1983, p. 94)

56. (Davis and Kenyon 1993, p. 58)

57. (Fano 1992, pp. 118–119)

58. (Arthur 1996) describes this pattern in detail for ICR's veteran debater Duane Gish, revealing a "morass of errors, omissions, misquotes, old data, distortions,

and non sequiturs" (p. 89) in his arguments that various scientists have publicly corrected in his presence, but that he continues to repeat year after year.

59. (Behe 1996, p. 5)

60. (Behe 1996, p. 15)

61. (Behe 1996, p. 13–14)

62. (Behe 1996, p. 141)

63. (Fano 1992, pp. 118–119)

64. (Comment on CNN Interactive WWW, "Should Creationism be taught in school?" November 7, 1995)

65. (Comment on CNN Interactive WWW, "Should Creationism be taught in school?" November 7, 1995)

66. (Frair and Davis 1983, p. 11)

67. We might note here as a warning that intelligent-design creationists have adopted the term "theory" to give an aura of scientific credibility to their view, calling their view "intelligent-design theory" and themselves "intelligent-design theorists." We should not be misled by this sort of self-puffery; as we will see, this is a theory in name only.

68. (Lakatos 1980)

69. (Johnson 1997, p. 66)

70. (Frair and Davis 1983, p. 21f)

71. (Frair and Davis, 1983, pp. 22–23)

Chapter 4

1. (Parrish 1988)

2. (van Till 1986, p. 201)

3. (Bocarsly et al. 1993)

4. (Silberman 1993, p. 9)

5. (Provine 1990, p. 20)

6. (Johnson 1990, p. 1)

7. One of Johnson's main complaints against science, which he illustrates by reference to the National Academy of Science's "friend of the court" brief in the Arkansas trial, is that it does not allow creationism to use merely negative arguments (1991, p. 8).

8. As we have seen, politically organized creationists are mostly Christian literalists who believe, for example, in biblical inerrancy, in special Creation by God of all biological kinds, and in a historical global Flood. The Bible-Science Association, the Creation-Science Fellowship, Inc., and the Creation Research Society, among others, all require that one sign a statement of belief in such principles to qualify for membership.

9. (Johnson 1991, p. 4)

10. (Johnson 1991, p. 113)

11. (Johnson 1990, p. 13)

12. (Culliton 1989) reports that only *half* of Americans accepted the statement that "Human beings as we know them today developed from earlier species of animals." Of these many will agree that this evolutionary devolopment was guided by God.

13. (Johnson 1990, p. 2)

14. Johnson notes the idea that the different races of human beings are all descendants of the original ancestral pair—Adam and Eve (Johnson 1991, p. 68). He calls this an example of allowable microevolution, but clearly the Creationist notion of diversification "within the limits" of created "kinds" is not the same as microevolution on the biological model. Furthermore, it is doubtful that creation-scientists, given their theological beliefs, could consistently accept that races differentiated on the basis of natural selection of random genetic variation.

15. (Johnson 1990, p. 2)

16. (Johnson 1991, p. 4)

17. (Johnson 1991, p. 4)

18. (Johnson 1990, p. 6)

19. The standard creationist arguments that Johnson brings out have previously been addressed by scientists and philosophers of science such as Philip Kitcher (1982), Michael Ruse (1982), Arthur N. Strahler (1987), and Tim Berra (1990).

20. (Johnson 1990, p. 13)

21. (Johnson 1990, p. 17)

22. (Johnson 1990, p. 8)

23. (Johnson 1990, p. 8)

24. (Johnson 1990, p. 13)

25. (Johnson 1990, p. 15)

26. To recognize the enrichment of the scientific ontology beyond the classical form of materialism, it is now more common to speak of "physicalism", where the reference is to the ontology of current physics, or sometimes "physicalistic materialism." Admittedly, some scientists and philosophers continue to use simply "materialism," but it is understood in the broader sense of the *fabric* of the universe, which includes space-time and electromagnetic fields and so on, rather than in the old sense of mere *matter*. Johnson, however, explicitly links naturalism and the old mechanistic materialism throughout his works, with the rhetorical effect of conflating naturalistic evolution with atheistic materialism.

27. (Johnson 1991, pp. 114–115)

28. (Johnson 1991, p. 115)

29. (Johnson 1991, p. 114)

30. Johnson provides no bibliographic references for any of these terms so we cannot evaluate the specifics of the definitions he may have in mind.

31. Another example is this passage: "The Soviet Cosmonaut who announced upon landing that he had been to the heavens and had not seen God was expressing crudely the basic philosophical premise that underlies Darwinism. Because we cannot examine God in our telescopes or under our microscopes, God is unreal. It is meaningless to say that some entity exists if in principle we can never have knowledge of that entity" (Johnson 1990, p. 14). Not only does Johnson once again link naturalism to atheism here by way of communism, but he explicitly characterizes it in positivist terms, with the reference to the verifiability criterion of meaning.

32. (Johnson 1990, p. 3)

33. (Simpson 1970, p. 61)

34. (Gould 1965, p. 226)

35. (Johnson 1990, p. 15)

36. This is not to say, however, that things we now think of as supernatural necessarily are so. It could turn out, for example, that ghosts exist but that unlike our fictional view of them, they are subject to natural law. In such a case we would have learned something new about the natural world (which may require revising current theories), and would not have truly found anything supernatural.

37. I should note a possible red herring on this point. Creationists often erroneously claim that evolutionary theory itself is not disconfirmable, and so they charge that it is in the same boat as their view. To his credit, Johnson does not make this error. He understands that the theory is disconfirmable, and all of his negative argumentation purports to show that in fact it has already been disconfirmed. Johnson's positive claim is that biologists' philosophical prejudices prevent them from recognizing the disconfirming evidence.

38. (Johnson 1990, p. 7)

39. (Johnson 1990, p. 9)

40. (Johnson 1990, p. 14)

41. (Johnson 1990, p. 14)

42. (Johnson 1990, p. 13)

43. (Johnson 1991, p. 31)

44. (Johnson 1991, p. 113)

45. (Johnson 1996, p. 561)

46. (Johnson 1996, p. 561)

47. (Johnson 1995c, p. 107)

48. (Johnson 1995c, p. 107)

49. (Johnson 1995c, p. 108)

50. (Plantinga 1997, p. 143)

51. (Plantinga 1997, p. 143–144)

52. (Plantinga 1997, p. 154)
53. (Plantinga 1997, p. 149)
54. (Johnson 1995c, p. 105)
55. (Newton 1962, p. 399)
56. (Burtt 1925, p. 215)
57. (Kuhn 1962, p. 148)
58. (Johnson 1997, p. 118)
59. (Johnson 1993 (1991), p. 123)
60. (Johnson, 1995. Open letter to John W. Burgeson published on the Internet.)
61. Johnson quoted in (Silberman 1993, p. 4)
62. Johnson's advocate John W. Burgeson mentioned this fact in a recent Internet post.
63. (Johnson 1993 (1991), p. 123)
64. (Johnson 1993 (1991), p. 134)
65. (Campbell 1997b)
66. (Johnson 1997, chap. 8)
67. (Johnson 1997, pp. 118–119)

Chapter 5

1. (Mitchell 1987, p. 127)
2. (Mitchell 1987, p 85)
3. (Baugh 1983)
4. (Morris 1985, p. 122)
5. (Cole, Godfrey, and Schafersman 1985; Godfrey 1985)
6. (Godfrey 1985, p. 36)
7. (Godfrey 1985, p. 36)
8. (Hastings 1985; Hastings 1986; Hastings 1987)
9. (Kuban 1986, pp. 13–16)
10. (Vol. 40, No. 3, p. 151)
11. Ham quoted in (Thomas 1996)
12. The scent of Paluxy is not entirely gone from the Trilogy, however, for page 332 of its second volume *Science and Creation,* refers the reader back to pages in Henry Morris's *The Biblical Basis of Modern Science* that promote the tracks (Morris 1984, p. 332).
13. (Morris and Morris 1996a, pp. 120–122)
14. (Shore 1984)

15. (Strahler 1987, p. 145)

16. (Awbrey 1983, p. 28)

17. (Van Till, Young, and Menninga 1988, p. 82)

18. This claim appears on the back cover of each volume of the *Trilogy*.

19. (Morris and Morris 1996a, p. 332)

20. (Morris and Morris 1996a, p. 317)

21. (Morris and Morris 1996a, p. 319)

22. (Morris and Morris 1996a, p. 319)

23. (Milne 1984, p. 3)

24. (Morris 1985, p. 168)

25. (Monroe 1986, p. 45)

26. For those who care to delve into some of the scientific details, Arthur Strahler's 552–page *Science and Earth History* (Strahler 1987) and the publications of the National Center for Science Education <www.natcenscied.org> are invaluable resources.

27. (Berra 1990, pp. 125–126)

28. (Davis and Kenyon 1993, p. vii)

29. (Heeren 1995, p. 169)

30. (Davis and Kenyon 1993, p. 126)

31. (Drake and Sobel 1992, p. 37)

32. (Drake and Sobel 1992, p. 88)

33. (Heeren 1995, pp. 36–37)

34. (Heeren 1995, p. 54)

35. (Dembski 1994, p. 83)

36. (Heeren 1995, p. 27)

37. (Raël 1986 (1974), p. 103)

38. (Terrusse 1996)

39. (Terrusse 1996)

40. (Terrusse 1996)

41. (Brand 1995)

42. (Raël 1986 (1974), p. 103)

43. A year after writing this, I learned from Tom McIver that at least one creationist holds a very similar sort of view. A. E. Wilder-Smith claims in *The Reliability of the Bible*, that God literally "cloned" Eve from Adam. He writes that "The entire report [in Genesis] reads exactly like a historical description of surgery under normal physiological conditions for surgery"(Wilder-Smith 1983, p. 55) and that in the operation God removed a cell from Adam's rib, deleted the Y chromosome

and doubled the remaining X to produce a female. When it comes to creationism, "Truth" is indeed stranger than fiction.

44. (von Däniken 1968, p. 55)

45. (von Däniken 1968, p. 57)

46. (Raël 1986 (1974), pp. 134–135)

47. (Raël 1986 (1974), p. 136)

48. (Flew 1966, pp. 85–88)

49. (Paley 1805, p. 241)

50. (Paley 1805, p. 244)

51. (Paley 1805, p. 244)

52. (Paley 1805, p. 245)

53. (Darwin 1996 (1859), p. 152)

54. (Darwin 1996 (1859), p. 153)

55. (Dawkins 1986; Dawkins 1996, chap. 5)

56. (Gould 1980, pp. 20–21)

57. (Nelson 1996, p. 514)

58. (Temple 1884, p. 117)

59. (Frair and Davis 1983, p. 31)

60. (Behe 1996, p. 223)

61. (Johnson 1997, p. 71)

62. (Johnson 1997, p. 72)

63. Quoted in (Flew 1966, p. 59)

64. (Geisler 1986, p. 39)

65. (Geisler 1986, p. 40)

66. (Geisler 1990, p. 247)

67. (Geisler 1990, p. 246)

68. Although Geisler appears to be the first to sketch the whole argument, it is only fair to point out that he picked up the key notion of specified complexity of DNA from Charles Thaxton, et. al. (Thaxton, Bradley, and Olsen 1984, p. 130), and that Thaxton was academic editor of *Pandas*. Thaxton in turn credits (Yockey 1977).

69. Quoted in (Gilkey 1985, pp. 76–77)

70. (http://www.seti-inst.edu/faq.html)

71. (Woodward 1996)

72. (Dembski 1997, p. 189)

73. Here I use Dembski's terminology of "complex *specified* information," but with a caveat to the reader that the term is subtly question-begging. Certainly, the information in biomolecules is *specific*, but, unlike his archery or other examples,

we cannot say that it is specified (which implies someone who was its specifier) in the sense that is needed for an inference to design.

74. (Dembski 1997, p. 189)

75. (Dembski 1997, p 186)

76. (Hoyle and Wickramasinghe 1981)

77. (Yockey 1981)

78. (Dawkins 1986, pp. 46–50)

79. (Sober 1993, pp. 37–38)

80. (Behe 1996, p. 221)

81. (Behe 1996, p. 221)

82. (Berra 1990)

83. (Johnson 1997, p. 63)

84. (Johnson 1997, p. 63)

85. (Dawkins 1986, p. 50)

86. (Dawkins 1986, p. 60)

87. (Dawkins 1986, p. 62)

88. (Johnson 1997, p. 80)

89. (Johnson 1997, p. 63)

90. (Darwin 1996 (1859), p. 154)

91. (Woodward 1996, p. 16)

92. (Behe 1996, p. 39)

93. (Behe 1996, p. 193)

94. (Behe 1996, p. 215)

95. (Behe 1996, p. 39)

96. (Behe 1996, p. 42)

97. (Behe 1996)

98. In the case of the mousetrap, its function really is specified and not just specific.

99. (Behe 1996, p. 216)

100. Quoted in (Sobel 1995)

101. (Orr 1997)

102. (Behe 1996, pp. 232–233)

103. (Davis and Kenyon 1993, p. viii)

104. (Gilchrist 1997, p. 14)

105. (Gilchrist 1997, p. 15)

106. (Behe 1996, p. 186)

107. (Gilkey 1985, p. 82)

Chapter 6

1. Jerry Falwell speaking on ABC local affiliate program "AIDS: The Anatomy of a Crisis." Falwell opened the program by citing Galatians: "When you violate moral, health, and hygiene laws, you reap the whirlwind. You cannot shake your fist in God's face and get by with it" (Shilts 1987, p. 347). Shilts cites similar comments made by other Fundamentalist and Evangelical leaders.

2. Quoted in (Toulmin and Goodfield 1962, p. 192)

3. (Newton 1962, p. 399)

4. (Sedgwick 1831, p. 313)

5. (Kottler 1974)

6. Quoted in (Heeren 1995, p. 129)

7. (Hempel 1964)

8. (Kitcher 1989).

9. (Salmon 1984)

10. (van Fraassen 1980)

11. (Whewell 1894, p. 151)

12. (Johnson 1991, pp. 14, 113)

13. (Johnson 1991, p. 113)

14. (Johnson 1991, p. 4)

15. (Johnson 1991, p. 71)

16. (Johnson 1991, p. 67)

17. (Frair and Davis 1983, p. 14)

18. (Davis and Kenyon 1993, p. 133)

19. (Behe 1996, p. 227)

20. (Behe 1996, p. 233)

21. (Johnson 1995c, p. 39)

22. (Johnson 1995c, p. 12)

23. (Johnson 1995c, p. 31)

24. (Shumaker 1972, p. 67)

25. (Shumaker 1972, p. 78)

26. (Shumaker 1972, p. 79)

27. (Johnson 1995c, p. 92)

28. (Johnson 1991, p. 31)

29. (Johnson 1994a; Johnson 1995a)

30. (Johnson 1994b; 1995b)

31. (Woodmorappe 1996)

32. (Johnson 1995c, p. 189)
.33. (Johnson 1995c, p. 204)
34. (Johnson 1995c, p. 110)
35. (Johnson 1995c, p. 110)
36. (Johnson 1997, p, 71)
37. (Geisler 1990, p. 247)

Chapter 7

1. Intelligent-design creationist Jonathan Wells brought up Nazism as one of the pernicious results of evolutionary theory, following one of my talks at a 1997 conference.
2. (Morris 1984, p. 107)
3. (Morris 1984, p. 109)
4. (Morris 1984, p. 109)
5. (Morris 1984, p. 110)
6. (Ammerman 1987, p. 52)
7. Quoted in (Desmond and Moore 1991, p. 449).
8. (Ankerberg and Weldon 1993, p. 37)
9. (Ross 1993, chap. 1)
10. (Johnson 1995c, p. 197)
11. (Johnson 1995c, p. 7)
12. (Johnson 1995c, p. 38)
13. (Johnson 1995c, p. 38)
14. (Johnson 1995c, p. 14)
15. (Anonymous 1996)
16. (Johnson 1995c, p. 116)
17. (Johnson 1995c, p. 22)
18. (Johnson 1995c, p. 41)
19. (Johnson 1995c, p. 41)
20. (Johnson 1995c, p. 144)
21. (Johnson 1995c, p. 141)
22. (Johnson 1995c, p. 45)
23. (Johnson 1995c, p. 197)
24. Darwin quoted in (Desmond and Moore, 1991, p. 122)
25. (Johnson 1995c, p. 116)
26. (Johnson 1995c, p. l 12)

27. (Johnson 1995c, p. 17)
28. (Johnson 1995c, p. 40)
29. (Johnson 1995c, p. 112)
30. (Johnson 1995c, p. 126)
31. (Provine 1990)
32. (Johnson 1995c, p. 49)
33. (Johnson 1995c, p. 40)
34. (Johnson 1995c, p. 41)
35. (Johnson 1995c, p. 139)
36. (Johnson 1995c, p. 139)
37. (Johnson 1995c, pp. 22, 161)
38. (Johnson 1995c, p. 32)
39. (Johnson 1995c, p. 157)
40. (Johnson 1995c, p. 157)
41. (Johnson 1995c, p. 39)
42. (Johnson 1995c, p. 39)
43. (Johnson 1995c, p. 26)
44. (Johnson 1995c, p. 196)
45. (Searle 1990).
46. (Johnson 1995c, p. 128)
47. (Johnson 1995c, p. 143)
48. (Johnson 1995c, p. 12)
49. (Johnson 1995c, p. 13)
50. (Heeren 1995, p. 230)
51. (Ross 1993)
52. (Johnson 1995c, p. 7)
53. (Johnson 1995c, p. 197)
54. (Johnson 1995c, p. 97, 211)
55. (Johnson 1995c, p. 97)
56. (Johnson 1995c, p. 194)
57. (Johnson 1995c, p. 204)
58. (Hamilton and Cairns, 1961, pp. 169–185)
59. (Desmond and Moore 1991)
60. (Gish 1993, p. vi)
61. (Johnson 1995c, p. 77)
62. (Johnson 1995c, p. 50)

63. (Johnson 1995c, p. 39)
64. (Johnson 1995c, p. 40)
65. (Johnson 1995c, p. 7)
66. (Johnson 1995c, p. 7)
67. (Morris and Rajca 1995)
68. (Jukes 1991, p. 205)

Chapter 8

1. (Special to the *New York Times* 1995)
2. (Brown 1996)
3. Quoted in (Applebome 1996, p. 12)
4. (Mossner 1967, p. 334)
5. (Erdoes and Ortiz 1984, p. 86)
6. (Tsunoda, Bary, and Keene 1958, p. 26)
7. (Erdoes and Ortiz 1984, p. 47)
8. (Longino 1990, p. 214)
9. Quoted in (Delattre 1988, p. 85). The textbook he quotes was published in 1986 by Accelerated Christian Education, Inc. in Lewisville, Texas as part of their PACE series.
10. (Delattre 1988, p. 86)
11. I discussed (Gilchrist 1997) in chapter 5. A previous literature search for the scientific basis of creation-science that was published in the *Quarterly Review of Biology* (Scott and Cole 1985) had also come up empty.
12. (Delattre 1988, p. 88)
13. (Johnson 1997, p. 89)
14. (Johnson 1997, p. 113)
15. Quoted in (Barbero 1993)
16. (Johnson 1997, p. 54)
17. (Johnson 1997, p. 54)
18. (Delattre 1988, p. 96–97)
19. (Delattre 1988, p. 97)
20. (Johnson 1997, p. 59)
21. (Johnson 1997, p. 58)
22. (Johnson 1997, p. 94)
23. (Johnson 1997, p. 57)
24. (Johnson 1997, p. 116)

25. (Delattre 1988, p. 93–94)
26. (Johnson 1997, p. 20)
27. (Johnson 1997, p. 65)
28. (Johnson 1997, p. 65)
29. (Johnson 1995c, p. 28)
30. (Johnson 1997, p. 28)
31. (Johnson 1997, p. 35)
32. (Johnson 1997, p. 35)
33. (Johnson 1995c, pp. 33–34)
34. (Johnson 1995c, p. 22)
35. (Johnson 1997, p. 22)
36. (Johnson 1995c, p. 26)
37. (Johnson 1997, p. 119)
38. (Root-Bernstein 1995, p. 41)
39. (Scott 1994, p. 13)
40. (Root-Bernstein 1995, p. 41)
41. See (Ross 1962) for an interesting history of the term.
42. (Root-Bernstein 1995, p. 38)
43. (Root-Bernstein 1995, p. 41)
44. (Delattre 1988, p. 87)
45. (Koll and Livingstone 1994 (1874))
46. (Johnson 1997, p. 113)
47. (Johnson 1997, p. 113)
48. (Johnson 1997, p. 117)
49. (Johnson 1997, p. 117)
50. (Johnson 1997, p. 113)
51. (Johnson 1997, p. 118)
52. (Johnson 1997, p. 54)
53. (Johnson 1997, p. 106)
54. (Meyer 1957, p. 11)

References

Adams, James L. 1974. *Conceptual Blockbusting: A Guide to Better Ideas.* San Francisco, CA: W. H. Freeman and Company.

Ammerman, Nancy Tatom. 1987. *Bible Believers: Fundamentalists in the Modern World.* New Brunswick, NJ and London: Rutgers University Press.

Ankerberg, John, and John Weldon. 1993. *The Facts on Creation vs. Evolution, The Anker Series.* Eugene, OR: Harvest House Publishers.

Ankerberg, John, and John Weldon. 1994. Rational Inquiry & The Force of Scientific Data: Are New Horizons Emerging? In *The Creation Hypothesis: Scientific Evidence for an Intelligent Designer,* edited by J. P. Moreland. Downers Grove: InterVarsity Press.

Anonymous. 1982. A High Powered Battery of Lawyers and Scientists Challenges Arkansas' "Creation-Science" Law. *Science* (January).

Anonymous. 1996. Creation/Evolution—What Are the Issues for the Church. In *Christian Answers.* http://www.ChristianAnswers.net/ac024.html.

Applebome, Peter. 1996. 70 Years after Scopes Trial, Creation Debate Lives. *New York Times,* March 10, 1996, 1–12.

Archer, Gleason. 1982. *Encyclopedia of Bible Difficulties.* Grand Rapids, MI: Zondervan.

Arthur, Joyce. 1996. Creationism: Bad Science or Immoral Pseudoscience? *Skeptic* 4 (4):88–93.

Atwood, E. Bagby. 1975. *The Regional Vocabulary of Texas.* Austin, TX: University of Texas Press.

Awbrey, Frank T. 1983. Space Dust, the Moon's Surface, and the Age of the Cosmos. *Creation/Evolution* 4 (3):21–29.

Barbero, Yves. 1993. Interview with Phillip Johnson. *California Committee of Correspondences Newsletter* (Third Quarter).

Baugh, Carl. 1983. *Latest Human and Dinosaur Tracks.* (Audiotape.) Minneapolis, MN: Bible-Science Association Tape of the Month.

Behe, Michael J. 1996. *Darwin's Black Box: The Biochemical Challenge to Evolution*. New York: The Free Press.

Berra, Tim M. 1990. *Evolution and the Myth of Creationism*. Stanford: Stanford University Press.

Bird, Wendell R. 1991 (1987). *The Origin of Species Revisited: The Theories of Evolution and of Abrupt Appearance*. 2 vols. New York: Philosophical Library, Inc.

Bocarsly, Andrew, et al. 1993. Ad Hoc Origins Committee's Letter to Colleagues.

Brand, Stuart. 1995. Interview: "The Physicist." *Wired*, 152–155.

Brewster, Edwin Tenney. 1927. *Creation: A History of Non-Evolutionary Theories*. Indianapolis, IN: The Bobbs-Merril Company.

Brown, Vicki. 1996. 71 Years after "Monkey" Trial, Evolution Flares in Tennessee. *Austin American-Statesman*, March 5, 1996, 1–6.

Brown, Walter. 1996 (1986). *In the Beginning: Compelling Evidence for Creation and the Flood*. Phoenix, AZ: Center for Scientific Creation.

Burtt, E. A. 1925. *The Metaphysical Foundations of Modern Science*. New York: Harcourt, Brace & Co.

Campbell, John Angus. 1997a. Reflections on the Mere Creation Conference. *Origins & Design* 18 (1).

Campbell, John Angus. 1997b. Theism, Naturalism, and Persuasive Design: A Rhetorical Analysis of Darwin's *Origin*. Paper read at Naturalism, Theism, and the Scientific Enterprise Conference, at Austin, Texas.

Cannon, Garland. 1991. Jones's "Spring from Some Common Source: 1786–1986." In *Sprung from Some Common Source: Investigations Into the Prehistory of Languages*, edited by S. M. Lamb and E. D. Mitchell. Stanford, CA: Stanford University Press.

Cavalli-Sforza, Luigi Luca, and Francesco Cavalli-Sforza. 1995. *The Great Human Diasporas: The History of Diversity and Evolution*. Reading, MA: Addison-Wesley Publishing Co.

Cavalli-Sforza, Luigi Luca, E. Minch, and Joanna Mountain. 1992. Coevolution of genes and languages revisited. *Proceedings of the National Academy of Sciences* 89:5620–5622.

Cavalli-Sforza, Luigi Luca, Alberto Piazza, Paolo Menozzi, and Joanna Mountain. 1988. Reconstruction of Human Evolution: Bringing Together Genetics, Archeology, and Linguistics. *Proceedings of the National Academy of Sciences* 85:6002–6006.

Chick, Jack T. 1992. *Big Daddy?* Chino, CA: Chick Publications.

Chomsky, Noam. 1975. *Reflections on Language*. New York: Pantheon Books.

Cole, J. R., L. R. Godfrey, and S. D. Schaferman. 1985. Mantracks? The Fossils Say *NO! Creation/Evolution* 5 (1):37–45.

Crick, Francis. 1968. The Origin of the Genetic Code. *Journal of Molecular Biology* 38:367–379.

Cronin, Helena. 1991. *The Ant and the Peacock*. Cambridge: Cambridge University Press.

Culliton, Barbara J. 1989. The Dismal State of Scientific Literacy. *Science* 243 (February):600.

Darwin, Charles. 1859, 1871. *The Origin of Species and the Descent of Man*. New York: Random House.

Darwin, Charles. 1955 (1872). *Expression of the Emotions in Man and Animals*. New York: The Philosophical Library.

Darwin, Charles. 1964 (1859). *On the Origin of Species, Facsimile of the First Edition*. Cambridge, MA: Harvard University Press.

Darwin, Charles. [1969]. *The Autobiography of Charles Darwin*. Edited by N. Barlow. New York: Norton.

Darwin, Charles. 1989 (1877). *The Descent of Man, and Selection in Relation to Sex, Part One*. Edited by P. H. Barrett and R. B. Freeman. Second, revised and augmented ed. Vol. 21, *The Works of Charles Darwin*. New York: New York University Press.

Darwin, Charles. 1996 (1859). *The Origin of Species*. Oxford: Oxford University Press.

Davis, Percival, and Dean H. Kenyon. 1993. *Of Pandas and People*. Dallas, Texas: Haughton Publishing Co.

Dawkins, Richard. 1976. *The Selfish Gene*. Oxford: Oxford University Press.

Dawkins, Richard. 1986. *The Blind Watchmaker*. New York and London: W. W. Norton & Company.

Dawkins, Richard. 1996. *Climbing Mount Improbable*. New York and London: W. W. Norton & Company.

de Bono, Edward. 1970. *Lateral Thinking: Creativity Step by Step*. New York: Harper & Row.

de Duve, Christian. 1995. *Vital Dust: Life as a Cosmic Imperative*. New York: Basic Books.

Delattre, Edwin J. 1988. *Education and the Public Trust: The Imperative for Common Purposes*. Lanham, MD: Ethics and Public Policy Center.

Dembski, William A. 1994. The Incompleteness of Scientific Naturalism. In *Darwinism: Science or Philosophy*, edited by J. Buell and V. Hearn. Richardson, TX: Foundation for Thought and Ethics.

Dembski, William A. 1995. What Every Theologian Should Know about Creation, Evolution, and Design. *Center for Interdisciplinary Studies Transactions* 3 (2):1–8.

Dembski, William A. 1997. Intelligent Design as a Theory of Information. *Perspectives on Science and Christian Faith* 49 3 (September):180–190.

Denton, Michael. 1985. *Evolution: A Theory in Crisis*: Adler & Adler.

Desmond, Adrian, and James Moore. 1991. *Darwin: The Life of a Tormented Evolutionist*. New York and London: W. W. Norton & Company.

Drake, Frank, and Dava Sobel. 1992. *Is Anyone Out There? The Scientific Search for Extraterrestrial Intelligence*. London: Souvenir Press.

Duvall, Jed. 1995. School Board Tackles Creationism Debate. *CNN Interactive (WWW)*, November 5.

Eco, Umberto. 1995. *The Search for the Perfect Language*. Translated by James Fentress. In *The Making of Europe*, edited by J. L. Goff. Oxford: Blackwell.

Erdoes, Richard, and Alfonso Ortiz, eds. 1984. *American Indian Myths and Legends*. New York: Pantheon Books.

Fano, Giorgio. 1992. *The Origins and Nature of Language*. Translated by Susan Petrilli. Bloomington and Indianapolis: Indiana University Press.

Fisher, John H., and Diane Bornstein. 1974. *In Forme of Speche Is Chaunge: Readings in the History of the English Language*. Englewood Cliffs, NJ: Prentice-Hall.

Flam, Faye. 1994a. Co-opting a Blind Watchmaker. *Science* 265 (August):1032–1033.

Flam, Faye. 1994b. Ecologist Plans to Let Cyberlife Run Wild in Internet Reserve. *Science* 264 (20 May):1085.

Flew, Antony. 1966. *God and Philosophy*. London: Hutchinson & Co.

Fox, Thomas D. 1987. Natural Variation in the Genetic Code. *Annual Review of Genetics* 21:67–91.

Frair, Wayne, and Percival Davis. 1983. *A Case for Creation*. 3rd ed. Chicago: Moody Press.

Futuyma, Douglas J. 1986. *Evolutionary Biology*. 2nd ed. Sunderland, MA: Sinauer Associates, Inc.

Geisler, Normal L. 1986. What Mount Rushmore and DNA Have in Common. *Creation/Evolution* XVII:39–40.

Geisler, Normal L. 1990. Of Pandas and People: The Central Questions of Biological Origins. *Perspectives on Science & Christian Faith* 42 (4):246–248.

Gilchrist, George W. 1997. The Elusive Scientific Basis of Intelligent Design Theory. *Reports of the National Center for Science Education* 17 (3):14–15.

Gilkey, Langdon. 1985. *Creationism on Trial: Evolution and God at Little Rock*. Minneapolis, MN: Winston Press.

Gish, Duane. 1995. *Evolution: The Fossils Still Say NO!* El Cajon, CA: Institute for Creation Research.

Gish, Duane T. 1993. *Creationist Scientists Answer their Critics*. El Cajon, CA: Institute for Creation Research.

Godfrey, Laurie R. 1985. Foot Notes of an Anatomist. *Creation/Evolution* 5 (1):16–36.

Goodall, Jane. 1986. *The Chimpanzees of Gombe: Patterns of Behavior*. Cambridge, MA: The Belknap Press of Harvard University Press.

Gosse, Philip Henry. 1857. *Omphalos: An Attempt to Untie the Geological Knot*. London: Van Voorst.

Gould, Stephen Jay. 1965. Is Uniformitarianism Necessary? *American Journal of Science* 265 (March):223–228.

Gould, Stephen Jay. 1980. *The Panda's Thumb*. New York: W. W. Norton.

Greenberg, Joseph H. 1991. Some Problems of Indo-European in Historical Perspective. In *Spring from Some Common Source: Investigations into the Prehistory of Languages*, edited by S. M. Lamb and E. D. Mitchell. Stanford, CA: Stanford Univ. Press.

Hall, Brian K., ed. 1994. *Homology: The Hierarchical Basis of Comparative Biology*. San Diego: Academic Press.

Harman, Gilbert. 1965. The Inference to the Best Explanation. *The Philosophical Review* 74 (1):88–95.

Hastings, Ronnie J. 1985. Tracking Those Incredible Creationists. *Creation/Evolution* 5 (1):5–15.

Hastings, Ronnie J. 1986. Tracking Those Incredible Creationists—The Trail Continues. *Creation/Evolution* 6 (1):19–27.

Hastings, Ronnie J. 1987. Tracking Those Incredible Creationists—The Trail Goes On. *Creation/Evolution* 7 (2):30–42.

Heeren, Fred. 1995. *Show Me God: What the Message from Space Is Telling Us About God*. Vol. 1, *Wonders That Witness to the Bible's Truth*. Wheeling, IL: Searchlight Publications.

Hempel, Carl. 1964. Studies in the Logic of Explanation. In *Aspects of Scientific Explanation*. New York and London: The Free Press.

Hoyle, Fred, and Chandra Wickramasinghe. 1981. *Evolution from Space*. London: J. M. Dent and Sons.

ICR. 1997. Radioisotopes and the Age of the Earth. *Acts & Facts*, July, 5.

Johnson, Phillip E. 1990. *Evolution as Dogma: The Establishment of Naturalism*. Dallas, TX: Haughton Publishing Company.

Johnson, Phillip E. 1991. *Darwin on Trial*. 1st ed. Washington, D.C.: Regnery Gateway.

Johnson, Phillip E. 1992. *Darwinism on Trial*. (Videotape.) Pasadena, CA: Reasons to Believe.

Johnson, Phillip E. 1993 (1991). *Darwin on Trial*. 2nd ed. Washington, D.C.: Regnery Gateway.

Johnson, Phillip. 1994a. A.I.D.S. and the Dog That Didn't Bark. *Insight*, 24–26.

Johnson, Phillip E., ed. 1994b. *Cases and Materials on Criminal Procedure*. 2nd ed. St. Paul, MN: West Publishing Company.

Johnson, Phillip. 1995a. The Thinking Problem in HIV-Science. In *A.I.D.S.: Infectious or Not*, edited by P. Duesberg. Dordrecht: Kluwer.

Johnson, Phillip E., ed. 1995b. *Criminal Law: Cases and Materials*. 5th ed. St. Paul: West Publishing Company.

Johnson, Phillip E. 1995c. *Reason in the Balance: The Case against Naturalism in Science, Law, and Education*. Downers Grove, IL: InterVarsity Press.

Johnson, Phillip. 1996. Response to Pennock. *Biology and Philosophy* 11:561–563.

Johnson, Phillip E. 1997. *Defeating Darwinism*. Downers Grove, IL: InterVarsity Press.

Jukes, Thomas H. 1991. Random Walking. *Journal of Molecular Evolution* 33:205–206.

Kauffman, Stuart. 1995. *At Home in the Universe: The Search for the Laws of Self-Organization and Complexity*. New York: Oxford University Press.

Kauffman, Stuart A. 1989. *The Origins of Order: Self-Organization and Selection in Evolution*. Oxford: Oxford University Press.

Kitcher, Philip. 1982. *Abusing Science: The Case against Creationism*. Cambridge, MA: The MIT Press.

Kitcher, Philip. 1989. Explanatory Unification and the Causal Structure of the World. In *Scientific Explanation*, edited by P. Kitcher and W. C. Salmon. Minneapolis, MN: University of Minnesota Press.

Kitcher, Philip. 1993. *The Advancement of Science: Science without Legend, Objectivity without Illusions*. New York: Oxford Univ. Press.

Koberg, Don, and Jim Bagnall. 1974. *The Universal Traveler: A Soft-Systems Guidebook to: Creativity, Problem-Solving, and the Process of Design*. Los Altos, CA: William Kaufmann, Inc.

Koll, Mark A., and David N. Livingstone, eds. 1994 (1874). *Charles Hodge: What Is Darwinism? and Other Writings on Science and Religion*. Grand Rapids, MI: Maker Book House.

Kottler, Malcolm Jay. 1974. Alfred Russell Wallace, the Origin of Man, and Spiritualism. *ISIS* 65 (227):144–192.

Kuban, Glen J. 1986. A Summary of the Taylor Site Evidence. *Creation/Evolution* 6 (1):10–18.

Kuhn, Thomas. 1962. *The Structure of Scientific Revolutions*. Chicago: University of Chicago Press.

La Follette, Marcel Chotkowski, ed. 1983. *Creationism, Science, and the Law: The Arkansas Case*. Cambridge, MA: The MIT Press.

Lakatos, Imre. 1980. Falsification and the Methodology of Scientific Research Programmes. In *Imre Lakatos, Philosophical papers*, edited by J. Worrall and G. Currie. Cambridge: Cambridge University Press.

Landsberg, Marge E., ed. 1988. *The Genesis of Language: A Different Judgement of Evidence*. Berlin: Mouton de Gruyter.

Linhart, Yan B. 1997. The Teaching of Evolution—We Need to Do Better. *BioScience* 47 (6):385–91.

Lipton, Peter. 1991. *Inference to the Best Explanation*. New York and London: Routledge.

Livingstone, David N. 1987. *Darwin's Forgotten Defenders: The Encounter Between Evangelical Theology and Evolutionary Thought*. Edinburgh: Scottish Academy Press.

Longino, Helen E. 1990. *Science as Social Knowledge: Values and Objectivity in Scientific Inquiry*. Princeton, NJ: Princeton University Press.

Lothers Jr., John E. 1995. Biology Teachers' Views on Evolution, Possible Distinctions of Theistic Views. *Perspectives on Science and Christian Faith* 47 (3):177–185.

Luther, Martin. 1958. *Lectures on Genesis*. Edited by J. Pelikan. Vol. 1, *Luther's Words*. Saint Louis, MO: Concordia Publishing House.

Lyell, Charles. 1881. *Life, Letters and Journals of Sir Charles Lyell*. Edited by M. Lyell. 2 vols. London: John Murray.

MacBeth, Norman. 1971. *Darwin Retried*. Boston, MA: Gambit.

Malthus, Thomas R. 1926 (1826). *An Essay on the Principle of Population, as It Affects the Future Improvement of Society, With Remarks on the Speculations of Mr. Godwin, M. Condorcet, and Other Writers*. 6th ed. London: Macmillan.

March, Frank Lewis. 1944. *Evolution, Creation, and Science*. Washington, D.C.: Review and Herald Publishing Assn.

Mather, Cotton. 1968 (1721). *The Christian Philosopher: A Collection of the Best Discoveries in Nature, With Religious Improvements*. Gainesville, FL: Scholars' Facsimiles & Reprints.

Matsumura, Molleen, ed. 1995. *Voices for Evolution*. Revised ed. Berkeley, CA: National Center for Science Education.

Mayr, E., and W. Provine. 1980. *The Evolutionary Synthesis*. Cambridge: Harvard Univ. Press.

McIver, Tom. 1992 (1988). *Anti-Evolution: An Annotated Bibliography*. Jefferson, NC: McFarland & Co.

McMasters, John H. 1989. The Flight of the Bumblebee and Related Myths of Entomological Engineering. *American Scientist* 77 (March–April):164–169.

Mengham, Rod. 1993. *The Descent of Language: Writing in Praise of Babel*. London: Bloomsbury.

Meyer, Agnes. 1957. *Education for a New Morality*. New York: Macmillan.

Milne, David H. 1984. Creationists, Population Growth, Bunnies, and the Great Pyramid. *Creation/Evolution* 4 (4):1–5.

Mitchell, Stephen. 1987. *The Book of Job*. Revised ed. San Francisco, CA: North Point Press.

Monroe, James S. 1986. More on Population Growth and Creationism. *Creation/ Evolution* 6 (2):44–46.

Moore, James. 1994. *The Darwin Legend: Are Reports of His Deathbed Conversion True?* Grand Rapids, MI: Baker Books.

Moreland, J. P., ed. 1994. *The Creation Hypothesis: Scientific Evidence for an Intelligent Designer*. Downers Grove, IL: InterVarsity Press.

Morris, Henry. 1984. *The Biblical Basis of Modern Science*. Grand Rapids, MI: Baker Book House.

Morris, Henry M. 1985. *Scientific Creationism*. 2nd ed. Green Forest, AR: Master Books.

Morris, Henry M., and John D. Morris. 1996a. *The Modern Creation Trilogy: Science and Creation*. 3 vols. Vol. 2. Green Forest, AR: Master Books.

Morris, Henry M., and John D. Morris. 1996b. *The Modern Creation Trilogy: Scripture and Creation*. 3 vols. Vol. 1. Green Forest, AR: Master Books.

Morris, John, and John Rajca. 1995. *A Walk Through History*. (Videotape.) Santee, CA: Institute for Creation Research.

Morris, John D. 1980. *Tracking Those Incredible Dinosaurs and the People Who Knew Them*. San Diego, CA: Creation Life Publishers.

Morton, Glenn R. 1995. *Foundation, Fall, and Flood: A Harmonization of Genesis and Science*. Dallas, TX: DMD Publishing Co.

Mossner, Ernest Campbell. 1967. Deism. In *The Encyclopedia of Philosophy*, edited by P. Edwards. New York: Macmillan.

Nelson, Byron C. 1948. *Before Abraham: Prehistoric Man in Biblical Light*. Minneapolis, MN: Augsburg.

Nelson, Byron C. 1952 (1926). *After Its Kind*. Minneapolis: Bethany.

Nelson, Byron C. 1968 (1931). *The Deluge Story in Stone: A History of the Flood Theory of Geology*. Minneapolis, MN: Bethhany Fellowship.

Nelson, Jill. 1997. Creationism: The Debate Is Still Evolving. *USA Weekend*, April 18–20, 12.

Nelson, Paul A. 1996. The Role of Theology in Current Evolutionary Reasoning. *Biology & Philosophy* 11 (4):493–517.

Newton, Isaac. [1962]. *Mathematical Principles of Natural Philosophy and His System of the World*. Translated by Andrew Motte. Revised by Florian Cajori. Berkeley, CA: University of California Press.

Nordenskiöld, Erik. 1928. *The History of Biology*. Translated by Leonard Bucknall Eyre. New York: Alfred A. Knopf.

Numbers, Ronald L. 1992. *The Creationists: The Evolution Of Scientific Creationism*. New York: Alfred A. Knopf.

Oller Jr., John W., and John L. Omdahl. 1994. Origin of the Human Language Capacity: In Whose Image? In *The Creation Hypothesis*, edited by J. P. Moreland. Downers Grove, IL: InterVarsity Press.

Orr, H. Allen. 1997. Darwin vs. Intelligent Design (Again). In *The Boston Review*. http://www-polisci.mit.edu/bostonreview/BR21.6/orr.html.

Osawa, Syozo. 1995. *Evolution of the Genetic Code*. Oxford: Oxford University Press.

Osborn, Alex. 1953. *Applied Imagination*. New York: Charles Scribner's Sons.

Overton, William R. 1982. United States District Court Opinion: *McLean v. Arkansas*. In *But Is It Science? The Philosophical Question in the Creation/Evolution Controversy*, edited by M. Ruse. Buffalo, New York: Prometheus Books.

Owen, Richard. 1849. *On the Nature of Limbs. A Discourse Delivered on Friday, February 9, at an Evening Meeting of the Royal Institution of Great Britain*. London: John van Voorst.

Paley, William. 1805. *Natural Theology*. Late London ed. New York: American Tract Society.

Parrish, Fred. 1988. How I Was Suckered into a Debate with a Creationist—And Won! *Creation/Evolution*.

Peacocke, Arthur. 1979. *Creation and the World of Science*. Oxford: Clarendon Press.

Peacocke, Arthur. 1993. *Theology for a Scientific Age: Being and Becoming—Natural, Divine and Human*. 2nd enlarged ed. London: SCM Press.

Peacocke, Arthur. 1997. Welcoming the "Disguised Friend"—Darwinism and Divinity. In *The Idreos Lectures*. Oxford: Harris Manchester College.

Pei, Mario. 1965. *The Story of Language*. Revised ed. Philadelphia and New York: J. B. Lippincott Company.

Pennock, Robert T. 1991. Marshall Nirenberg invents an experimental technique that cracks the genetic code. In *Great Events from History II: Science and Technology Series*. Edited by Frank N. Magill. Englewood Cliffs, NJ: Salem Press.

Pennock, Robert T. 1995a. Epistemic and Ontic Theories of Explanation and Confirmation. *Philosophy of Science (Japan)* 28:31–45.

Pennock, Robert T. 1995b. Moral Darwinism: Ethical Evidence for the Descent of Man. *Biology and Philosophy* 10.

Pennock, Robert T. 1996a. Naturalism, Creationism and the Meaning of Life: The Case of Phillip Johnson Revisited. *Creation/Evolution* 16(2) (39):10–30.

Pennock, Robert T. 1996b. Naturalism, Evidence and Creationism: The Case of Phillip Johnson. *Biology and Philosophy* 11 (4):543–559.

Pennock, Robert T. 1996c. Reply: Johnson's Reason in the Balance. *Biology and Philosophy* 11 (4):565–568.

Pennock, Robert T. 1998. The Prospects for a Theistic Science. *Perspectives on Science and Christian Faith* 50 (3):205—209.

Pennock, Robert T. In press. Creationism in the Science Classroom: Private Faith vs. Public Knowledge, edited by K. Burkam.

Pinker, Steven. 1994. *The Language Instinct*. New York: W. Morrow and Co.

Plantinga, Alvin. 1997. Methodological Naturalism? *Perspectives on Science and Christian Faith* 49 (3):143–154.

Popper, Karl R. 1957. *The Poverty of Historicism*. Boston: The Beacon Press.

Popper, Karl R. 1968 (1962). *Conjectures and Refutations: The Growth of Scientific Knowledge*. New York: Harper and Row.

Popper, Karl R. 1978. Natural Selection and the Emergence of Mind. *Dialectica* 32:339–355.

Price, George McCready. 1902. *Outlines of Modern Christianity and Modern Science*. Oakland, CA: Pacific Press.

Provine, William B. 1990. Response to Phillip Johnson. In *Evolution as Dogma: The Establishment of Naturalism*, edited by P. E. Johnson. Haughton Publishing Company.

Raël, Claude Vorilhon. 1986 (1974). The Book Which Tells the Truth. In *The Message Given to Me By Extra-Terrestrials*. Tokyo: AOM Corporation.

Raff, Rudolf A. 1996. *The Shape of Life: Genes, Development, and the Evolution of Animal Form*. Chicago and London: The University of Chicago Press.

Roose, and Gottlieb. 1976. Genetic and Biochemical Consequences of Polyploidy in Tragopogon. *Evolution* 30:818–830.

Root-Bernstein, Robert S. 1995. Darwin's Rib. *Discover* (September):38–41.

Ross, Hugh. 1993. *The Creator and the Cosmos*. Colorado Springs, CO: NavPress.

Ross, Hugh. 1994. *Creation and Time: A Biblical and Scientific Perspective on the Creation-Date Controversy*. Colorado Springs, CO: NavPress.

Ross, Sydney. 1962. Scientist: The Story of a Word. *Annals of Science* 18 (2):65–85.

Ruse, Michael. 1982. *Darwinism Defended*. Reading, MA: Addison-Wesley.

Safire, William. 1993. On Language. *The New York Times Magazine*, October 31.

Salmon, Wesley C. 1966. *Foundations of Scientific Inference*. Pittsburgh, PA: Univ. of Pittsburgh Press.

Salmon, Wesley C. 1984. *Scientific Explanation and the Causal Structure of the World*. Princeton, NJ: Princeton University Press.

Schleicher, August. 1983 (1863). Darwinism Tested by the Science of Language. In *Linguistics and Evolutionary Theory: Three Essays by August Schleicher, Ernst Haeckel, and Wilhelm Bleek*, edited by K. Koerner. Amsterdam/Philadelphia: John Benjamins Pub. Co.

Schrödinger, Erwin. 1967 (1944). *What is Life?* Cambridge: Cambridge University Press.

Scofield, C. I. 1917. *Reference Bible.* New York: Oxford University Press.

Scopes, John T., and James Presley. 1967. *Center of the Storm: Memoirs of John T. Scopes.* New York: Holt, Rinehart and Winston.

Scott, Eugenie C. 1994. The Struggle for the Schools. *Natural History,* July 1994, 10–13.

Scott, E. C., and H. P. Cole. 1985. The elusive basis of creation "science." *Quarterly Review of Biology* 60:21–30.

Searle, John. 1990. Is the Brain's Mind a Computer Program? *Scientific American* 262 (1):26–31.

Sedgwick, Adam. 1831. Address of the President (1831). *Proceedings of the Geological Society of London 1826–1833* 1.

Sedgwick, Adam. 1988 (1860). Objections to Mr. Darwin's Theory of the Origin of Species. In *But Is It Science?: The Philosophical Question in the Creation/Evolution Controversy,* edited by M. Ruse. Buffalo, New York: Prometheus Books.

Shilts, Randy. 1987. *And the Band Played On: Politics, People, and the AIDS Epidemic.* New York: St. Martin's Press.

Shore, Steven N. 1984. Footprints in the Dust: The Lunar Surface and Creationism. *Creation/Evolution* 4 (4):32–35.

Shumaker, Wayne. 1972. *The Occult Sciences in the Renaissance: A Study in Intellectual Patterns.* Berkeley and Los Angeles: University of California Press.

Sibley, C. G., and J. E. Ahlquist. 1987. DNA Hybridization Evidence of Hominoid Phylogeny: Results from an Expanded Data Set. *Journal of Molecular Evolution* 26:99–121.

Sibley, C. G., J. A. Comstock, and J. E. Ahlquist. 1990. DNA Hybridization Evidence of Hominoid Phylogeny: A Reanalysis of the Data. *Journal of Molecular Evolution* 30:202–236.

Silberman, Gil. 1993. Phil Johnson's Little Hobby. *The Boalt Hall Cross-Examiner* 6 (2):1,4,9–10.

Simpson, George Gaylord. 1970. Uniformitarianism: An Inquiry into Principle, Theory, and Method in Geohistory and Biohistory. In *Essays in Evolution and Genetics in Honor of Theodosius Dobzhansky,* edited by M. K. Kecht and W. C. Steere. New York: Appleton-Century-Crofts.

Sobel, Dava. 1995. *Longitude: The True Story of a Lone Genius Who Solved the Greatest Scientific Problem of His Time.* New York: Penguin Books.

Sober, Elliott. 1993. *Philosophy of Biology.* Edited by N. Daniels and K. Lehrer, *Dimensions of Philosophy Series.* Boulder, CO: Westview Press.

Soltis, Douglas, and Pamela Soltis. 1989. Allopolyploid Speciation in Tragopogon: Insights from Chloroplast DNA. *American Journal of Botany* 76 (8):1119–1124.

Special to the *New York Times.* 1995. Town Divided Over Creationism in Schools. *The New York Times,* February 13, A7.

Spong, John Shelby. 1991. *Rescuing the Bible from Fundamentalism*. San Francisco, CA: Harper.

Strahler, Arthur N. 1987. *Science and Earth History: The Evolution/Creation Controversy*. Buffalo, NY: Prometheus Books.

Temple, Rev. Frederick. 1884. *The Relations between Religion and Science*. New York: Macmillan.

Terrusse, Marcel. 1996. Obscurantism and the Neo-Darwinian Myth. http://www.rael.org/broch01.html#50.

Thaxton, Charles B., Walter L. Bradley, and Roger L. Olsen. 1984. *The Mystery of Life's Origin: Reassessing Current Theories*. New York: Philosophical Library.

Thomas, Dave. 1996. NBC's "Origins" Show. *National Center for Science Education Reports* 16(1):7.

Toulmin, S., and J. Goodfield. 1962. *The Architecture of Matter*. New York: Harper & Row.

Travis, John. 1991. Electronic Ecosystem: Evolving "Life" Flourishes and Surprises in a Novel Electronic World. *Science News* 140(6):88–90.

Tsunoda, Ryusaku, Wm. Theodore de Bary, and Donald Keene, eds. 1958. *Sources of Japanese Tradition*. Edited by W. T. D. Bary. 2 vols. New York and London: Columbia University Press.

Van Bebber, Mark, and Paul S. Taylor. 1995. *Creation and Time: A Report on the Progressive Creationist Book by Hugh Ross*. 2nd ed. Mesa, AZ: Eden Communications.

van Fraassen, Bas C. 1980. *The Scientific Image*. London: Oxford University Press.

Van Till, Howard J. 1986. *The Fourth Day: What the Bible and the Heavens Are Telling Us about the Creation*. Grand Rapids, MI: William B. Eerdmans Publishing Co.

Van Till, Howard J., Davis A. Young, and Clarence Menninga. 1988. *Science Held Hostage—What's Wrong with Creation Science AND Evolutionism*. Downers Grove, IL: InterVarsity Press.

Vardiman, Larry. 1997. The First Young-Earth Conference on Radioisotopes. *Impact*, August, i–iv.

von Däniken, Erich. 1968. *Chariots of the Gods: Unsolved Mysteries of the Past*. New York: G. P. Putnam's Sons.

Weber, Bruce H., David J. Depew, and James D. Smith. 1988. *Entropy, Information, and Evolution: New Perspectives on Physical and Biological Evolution*. Cambridge, MA: The MIT Press.

Weiner, Jonathan. 1994. *The Beak of the Finch: A Story of Evolution in Our Time*. New York: W. A. Knopf.

Whewell, William. 1894. *History of the Inductive Sciences*. 3rd ed. 2 vols. New York: D. Appleton and Company.

Wicken, Jeffrey S. 1987. *Evolution, Thermodynamics, and Information: Extending the Darwinian Program.* New York: Oxford University Press.

Wilder-Smith, A. E. 1983. *The Reliability of the Bible.* San Diego: Master Books.

Woodmorappe, John. 1996. *Noah's Ark: A Feasibility Study.* El Cajon, CA: Institute for Creation Research.

Woodward, Tom. 1996. Meeting Darwin's Wager: How Biochemist Michael Behe Uses a Mousetrap to Challenge Evolutionary Theory. *Christianity Today*, April 28, 15–21.

Wray, Gregory A., Jeffrey S. Levinton, and Leo H. Shapiro. 1996. Molecular Evidence for Deep Precambrian Divergences Among Metazoan Phyla. *Science* 274 (October):568–73.

Yockey, Hubert. 1977. A Calculation of the Probability of Spontaneous Biogenesis by Information Theory. *Journal of Theoretical Biology* 67 (J):377.

Yockey, Hubert. 1981. Self-Organization Origin of Life Scenarios and Information Theory. *Journal of Theoretical Biology* 91 (J):13–6.

Index

history of development of, 59–78,
81, 82, 84–90, 142
hypotheses that comprise, 55–59,
104, 144, 160, 161, 169, 332,
334, 372
interconnected with other sciences,
90, 146, 340
practical utility of, 151, 353
a progressive research program, 38,
82, 86, 171
punctuated equilibrium model, 58,
156
tool of the devil? 277, 310,
309–311, 374, 377
"tree of life" concept, 56, 58, 78,
98, 110, 142, 143, 144, 150,
153, 237, 333, 368
as unifying explanatory framework,
43, 54, 55, 59, 340
Existentialism, 313, 330
Explanation. *See also* Creationist
explanation; Scientific
explanation; Supernatural
explanation
causal, 56, 57, 67, 102, 156, 157,
195, 205, 223, 256, 265, 281,
286, 288
functional, 256
Kitcher's model of, 288
Salmon's model of, 288
van Fraassen's model of, 289
Extinction, 107
of languages, 135
of species, 62, 72, 113, 135, 223
theological concerns about, 135
Extraterrestrial Intelligent-Design
(ET-ID), 234, 235, 249, 275,
304. *See also* Intelligent-design
creationism (IDC); SETI analogy
biblical evidence for? 240
scientific evidence of? 241, 252,
274, 275, 368

Factors, Mendelian, 85
Faith. *See* Religious faith
Fisher, Sir Ronald, 142

Fitness, 97, 101, 102, 106, 107
misunderstandings of, 97, 99–100
a tautology? 99, 100–102
Flood Geology, 13, 16, 48, 61, 62,
74, 76, 83, 84, 282
Fossils, 4, 73, 75, 77, 135, 152
creationist views about, 48, 73, 76,
77, 153, 236, 263, 370
discovered by Cuvier, 72
discovered by Darwin, 62, 63, 64
deficiencies of, 58, 72, 109, 153,
154, 264
early creationist views of, 24, 72
as evidence of creationism? (*see*
Paluxy "mantracks")
as evidence of evolution, 35, 58, 63,
64, 72, 73, 77, 110, 111, 153,
154, 261
incompatible with evolution? 48,
75, 158, 163, 188
of languages, 142, 152
problems with creationist views of,
76, 77, 78, 111, 152, 154, 163
transitional (*see* Intermediate forms)
Founder Principle, 124
Fowler, State Senator David, 344
Frair, Wayne, 13, 14, 162, 173, 178,
179, 248, 292
Functionality, 94, 107
biological, 57, 87, 91, 92, 97, 106,
256, 255–256
creationist arguments about, 92, 94,
96, 245, 255, 260, 264, 265,
266, 267
creationist notions of, 78, 162, 244,
268, 293
problems with creationist views of,
255, 256, 261, 263, 266, 267,
268, 269, 270, 274
scientific notion of, 255–256
Fundamentalism, 8, 12, 26, 31, 35,
37, 39, 49, 310, 374
and Evangelicals, 8–9

Galileo, 39, 40, 134, 155, 156, 287,
339